Terra Thrive

John Joseph Puthenkalam S.J
Joseph Chacko Chennattuserry
Lija Jacob • Haemi Park • S. Vijayalakshmi
Editors

Terra Thrive

Achieving SDG 15: Life on Land for Future Generations

Springer

Editors
John Joseph Puthenkalam S.J
Graduate School of Global Environmental Studies
Sophia University
Tokyo, Japan

Lija Jacob
School of Sciences
Christ University
Bengaluru, Karnataka, India

S. Vijayalakshmi
School of Sciences
Christ University
Bengaluru, Karnataka, India

Joseph Chacko Chennattuserry
Vice Chancellor
Christ University
Bengaluru, Karnataka, India

Haemi Park
Graduate School of Global Environmental Studies
Sophia University
Tokyo, Japan

ISBN 978-3-032-02737-5 ISBN 978-3-032-02738-2 (eBook)
https://doi.org/10.1007/978-3-032-02738-2

This Springer imprint is published by the registered company Springer Nature Switzerland AG
The registered company address is: Gewerbestrasse 11, 6330 Cham, Switzerland

Preface

The Earth's terrestrial ecosystems are facing extraordinary challenges, due to rapid deforestation and species extinction, habitat degradation, and climate change impacts. This book explores Sustainable Development Goal 15, "Life on Land" not merely for a country's commitment but as a critical blueprint for transforming our relationship with the natural world. As human activities increasingly disrupt ecosystems that took millions of years to evolve, the public must urgently shift from being conquerors of the land to becoming its stewards. This book will take the readers on a comprehensive journey through the rich tapestry of terrestrial life, celebrating the incredible diversity of flora and fauna while examining the sobering realities of environmental destruction. The vital role of keystone species in maintaining ecosystem stability, investigating the plight of endangered plants and animals across the globe, and showcase innovative conservation strategies that offer hope for the future are being explored here. This book recognizes that humans are not separate from terrestrial ecosystems but integral to them—our cities, farms, and infrastructure are part of the landscape, and our choices ripple through natural systems in ways both seen and unseen. Through examining sustainable land management, urban planning for biodiversity, climate change mitigation, and the power of environmental education, we provide both understanding and practical tools for action. The solutions exist, from carbon sequestration in forests to pollution control and green infrastructure, but they require our collective commitment to implement. This exploration bridges scientific knowledge with practical action, offering readers the wonder of discovery alongside the tools for meaningful change, because the future of life on land depends on our ability to transform awareness into advocacy and knowledge into action. This book is organized as follows:

Chapter "Introduction to SDG 15—Life on Land: Harmony in Habitat" introduces the SDG 15 and its importance in achieving the ecological balance on earth. A review of India's terrain was done, its surrounding areas, and its collaboration with Japan. This book presents a comprehensive review of the opportunities and potential threats that confront terrestrial ecosystems and to provide suggestions to how important it is to embrace sustainable behavior that promotes a peaceful coexistence between nature and humans.

Chapter "Exploring the Rich Tapestry of Land" discusses the picturesque scenery and landscapes, rich cultural heritage sites, the varied terrains, and environments, from the lands to the sea, from the hills to the plains following its own unique characteristics and stories.

Chapter "Ecosystem Restoration and Sustainable Land Practices Using AI, IoT and Remote Sensing Technologies" explores the importance of ecosystem services. It discusses the incorporation of novel technologies, such as artificial intelligence (AI), the Internet of Things (IoT), and remote sensing, and how they enable the monitoring, maintenance, and rehabilitation of ecosystems in a more effective manner than has previously been possible.

Chapter "Balancing Development and Conservation" explores the critical interplay between balancing development and conservation focusing on how Sustainable Development Goals (SDGs) create a blueprint for achieving equilibrium. Selected SDGs such as Sustainable Cities and Communities (SDG 11) and Life on Land (SDG 15) are examined to understand their role in aligning developmental goals with ecological sustainability. The chapter highlights various theoretical frameworks and case studies which emphasize the conflicts and synergies between development and conservation, through the lens of urbanization, agricultural practices, industrial growth, and community-led conservation efforts.

Chapter "Climate Change and the Future of Land Ecosystems: The Multifaceted Impact of Climate Change on Terrestrial Biodiversity, Vegetation, and Soil Health" describes that climate change is a strong trend that takes place on a global scale nowadays. This chapter discusses various impacts of the climate changes, such as the temperature increase and related changes in climate regimes, loss of species habitats, and their shift to survive.

Chapter "Educating Sustainable Practices" explores the role of educational institutions in creating values, skills, and behaviors through curriculum, policies, and different evaluation and pedagogical methods. By focusing on SDG 4 (Quality Education), this clearly explains how different pedagogies lead to achieving Sustainable Development Goals (SDGs). The chapter highlights various theoretical frameworks and case studies which emphasize methods and strategies an educational institution should adopt to create better minds and generation.

Chapter "Sowing Seeds of Change: Combating Deforestation Through Environmental Education in the Indo-Japan Context" emphasizes the importance of environmental education in the Indo-Japan context toward deforestation. This chapter, based on a comparative analysis of Japan and India, concretely examines the function of Environmental Education (EE) in combating deforestation. Despite their different socio-economic contexts, both countries have a strong commitment to forest protection.

Chapter "Into the Future: Nurturing Sustainable Life on Land" explores sustainable practice implementation for terrestrial ecosystem defense which matches the

Sustainable Development Goal 15 (SDG 15) "Life on Land" approach. This practical approach provides ways to deal with present-day habitat destruction and deforestation problems alongside biodiversity decline through localized scalable solutions.

Chapter "International Collaboration for Sustainable Development: Navigating Life on Land Across the Asian Landscape" examines a few initiatives—possibly most important ones to emerge through countries working to address at least five different sets of legal frameworks, mechanisms, and policies presently available in the world, i.e., the United Nations Sustainable Development Goals (SDGs), a few regional environmental agreements and cross-border conservation project—which are the focus of this chapter. The chapter also reiterates the role of local communities, indigenous knowledge, and technological innovations in the effectiveness of sustainability initiatives. It summarizes with a section on actions implemented via regional collaboration, policy alignment, and knowledge sharing, paving ways to mass awareness.

Chapter "Holistic Land Governance-Revolutions for Ecological Stability and Socio-Economic Growth" analyzes strategic directions and actionable recommendations to improve land management in accordance with global frameworks like Sustainable Development Goal 15, the Bonn Challenge, and the Convention on Biological Diversity. One of the main arguments of the chapter is the need to mainstream sustainable land management in national development strategies.

Chapter "Conservation of Terrestrial Ecosystem Through Carbon Cycle Monitoring and Management" initially discusses the carbon cycle. Then, the function of the terrestrial ecosystem is discussed. It also considers that terrestrial ecosystems are important in providing a foundation for sustainable development through the carbon cycle, and it is necessary to conserve them to cope with climate change. This chapter describes how remote sensing technology is utilized in the conservation of terrestrial ecosystems, encompassing almost all fields from scientific research to local community activities.

Chapter "Prevention of Biodiversity Loss" discusses the various factors that cause biodiversity loss, including natural and human causes. The various consequences faced by the ecosystem along with the global challenges such as health, climate change, and food security will be analyzed. It also explores various strategies that can be implemented for the conservation and restoration of biodiversity, such as establishing protected areas, restoring various habitats and species through recovery programs, and building ecological corridors.

Chapter "Sustainable Usage of Terrestrial Ecosystem" investigates the key aspects of SDG 15 in supporting the achievement of other SDGs and the broader 2030 Agenda. It examines policy pathways to enhance the implementation of SDG 15, highlighting its interconnectedness with global sustainability efforts. A case study is included to illustrate how insufficient progress on SDG 15 targets can increase the risk of future zoonotic diseases.

Chapter "Role of Humans Toward Biodiversity Protection" discusses biodiversity on Earth and how human activities degrade it. The authors conclude that by promoting biodiversity-friendly initiatives, restoring ecosystems, and raising public awareness, we must all work together to secure a future that supports both nature and humanity.

Chapter "Integrating Indigenous Knowledge and Practices into Land Management Systems" explores the necessary elements for maintaining sustainable land management and ecological resilience over the long term, including stronger legal protections, knowledge transfers between generations, and partnerships between communities, researchers and policymakers.

Chapter "Sustainable Consumption of Life on Land: Impact of "Cruelty-Free" and "Ethically Certified" Cosmetics Labels" focuses on determining whether attitudes and purchase intentions to consume ethically certified and cruelty-free cosmetic labels lead to sustainable consumption of life on land in India. Sustainable consumption has been identified as one of the means to create climate resilient policies. The chapter also measures the mediating effects of a lifestyle choice, consumer attitude, and consumer preference toward purchase intention, which results in sustainable consumption.

Bengaluru, Karnataka, India — Joseph Chacko Chennattuserry
Tokyo, Japan — John Joseph Puthenkalam S.J
Bengaluru, Karnataka, India — Lija Jacob
Bengaluru, Karnataka, India — S. Vijayalakshmi
Tokyo, Japan — Haemi Park

Acknowledgments

The editors extend their sincere gratitude to all contributing authors for their cooperation, dedication, and commitment to this publication. The editors also acknowledge the support and funding provided by the Jesuit Research Fund of S. J. House, Tokyo, which made the research included in this publication possible. We also express our heartfelt thanks to Dr. Fr. Joseph C. C., Vice Chancellor, CHRIST (Deemed to be University), India, and to the Leader of the SDGs Books Publication Project, Prof. John Joseph Puthenkalam S.J, Trustee and Board Member of Sophia University, Japan, for their continued guidance, encouragement, and unwavering support of the SDGs Book Project.

The editors further convey their deep appreciation to the Springer teams for their expert guidance and assistance at every stage of the publication process. We hope that this collective effort will contribute toward a more sustainable world grounded in quality education, leading to higher literacy, healthy longevity, and prosperity centered on well-being for all.

Joseph Chacko Chennattuserry
John Joseph Puthenkalam S.J
Lija Jacob
S. Vijayalakshmi
Haemi Park

Contents

Editors and Contributors

About the Editors

Joseph Chacko Chennattuserry is the Vice Chancellor of CHRIST University and brings a wealth of experience and expertise to the institution. His academic background is rooted in History and Education, providing him with a solid foundation for his leadership role. In addition to serving as Vice Chancellor, Fr. Joseph also holds the position of Director of Student Affairs and is a respected Professor within the Department of International Studies and History. Throughout his tenure, Fr. Joseph has been a driving force behind the educational innovations and practices implemented by the University. His visionary leadership has contributed to the institution's growth and reputation for excellence. Fr. Joseph's journey with CHRIST (Deemed to be University) began in 1996, marking the start of a long and fruitful collaboration. Over the years, he has held various leadership positions, including Pro-Vice-Chancellor and Director of Student Services. He has also played a key role in directing important departments such as the Centre for Counselling and Health Services, the Department of Sociology, and the Department of Social Work. Fr. Joseph's influence extends beyond the main campus of CHRIST, as he was instrumental in establishing the Pune Lavasa campus and overseeing the development of the Delhi NCR campus, further expanding the University's reach and impact. In addition to his administrative and academic responsibilities, Fr. Joseph is the

Chief Editor of the prestigious international book project, *Encyclopedia of New Populism and its Responses in the 21st Century*, published by Springer. This role underscores his commitment to scholarly excellence and engagement with contemporary issues on a global scale.

John Joseph Puthenkalam S.J was born in 1956 in Kerala, India. He embarked on a journey of academic pursuit that transcended geographical boundaries. His intellectual voyage commenced with undergraduate studies in Economics and Philosophy at esteemed institutions in India. Seeking to broaden his horizons, he ventured to Japan in 1986, where he delved into the realm of Theology, earning a degree from Sophia University. Undeterred by disciplinary boundaries, Fr. John continued his academic odyssey, ultimately earning a Ph.D. in Economics from the University of Glasgow in the United Kingdom. Fueled by a passion for addressing the pressing issues facing developing nations, his research endeavors centered on crafting models that harmonize sustainable environmental practices with economic progress. Central to his approach were considerations of human development, the empowerment of marginalized communities, and a vision of globalization that prioritized real, tangible benefits for all. Throughout his distinguished career, Fr. John has authored numerous books and articles, cementing his reputation as a thought leader in his field. His contributions not only enrich academic discourse but also offer practical insights into fostering inclusive growth and environmental stewardship in an interconnected world.

Lija Jacob is a seasoned educator and researcher with over 20 years of experience in the field of teaching and academia. With a strong commitment to fostering learning and innovation, she has demonstrated effective leadership skills and a passion for both teaching and research. Dr. Jacob holds a Doctor of Philosophy (PhD) degree from Bharathiar University, Coimbatore, along with a Master of Philosophy (MPhil-CS) from Madurai Kamaraj University, Madurai. Currently serving as an Associate Professor and Head in the Department of Data Science at CHRIST (Deemed to be University), Pune Lavasa Campus, Pune, Dr. Jacob has been instrumental in shaping the academic landscape of the institution. Her responsibilities include curriculum development, research coordination, and technical program chairmanship for various conferences. Prior to her current role, Dr. Jacob worked as an Assistant Professor [Senior] in the Department of Data Science at CHRIST (Deemed to be University), Pune Lavasa Campus, Pune, and also held teaching positions at Saintgits College of Engineering, Pathamuttom, Kottayam, Lourdes Matha College of Science and Technology under Kerala University, and BCM College for Women under Mahatma Gandhi University, Kottayam. Dr. Jacob's research interests span across machine learning, deep learning, data science, and computational security, as evidenced by her numerous international publications and presentations. She has also secured project funding from prestigious institutions like the Kerala State Council for Science, Technology and Environment (KSCSTE) for her research endeavors. Beyond academia, Dr. Jacob is actively involved in consultancy projects, faculty development programs, and has received recognition for her contributions, including awards and prizes for her innovative work. She is a certified member of Google Cloud Platform (GCP) Associate Cloud Engineer course and has chaired technical sessions in various international conferences.

S. Vijayalakshmi is an Associate Professor in the School of Sciences, at Christ University, Pune Lavasa Campus, where she serves in multiple academic portfolios. Her research expertise encompasses Image Processing and Internet of Things (IoT), areas in which she has made significant contributions through publications in numerous international and national conferences as well as reputed journals. Dr. S. Vijayalakshmi has authored several book chapters and currently has multiple books in development. As an active member of academic bodies such as IFERP, IAENG, and IEEE, she remains engaged with the broader research community while dedicating considerable effort to guiding students in both research and academic pursuits. Her commitment to advancing knowledge in data science and emerging technologies, combined with her passion for student mentorship, makes her a valuable contributor to both academia and the field of technological innovation.

Haemi Park completed her doctoral studies in the Department of Civil Engineering, University of Tokyo in 2015. Her major is in Environmental Remote Sensing, especially carbon and water cycle modeling in land area. Between 2015 and 2018, she was a postdoc of the Intelligent Remote sensing and geospatial Information Science (IRIS) Lab in Ulsan National Institute of Science and Technology in Republic of Korea. She went back to UT–Institute of Industrial Science (IIS) and worked as a postdoc in 2019. From 2020 to 2022, she has been working at Japan Aerospace Exploration Agency—Earth Observation Research Center (JAXA–EORC) as a researcher. Since 2022, she is working at Graduate School of Global Environmental Studies of Sophia University, Japan, as an Assistant Professor. The research interests include the carbon/ water balance in wetlands, soil moisture, sustainability of forests, and effects of human activity.

Contributors

Vandana R. Bais Department of Computer Science, Christ College, Pune, Maharashtra, India

C. Balaji SRM Institute of Science and Technology, Trichy, India

Salu Dsouza School of Law, CHRIST University, Bengaluru, India

Nilima Gaikwad Department of Computer Science, School of Sciences, Christ University, Bengaluru, Karnataka, India

T. Genish Arumugam Pillai Seethai Ammal College, Tiruppattur, India

R. Gunavathi Christ University, Bengaluru, India

Sharad Gupta Christ University, Bengaluru, Karnataka, India

John Joseph Puthenkalam S.J Sophia University, Tokyo, Japan

Lekha J School of Sciences, Christ University, Bengaluru, Karnataka, India

Resham Lohani Christ University, Bangalore, India

K. Malathi Department of Economics, University of Madras, Chennai, India

G. Manivasagam Jain (Deemed to be University), Bangalore, India

R. Nithya KPR College of Arts Science and Research, Coimbatore, India

Tiny S. Palathara CHRIST Deemed to be University, Pune, India

Haemi Park Graduate School of Global Environmental Studies, Sophia University, Tokyo, Japan

B. Meena Preethi Department of Software Systems, Sri Krishna Arts and Science College, Coimbatore, Tamil Nadu, India

Manjeri Subin Sunder Raj School of Law, CHRIST University, Bangalore, Karnataka, India

Mabel Sarpong-Kusi Sophia University, Tokyo, Japan

M. Savithri Christ University, Bengaluru, India

Anto Sebastain School of Law, CHRIST (Deemed to be University), Pune Lavasa, Maharashtra, India

Sharaf Ali V Liwa University, Al-Ain Campus, Abu Dhabi, United Arab Emirates

Manjari Sharma Christ University, Bengaluru, Karnataka, India

V. Shobana Dr. N.G.P. Arts and Science, Coimbatore, India

K. Thenmozhi Dr. N.G.P. Arts and Science, Coimbatore, India

Sharon Susan Thomas Department of Computer Science, Christ University, Bengaluru, Karnataka, India

Amrutha Rose J. Valavi School of Law, CHRIST (Deemed to be University), Pune Lavasa, Maharashtra, India

S. Vijayalakshmi Department of Data Science, School of Sciences, Christ University, Bengaluru, Karnataka, India

Department of Data Science, Christ (Deemed to be University), Pune Lavasa, Maharashtra, India

Ravi Kumar Yadav Christ University, Lavasa, Maharashtra, India

Shraddha Yadav Christ University, Lavasa, Maharashtra, India

Lihui Zhou Sophia University, Tokyo, Japan

Introduction to SDG 15—Life on Land: Harmony in Habitat

Mabel Sarpong-Kusi and John Joseph Puthenkalam S.J

1 Introduction to *Terra Thrive*

The Sustainable Development Goal 15 (SDG 15) has as its central goal to preserve and sustain our earthly resources, and this has served as an important component or aspect of the United Nations 2030 Agenda toward Sustainable Development of our nations. In these present days, where environmental sustainability has become a pressing concern, this SDG 15 serves as a guide in helping to solve these problems facing our land-based resources and intend to protect, restore, and promote the usage of terrestrial resources sustainably. Globally, the lives of billions of people depend on the essential provision of vital resources such as food, water, and clean air provided by our earthly resources of which SDG 15 seeks to protect and restore these terrestrial ecosystems knowing the important connections between human life and the health of our natural environment for the continuous provisions of these services. Taking a deeper look into this all-important goal is what this book *Terra Thrive* seeks to provide to its readers. In doing this, readers will be provided with basic understanding of the effects of our (human) interactions with natural resources and vice versa as we co-exist on this planet earth. By laying emphasis on environmental sustainability, this book looks at the vivacious landscape of India and its surrounding countries, not forgetting their collaborations with Japan. *Terra Thrive* presents the complexity of the human-environmental relationship while stressing on the essentialities of sustainable practices, protecting forest, and conserving biodiversity (Karr, 2012). By deeply exploring this theme, *Terra Thrive* will motivate and encourage readers to be more proactive toward the maintenance of our environment. The book highlights the importance of the enhancement of environmental practices and provides the information toward conservation efforts and good environmental

M. Sarpong-Kusi (✉) · J. J. Puthenkalam S.J
Sophia University, Tokyo, Japan
e-mail: j-puthen@sophia.ac.jp

J. J. Puthenkalam S.J et al. (eds.), *Terra Thrive*,
https://doi.org/10.1007/978-3-032-02738-2_1

practices. The successful initiatives and real-world examples that readers will encounter will give them hope, serve as tools, and offer insights for the creation of policies, all of which will ultimately contribute to a sustainable future for both current and future generations (Tully, 2019; Saleh et al., 2024).

2 Overview of SDG 15

Deforestation can be explained as the destruction of forest or land by cutting down trees for the purpose of different developmental projects like mining, farming, fuel, and other urban expansion and development projects. These activities are mostly caused by humans with a few being caused by natural events like volcanic eruptions and wildfires (Antonelli et al., 2020; Ritchie & Roser, 2024). The conservation and preservation of terrestrial ecosystems like mountains, forests, wetlands, and drylands remains the focus of SDG 15. This SDG primarily aims to conserve these earthly resources (15.1 and 15.2) and provide guidelines that will help to avert deforestation and restoration of damaged lands. SDG 15 lays emphasis on the continuous and active restoration, conservation, and preservation initiatives (SDG 15.3, 15.4, and 15.5) geared toward the maintenance of these ecosystems and protection of biodiversity. To ensure the fair and equitable distribution of the benefits that we derive from terrestrial resources, SDG 15 highlights the importance of this form of distribution which includes medicinal plants distribution among stakeholders laying emphasis on the local and indigenous communities (SDG 15.6). The strategy continues to hammer on the stoppage of illegal activities like plant and animal trafficking and poaching which together go to endanger biodiversity (SDG 15.7). Also, two of the SDG 15's objectives (15.8 and 15.9) work on the security of disrupted ecosystems and native species threatened by invasive alien species. To put a stop to this, SDG 15 encourages the incorporation of these into national and local development planning processes to address these threads. For the protection of these natural resources, this SDG requires that biodiversity issues be incorporated into developmental projects and policy formulations.

To achieve the general objectives of SDG 15, three main targets have been outlined for implementation which serve as a necessary structure and support for the achievement of these goals. These three implementation targets are: increasing financing for biodiversity conservation (15 A), promoting sustainable forest management (15 B), and stepping up measures to stop poaching and protected species trafficking (15 C). Paying attention to these three strategies are important because it will help to achieve this overarching goal of restoring, maintaining, and promoting the sustainable usage of terrestrial ecosystems to ensure a balanced and resilient environment for the present and future generation.

Protect and Restore Ecosystems (15.3, 15.4, and 15.5) The aim of the above SDG is to address the loss of biodiversity, and fight desertification. The conservation of freshwater and terrestrial ecosystems are the focuses of several objectives in

SDG 15. The achievement of these targets is affected by human activities like: Agriculture, a primary driver of deforestation that accounts for over 70% of the world's deforestation rates due to land expansion for crops and animals. Logging is another significant cause of deforestation, as it involves the commercial extraction of wood, paper, and other wood products from forests in meeting the needs of humankind. Building roads, highways, and other social amenities as infrastructure development, which frequently necessitates the removal of huge areas of forest. Mining and mineral extraction are necessary to get a nation's mineral resources, and this often necessitates the destruction of waterways, forests, and other natural areas. Fire may also result in extensive deforestation, whether it is caused by natural or human-made disasters. Because forests serve as habitat for a large number of terrestrial species on Earth, when deforestation destroys ecosystems, species go extinct. Because trees absorb the greenhouse gas carbon dioxide, global warming is exacerbated when these trees are cut down. These trees and other vegetation cover help to prevent soil erosion by slowing down the movement of water through the soil. Without them, the likelihood of soil erosion increases, which can cause the loss of arable land and increased sedimentation in streams. Additionally, forests are essential for controlling water cycles, and their loss can affect rainfall patterns and water quality. The effects forests have on indigenous populations cannot be underestimated because forests are important to many indigenous communities, serving as a source of identity, culture, and means of subsistence. The degradation of these forests can cause disruption and upheaval in these communities (Olagunju, 2015; Duguma et al., 2019).

The negative impacts on forests can be lessened by remediation and corrective measures including implementing and encouraging sustainable logging methods. Reforestation and afforestation are also essential techniques that involve planting trees in areas where trees have been cut down and where there has been no tree coverage. Creating national parks, wildlife reserves, and other protected areas can help preserve forests. Enacting laws and regulations that restrict illegal logging and promote forest conservation can also be an effective measure. These measures cannot be achieved when the human factor is left unattended to and so because of this, involving local communities in forest management and conservation initiatives can lead to a sense of investment in the preservation of forests among those who rely on them. Also, immediate and consistent action from global, national, and local stakeholders is necessary to combat the significant environmental issue of deforestation and minimize its consequences. Since 1990, it is projected that different land users have destroyed more than 400 million hectares of forest (Barraclough & Ghimire, 1995; Antonelli et al., 2020; Ritchie & Roser, 2024). Climate stability, biodiversity, and indigenous communities are all under risk from deforestation. The actions of we humans as occupants of this land including habitat destruction, over-exploitation, pollution, not forgetting climate change are attributed to as sources of biodiversity loss of our times according to the Intergovernmental Science-Policy Platform on Biodiversity and Ecosystem Services (IPBES). Data indicate that about 1 million plant and animal species are in danger of extension due to human-induced activities with a few more being pushed into that condition in the nearest few decades. As humans depend on these resources for survival and continuous assistance, this threatening situation of biodiversity loss shows how important it is to welcome any action toward conservation and sustainability strategies meant to protect biodiversity. It also shows how important it is to maintain and restore ecosystems that give support to many creatures that call these resources Home. To avert deforestation, this STG 15 encourages an advocate for properly implementing laws against illegal logging, motivating and encouraging best and sustainable forestry practices while supporting with afforestation and reforestation strategies to help restore degraded forest lands. In doing these toward the maintenance of ecological balance, it will help to enable forests to continuously provide important services which include carbon sequestration, aquatic management, and home for man species (Husain et al., 2024; Monaco, 2024).

Combat Desertification and Recover Deteriorated Land and Soil (15.3) Desertification occurs when productive land turns into a desert because of drought, deforestation, or unsuitable agriculture. To counteract desertification, SDG 15 calls for sustainable land management techniques that improve soil fertility, stop erosion, and increase production in degraded areas. This includes practices such as crop rotation, agroforestry, and the use of organic fertilizers—all of which are emphasized in the SDG objectives as essential tactics for attaining sustainability. In addition to these the following are also there.

- ***Protection Against Desertification:*** Human life, water supply, and biodiversity restoration all are dependent upon mountain habitats. SDG 15 seeks to protect these habitats by halting soil erosion, protecting the unique plants and animals of mountainous ecosystems, and implementing sustainable water resource management. In addition to ensuring the survival of mountain communities and the supply of freshwater for people in the communities below, conservation activities in these locations contribute to the preservation of biodiversity.
- ***Protection of Biodiversity—Fauna:*** One surefire approach to preserve the diversity of protected species is to stop poaching and trafficking in them. This is because due to these activities, a greater number of species face extinction as a result of illegal poaching and wildlife trafficking which pose a greater threat to biodiversity. To find a way out of this challenge that confronts biodiversity, SDG 15 seeks to halt these activities that bring about biodiversity loss by encouraging the enforcement of stronger legislation on the protection of wildlife, enhanced and updated law-enforcement tools, and more collaborations at the international level against wildlife crimes. These initiatives and strategies will help us to achieve equilibrium between biodiversity and ecology which is made possible when endangered species and inhabitants are protected.
- ***Floral Restoration for Natural Habitat:*** Habitat loss has been a major causative agent toward the reduction in biodiversity. This issue prevents SDG 15 from achieving the goal of protecting and restoring natural resources including wetlands, grassland, and forest. To end this, there is the need for the creation of protected areas and its management; there is the need for action toward ecosystem restoration, and the acceptance of sustainable land use best practices that

help to reduce habitat fragmentation and many more which serve as examples of conservation strategies. That is when these are carried out and actions will help to safeguard our planet.

- ***Encourage Equitable and Fair Benefit Sharing (15.6)***: The need for collaborative effort in achieving this SDG target cannot be understated. This is because all governments, businesses, and individuals are needed in the achievement of this all-important goal. This target encompasses the preservation of the natural ecosystems, carbon footprint reduction, and education of the significance of biodiversity conservation. The goal also emphasizes the need for enhanced funding, updated and realistic policies and strategies focused on sustainable land management, ecosystem restoration, and conservation of natural capital. This way we can all enjoy the benefits that it has to offer (Husain et al., 2024; Whipple & Moss, 2024).
- ***Encouraging Sustainable Land Management (SLM)—(15.1 and 2):*** To achieve the provision of human population's needs, it is important and urgent to support actions toward promotion of sustainable land management (SLM). To preserve the land for a long-term health condition and production purposes for the up-and-coming generations, land use techniques are being implemented as sustainable measures by SLM. These approaches included ecological economic and social aspects into the management processes in addressing concerns like diversity laws, deforestation, and others. All these will ensure the promotion of food security and sustainable development.

2.1 *Achieving Development That Is Sustainable*

Countries and individuals throughout the world now recognize the need for sustainable development, making it a major concern. Adopting a balanced approach to growth and progress is becoming more and more important as we face these concerns. By striking a balance between environmental preservation and human advancement, sustainable development seeks to guarantee that future generations may live at the same level as we do. By promoting sustainable activities, we can address issues such as resource depletion, social injustice, and climate change. To take the right action to achieve the this, the following actions could be considered.

Biodiversity, A Critical Component of a Healthy Planet The planet depends on its biodiversity. These varieties include genetic variation within species, the complex nature of ecosystems they create not forgetting the diversity of the species. The support that biodiversity provides to human lives makes it an important resource. The availability and supply of food, fuel, and freshwater and the management of water filtration, disease control, flood prevention, and climate regulation are among other things that by diversity directly supports us. We can as well talk about the culture services like leisure that biodiversity supports us with, the art and spirituality as well as soil health, and water retention that helps in the promotion of agriculture production and food security are also what biodiversity supports us with. For example, biodiversity serves as a natural water filter from Westland that is rich in biodiversity that helps to lower the need for more costly water treatment facilities. Also, several trees or plant species help in water retention. A healthy ecosystem with abundant biodiversity can withstand events like natural catastrophes, human activities, and changes in the climate which constitute environmental destruction. For biodiversity to continue to be resilient, the ecosystem must be able to withstand shocks and stresses. This resilience allows ecosystems to better withstand shocks and stresses, maintaining their functionality and providing continuous ecosystem services.

This resilience allows ecosystems to better withstand shocks and stresses, maintaining their functionality and providing continuous ecosystem services. Therefore, it is hard to overstate how important biodiversity is to preserve the planet's health (Kopittke et al., 2024; Masso, 2024; Singh, 2024; Yüksel & Karaçuhalılar, 2024). Biodiversity serves as a comprehensive source of genetic resources that may solve emerging challenges. The medicinal value of biodiversity is reflected in many pharmaceuticals derived from wildlife and plant sources. For example, the rosy periwinkle, a native plant of Madagascar, has provided compounds used in cancer treatment. The financial importance of biodiversity is vast, although it is often underestimated. Agriculture, forestry, tourism, and fisheries are among the businesses that rely on ecosystem services, which have a substantial impact on world economies. Pollinators, including bees, are crucial for crop production, and their services are valued at billions of dollars annually. Despite its significance, human actions including habitat destruction, pollution, overexploitation, climate change, and invasive species pose a danger to biodiversity. These factors are driving species to extinction at an alarming rate, which undermines ecosystem functionality and reduces the standard of living for human populations. Therefore, protecting and promoting biodiversity is essential to attaining sustained growth and ensuring the well-being of people. As we face unprecedented environmental challenges, recognizing and appreciating biodiversity's contribution is essential to creating a sustainable and harmonious future for our planet (Han et al., 2024; Pla et al., 2024; Wang et al., 2024).

Preserving and Recovering Ecosystems Ecosystem protection and restoration are crucial measures for ensuring the well-being and functionality of the planet globally. Ecosystems that are resilient and in good health provide invaluable services, including soil fertility, water purification, climate regulation, and habitat for biodiversity. The provisioning, regulating, supporting, and cultural services that make up these ecosystem services are all essential for maintaining the earth's balance. Unfortunately, many ecosystems have been severely damaged by human activities, which threaten their ability to function properly. Examples of these activities include deforestation, pollution, urbanization, and overexploitation of resources. These stressors, in addition to climate change, have made it more challenging for landscapes to supply the required resources that communities need to survive (Emerson, 2024; Jurjonas et al., 2024; Nelson et al., 2024).

In order to stop the undesirable effect associated with climate change, it is important to have and to implement a comprehensive approach that embodies sustainable land practices, active participation of the local communities in conservation initiatives, and the recreation of restricted areas with the communities. A successful illustrated approach serving as an example of this strategy is exhibited by the European Union Natura 2000 conservation network acts. This approach covers over one-fifth of the EU's total land property, and this provides enough space that enables the species to operate and to adapt to the challenges they faced by climate change. For example, when deforested areas are reforested, wetlands and peatlands are restored, and there is a significant decrease in the other damaging environmental impacts, these will all help to reduce the effects of greenhouse gas emission on future generations significantly. Another practical approach toward solving the problem of biodiversity loss and climate change can be the implementation of nature-based solutions (NBS). These nature-based interventions can include, but are not limited to, measures for adapting to the negative effects of climate change, and the reduction of greenhouse gas, for biodiversity, and while doing this, remembering the social aspect of these issues. All these approaches, when implemented, have several benefits for the ecosystem (Li et al., 2024a, b; Rawat et al., 2024). We should embrace the implementation of SDG 15 and attach all forms of importance that will highly prioritize this SDG, if we want to guarantee the long-term sustainable use of our natural resources and the benefits they offer. In our quest to protect the climate and build-up more resilience to the changes, the best approach and effective action to take is to cooperate with nature.

Caring for and Conserving Our Forests According to available data, over 31% of earth's surface area can be said to be forested; terrestrial species also use over 80% of the surface of the land as Home. So, the capacity of forests to regulate carbon dioxide and release oxygen positions them as a critical component of climate stabilization. Looking at this situation from the global perspective, the livelihood of over 1.6 billion estimated people worldwide depend on the essential resources that are derived from forests like: fuel, timber, and other medicinal plants. To meet up with the demand for these essential services by the populace, it requires good forest practices to preserve this forest biodiversity. These issues are critical for the effective management of forest practices and include approaches that promote the sustainable use of forests. This requires maintaining their ecological integrity through protected areas created along with agroforestry system implementation and the proper practices of selective logging application methods. These forests provide countless benefits which span across ecological systems, and the economic and social benefits that come along with their usage.

These forests contribute greatly to the mitigation of global warming through the provision of carbon sinks that help to absolve significant harm from the environment. Apart from forests serving as habitat for several animals and humans, the health of soil and regulation of water supply are critical aspects of agriculture and for human consumption. Forests provide timber and non-timber resources to meet up with the demand for forest supplies, especially in rural areas, areas supporting millions of the populous livelihood. Through these forest products, businesses are generated. For example, in California, data indicate that about $16.8 billion of yearly revenue is generated and the same employs around 68,000 people, which helps to reduce the number of unemployment and provide increased income for the people's survival. Considering their cultural and social importance, forests provide spaces for relaxation, foster reflective thinking, and strengthen human connections with nature. These forests provide natural spaces that promote wellbeing which contribute to several health benefits in the society. To retain these numerous benefits, we derive from the forest, it is appropriate to maintain good forest management practices that contribute to balance resource extraction and conservation actions. Through these forest management practices, other forest assets are protected; these practices promote the selective and periodic harvesting of these resources and support the long-term health of our forest ecosystem. Through these forest conservation efforts, economic growth, ecosystem health protection, and improved quality of life are promoted globally. Measures put in place to prevent soil erosion reduce the effects of climate change, and protect wetlands and have swift effects on human wellbeing and economic stability.

2.2 *Barriers Hindering Conservation Efforts*

Despite the number of obstacles that conservation efforts are faced with, it is critical to protect the earth's biodiversity and natural resources. These efforts to preserve the environment face threats such as pollution, illicit wildlife trading, and habitat loss. The issue of poverty and development can negatively impact conservation efforts. It is therefore important to understand the pivotal roles these challenges play to be able to establish successful interventions for preserving our environmental heritage for the yet unborn generation. Striking a balance between environmental protection and human needs is very important, which may need difficult decisions and sacrifices for its accomplishment

Conservation of Habitats and Human Impact The natural environment has negatively been impacted by the actions of its human occupants reducing biodiversity and degrading ecosystems. These can be attributed to key contributing factors like pollution, deforestation, urbanization, industry, and agriculture as the main drivers of this phenomenon. The protection of habitats is important to the provision of ecological services, and promotion of the well-being of all living things that exist including humans. It is therefore important to use effective preservation techniques and identify human impact if we want to achieve environmental sustainability. Unfortunately, conservation initiatives face challenges like financial constraints, political support, the degradation of habitats, and impact of climate change (Nickel et al., 2020; Riggio et al., 2020). Collaborative efforts and multifaceted approaches are essential in addressing these complex and linked concerns. Identifying and addressing these challenges is critical for conserving diversity and ensuring resource sustainability for future generations in the long run. Some of these obstacles include but not limited to the following:

- ***Habitat Loss and Fragmentation:*** Development of infrastructure, agriculture, mining, and urbanization contribute significantly to habitat degradation, which poses one of the greatest threats to ecosystems.
- ***Change in the Climate:*** The environment is being altered due to the impact of climate change, which impacts the distribution and behavior of species. The loss of vulnerable species is due to the rapid changes because of these activities (SINGH, Wiens, 2016).
- ***Invasive Species:*** Non-native creatures introduced intentionally or accidentally can have unforeseen negative consequences, such as the displacement of native species and the threat to the ecological relationships that support aquatic ecosystems (Yadav et al., 2024).
- **Pollution:** Native fish and wildlife that rely on clean water are at risk from both point source and nonpoint source pollution. Long-term environmental degradation, disruption of food chains, and contaminated water can cause harm or even death to plants and animals, soil, and air (Mulder & Coppolillo, 2005; Meffe et al., 2012).
- ***Decreased Water Quantity:*** Ecosystems, habitats, and animals whose survival depends on the steady flow of rivers or streams are impacted by variations in water availability as the population expands and water needs increase (Dagnachew et al., 2024).

2.3 *Techniques for Effective Conservation*

Maintaining natural resources and guaranteeing the worldwide conservation of biodiversity is essential and requires effective conservation measures. As human activities continue to impact ecosystems, environmental protection and enhancement policies are becoming more and more essential. This manual delves into various practical conservation techniques, focusing on approaches that can reduce habitat loss, enhance the recovery of their species, and boost their ecological resilience as well. Getting a better understanding and executing these strategies can lead to a harmonious future between humans and its environment.

Creating Protected Areas The creation and administration of protected areas, including national parks, wildlife reserves, and marine sanctuary sites, is an important action to take. When these areas are set aside, it helps to protect ecological processes and provide habitats for animals and promote tourism and enhance scientific studies (Alles, 2012).

Management of Resources and the Sustainability of Land Use The management of fisheries, forestry, and agriculture is a sure way toward the mitigation and reduction of the negative effects of human activities on ecosystems. Promoting sustainable login, agroforestry, and proper conservation of agriculture can be a good example of these methods that will help to reduce habitat degradation as well as promoting biodiversity improvement (Li et al., 2024a, 2024b).

Educating the Community to Improve Participation The involvement of local communities in conservation efforts helps to develop stewardship and ensures that conservation initiatives are economically and culturally appropriate, providing the communities with information and skills as awareness creation can be done successfully through education and outreach programs. In doing this, it will help the community members to develop the passion for biodiversity conservation and motivate actions to safeguard our natural environment (Griswold et al., 2024).

Ecosystem Assessment and Impact Evaluation To continuously monitor species and inhabit for a timely intervention and adaptation measures, scientist must continue to conduct scientific research to help in the better understanding of ecosystems, identify timely threats, and establish efficient conservation strategies to help provide for timely interventions and adaptation management (Mori & Grover, 2024).

Environmental Policy Enforcement Protecting our environment and the species found within it requires that we enact strong environmental laws and regulations to protect these species and the ecosystems. The efficient and effective implementation of laws and regulations will help to enforce their compliance by the people and the sanction will prevent people from breaking these rules (Robinson & Reilly, 2024; Rockström et al., 2024).

Although conservation efforts face numerous challenges, success is achievable with calculated risk and teamwork. Applying the above strategies to address pollution, habitat loss, climate change, and other concerns can help to preserve biodiversity and ensure the sustainability of our planets' ecosystem.

2.4 The Importance of Ecological Land Management

The production of food, the supply of water, and biodiversity all depend on the productivity and health of our land. The aim of sustainable land management (SLM) is to strive to sustain productivity by reducing unsustainable activities like overgrazing, deforestation, and intensive agriculture, which can deplete soil, reduce production, and interrupt ecosystem services. Land degradation, rehabilitation of damaged lands, and ensuring land resources can meet with both natural and human demands can be handled by SLM strategies that help prevent these issues (Hyman & Leibowitz, 2000; Gillis et al., 2017). Doing all this aligns with SDG 15. This SDG 15 provides nations with the right guidelines on how to accomplish their Sustainable Development Goals. Included in this strategy are, but not limited to, the following:

Resilient Nature-Based Farming Practices Alternating preservative agricultural practices involves measures like crop cycles, maintaining soil cover, and limiting soil disruptions. These strategies strengthen soil health, water retention, and prevent erosion, resulting in improved agriculture output and sustainability.

Forest-Based Farming Management Agroforestry strategies involve the cultivation and using of trees and shrubs in farming environments to promote soil fertility, biodiversity, and carbon sequestration, among other benefits. Apart from the benefits we derive from trees like the provision of shades, wind breaks, and erosion management, this can also boost crop yields.

Ecological Pest Control Strategies An integrated sustainable strategy for pest control that combines chemical, biological, cultural, and physical approaches is the integrated pet management (IPM). By significantly reducing the use of chemical pesticides, this strategy reduces pollution, harm to non-target species, and enhances human health.

Management of Land and Water Strategies like landscaping, checking of reservoirs, and contour farming are effective ways for saving soil and land. When these strategies are employed, they help to promote sustainable agriculture and enhance water filtration, reduce soil erosion and increase nutrient retention for the promotion of sustainable agriculture.

2.5 *Sustainable Land Management (SLM) Benefits*

To achieve sustainable land management (SLM), it is important to adopt techniques that will ensure land resource health and production. SLM strategies consider environmental, economic, and social factors to maximize land use while maintaining future generation's ability to meet their own needs. This session discusses the benefits that we derive from SME such as biodiversity conservation, increased health, increased agriculture productivity, and sustainable livelihoods. Understanding this benefit is important for developing more resilient ecosystems and communities. Key factors among others include the following:

- ***Enhanced Food Security***: SLM techniques provide a stable and sustainable food supply, promoting food security for growing populations. They help to lower the rates of hunger and malnutrition, especially in poor nations.
- ***Climate Change Mitigation and Adaptation:*** The sequestration of carbon essential for mitigation. By making soils and ecosystems more resilient, SLM techniques enable these systems to better withstand harsh environmental fluctuations.
- ***Restoring Degraded Habitats Is Essential for Biodiversity Conservation***: This involves addressing the underlying reasons behind devastation, reintroducing suitable species, preserving habitats, enhancing ecosystem services, and mitigating agriculture's negative impacts on natural ecosystems to promote biodiversity (Memmott, 2009).

Sustainable land management (SLM) strategies enhance water resources by improving both quantity and quality. These strategies support agriculture and the environment by strengthening soil structure and retaining water, thereby sustaining livelihoods for all dependent habitats (Loucks, 2000).

2.6 *Obstacles and Solutions for SLM Promotion*

To overcome these obstacles, a comprehensive approach that includes these elements is necessary. Despite the many benefits that ecosystem service presents with the implementation and promotion of sustainable land management, they face many challenges in its implementation and promotion for sustainable management. Among the challenges include, but not limited to, the lack of knowledge, socioeconomic, difficulties, and technical and or policy in variation. To overcome these barriers, a complete strategy incorporating these factors is needed.

As lack of knowledge of the adequacy of it has been identified as a challenge to successfully implement SLM, there is the need to give training to key stakeholders like politicians, land managers, and farmers alike. This training program will empower participants with knowledge and skills to safeguard and improve land resources, promote environmental health, and sustainable agricultural production. These individuals can be trained on current issues about agroforestry, crop diversification, water management, and soil conservation. To maximize the benefits of these training programs, a comprehensive approach is needed. Agricultural extension services offer technical assistance and protocol training to local communities. The provision of these services connects academic institutions and industrial specialists, assuring the successful implementation of cutting-edge scientific discoveries. Presenting to the beneficiaries, successful projects present to them a real-life example which is effective for SLM approach and can serve as a significant resource. Initiatives like this can promote the widespread adaptation of successful sustainable best practices when these demonstrated project results are presented. Web-based communication will enable information sharing and experience exchange among stakeholders and knowledge sharing venues like conferences and seminars. Platforms like this allow farmers and land managers, scientists, and policymakers to have communication, debate challenges, and generate creative solutions. These projects promote a network of practices around sustainable land management, accelerating the adoption of effective solutions and expanding social support. Training politicians is important for the enactment of laws and policies that support sustainable land management; lawmakers may support sustainable land management through the provision of incentives, subsidies, and regulatory frameworks. It is imperative for policymakers to understand the benefits and drawbacks of sustainable initiatives. We need a comprehensive approach to education and knowledge sharing to sustain our resources and promote the general acceptance of sustainable land management (Palmer et al., 2009).

- Campaigning for sustainable land management (SLM) demands institutional and regulatory support, along with financial incentives. These are needed to promote sustainable practices across multiple sectors. Clear guidelines are necessary for maintaining its effectiveness.
- Promoting sustainable behaviors: Policies geared toward the promotion of sustainable practices may include grants, taxes breaks, financial rewards, subsidies, and other. Assisting stakeholders with financial incentives to embrace agroforestry, natural farming, and soil conservation practices would increase their attractiveness and accessibility. Reimbursing ecological services that include carbon sequestration and biodiversity conservation could financially benefit individuals that promote environmental sustainability.
- Institutional support and policy framework: When supportive legislations are created that promotes sustainability, secure land tenure, and supporting regulatory framework. They serve as an important aspect of successfully implementing sustainable land management. Therefore, strengthening the collaboration between stakeholders and institutions can increase the efficiency of sustainable land management actions (Harari et al., 2023; Haregeweyn et al., 2023).
- Creative solutions: Considering innovative processes like ecosystem service payments, green bonds, and agricultural subsidies can help attract financial resources for sustainable land management endeavors. Both the public-private partnership and international collaboration can help provide financial and technical support (Louman et al., 2020; Boufounou & Dellis, 2021).

To maintain and sustain land productivity and health, increase biodiversity, and improve climate resilience, sustainable land management has become essential. Using SLM strategies can achieve a balance between environmental sustainability and human demands, thereby promoting sustainable development and the well-being of generations yet to be born.

2.7 *Collaborative Efforts for a Greener Future*

To create an environmentally friendly, effective collaboration between and among all stakeholders is crucial. This collaboration is needed to resolve environmental challenges, carry out conservation initiatives, and promote healthy environmental behaviors. Governments' roles cannot be understated due to the importance of the functions of the roles play. There is the need for state governments to create policies and motivate behaviors as indicated by the international agreements such as the Paris Agreement and SDG 15 initiatives. These notwithstanding, they can also support green technology research, and promote environmentally advantageous behaviors. The business sectors also have significantly great influence on the environment and can bring about long-term transformation. Businesses may also bring on board their support for the achievement of Sustainable Development Goals (SDGs) by getting involved with collaborative initiatives like the United Nations global compact and public-private partnership that promote green technologies and sustainable practices. Active community participation and nongovernmental organizations (NGO) have a vital part to play in protecting the environment, and engaging local communities in conservation efforts is a sure way to go. Conservation initiatives that are community based promote accountability and ownership which aligns with SDG 15. Educational institutions and research ventures help in advancing our understanding of environmental challenges while providing novel approaches to solving these challenges. These partnerships between corporations, governments, and nongovernmental organizations allow for interdisciplinary research and innovative recommendations based on facts.

We need international collaborations in addressing environmental challenges that go beyond national boundaries. International organizations like the World Bank, IUCN, and UNEP make it possible for large scale conservation projects, policy harmonization, and information exchange. Not forgetting the important role of technological development that helps in reducing the negative environmental effects and optimizing resource efficiency, particularly in areas like waste management, renewable energy, and sustainable agriculture. Using analytical and digital tools also support environmental monitoring and decision making. Measures focused on advocacy and public awareness are necessary for motivating group actions and environmental responsibility. NGOs, grassroots organizations, and environmental activists are all important due to the essential roles they play in creating awareness of the important issues and advancing sustainable behavior. They promote buying of sustainable products, reducing waste, and conserving energy. With all stakeholders working together, we can bring together the resources, and the technical know-how of many stakeholders to establish and put in place practical and infeasible strategies for an adaptable and sustainable future, guaranteeing the survival of our planet for the next generations.

3 Modern Approaches to Protecting Biodiversity: Advancements in Technology

Biodiversity is important in providing functioning services to ecosystems, human welfare, and the earth's resistance to environmental challenges. However, human actions like resource abuse, pollution, habitat degradation, and climate change threaten biodiversity. To achieve successful conservation and address these challenges, innovative approaches that combine political, technology, and scientific inquiry with community engagement are all needed to help address these challenges. Technological breakthroughs present an unprecedented prospect for improved environmental monitoring, management, and preservation of biodiversity. Technologies that promote environmental conservation, strengthen land management, and enhance sustainable development include artificial intelligence (AI), data analytics, satellite photography, and remote sensing. Integrating technological initiatives with environmental policies promotes real-time monitoring, early environmental threat identification, and effective resource management. For instance, drones and satellite technologies enable precise mapping and tracking of land use changes, forest cover, and wildlife populations. Machine learning and artificial intelligence algorithms can analyze vast data sets to predict deforestation patterns, detect illegal logging, and assess ecosystem health. Technology also empowers communities by providing accessible platforms for collaboration, education, and data sharing. Online resources and mobile applications help local communities adopt sustainable practices, report environmental issues, and engage in conservation efforts. Leveraging technological advancements is crucial for achieving SDG 15. Technology enhances our capacity to protect and revitalize land-based environments, ensuring a sustainable future for all by utilizing the following:

- Remote sensing and GIS have become indispensable for environmental scientists, offering extensive data on ecosystem health, species distribution, and habitat changes.
- Biotelemetry and GPS tracking facilitate the study of animal behavior and movements, aiding in environmental initiatives and the establishment of protected areas.
- Environmental DNA (eDNA) analysis and DNA barcoding allow species identification and monitoring even in challenging environments, enhancing ecosystem understanding.
- Technologies like ML and AI have transformed environmental science by enabling early warning systems, species identification, and habitat suitability modeling.

Innovations in Policy and Governance Efficacious governance and policy innovations are crucial for addressing the complexities of conserving and managing land. To create an environment that is supportive to the expansion of sustainable development, it is important to develop a robust and efficient governance and policy frameworks. These may include the establishment of all-encompassing legal frameworks, implementing an integrated land use strategy, encouraging the adoption of participatory governance model, and the establishment of international cooperation frameworks. The enforcement of sustainable land use and environmental preservation necessitate a strong policy framework. These guarantees clear, legally binding, and internationally accepted regulations. Laws that offer incentives (rewards and punishment) for sustainable activities like fines for illicit logging or grants for sustainable farming can spark positive change at the local levels.

Integrated land-use planning is another important strategy, taking into accounts the economic, social, and environmental factors when planning for and deciding on land use. This strategy ensures the long-term sustainable use of our lands.

When trying to speed up ownership and promote accountability, collaborative government models are needed to achieve this objective. This may include the incorporation of local communities, indigenous population, and other stakeholders into the decision-making process. These procedures may lead to a more stronger and effective land management system.

Again, to guarantee that policies are adequately backed and tailored in the local community, it may be viable to effectively exploit local knowledge by considering collaborative governance framework and community-based management systems.

Many challenges connected to SDG 15, including biodiversity loss and climate change, have international influence. Because of this, solving these challenges through international collaboration and successful cooperation will be essential. These corporations between countries, NGOs, and international organizations will improve resource mobilization, information sharing, and promotion of best practices.

Groundbreaking approaches to governance and policy making are important to achieve SDG 15. The conservation and sustainable management of our terrestrial resources may be ensured by developing environment for sustainable management and through robust policies and inclusive governance, which will result in a more resilient and helpful environmental for the next generation.

3.1 The Importance of Payment for Ecosystem Services (PES) in Achieving SDG 15

To achieve Sustainable Development Goal 15 (SDG 15), which aims to protect, restore, and promote sustainable use of terrestrial ecosystems, efforts to pay for ecosystem services (PES) must be performed. PES programs pay communities and landowners to manage their lands in ways that are friendly to biodiversity and ecosystem services. These initiatives are critical for protecting and restoring habitats that are critical to the survival of many species. Landowners who participate in activities such as wetland restoration, tree planting, and protection of endangered species habitats must receive these payments (Ding et al., 2019; Gaglio et al., 2023). For example, PES-supported land management techniques assist riparian buffers and wetlands preservation to filter pollutants and control flow; thus, the same ecosystem intervention can be analyzed both in terms of improvements in ecosystem condition and ecosystem services, and it is also an intervention benefitting human populations in terms of improved water quality and quantity.

Studies reformulating PES programs show significant benefits for carbon sequestration through reforestation and forest conservation programs. Such measures are integral to maintaining the balance of the global climate system and mitigating the impacts of climate change. Such programs can provide landowners with financial benefits that serve as means to reduce rates of deforestation and combat land degradation—contributing directly to the achievement of Sustainable Development Goal 15 (SDG 15). By incorporating local communities into the decision-making process, these programs ensure that conservation approaches are grounded in Indigenous knowledge and respond to local needs, thus promoting community participation and ownership of conservation efforts. PES initiatives highlight the economic value of ecosystem services, encouraging policymakers to bring these aspects of land-use planning and policymaking into play. Such an approach not only helps in attaining

SDG 15 but also builds a more resilient and sustainable global ecosystem (Ansari et al., 2021; Thompson, 2021; Gaglio et al., 2023; Li et al., 2023).

Research and Innovation in Science—Aiming to Attain SDG 15 Sustainable Development Goal 15 is "Life on Land" which aims to ensure the conservation and sustainable use of terrestrial ecosystems. Major scientific and research improvements are also essential, as are a robust legislative framework and community involvement to reach this goal. Innovations in these fields promote biodiversity and ecosystem resiliency by equipping the people who need them with the skills and information to better understand, manage, and monitor natural resources. Improvements in technology for data collection and monitoring are revolutionizing our ability to track ecosystems. Using satellite imaging and remote sensing technologies can help in the real-time monitoring of deforestation, land degradation, and biodiversity loss across vast areas. These technologies are also very useful for collecting data that help inform policy and conservation decisions. Using drones and aerial surveys is a cost-effective method to collect data in hard-to-reach places, as they can create high-resolution pictures and video footage, monitor animal numbers, assess habitat conditions, and detect crimes, such as poaching and deforestation. Using Geographic Information Systems (GIS) technology allows for the integration of spatial data and analysis, mapping of ecosystems, pinpointing of hotspots for biodiversity, and land use arrangement with minimal environmental adverse effect (Moallemi et al., 2020; Adenle et al., 2023).

Genetic Research and Biotechnology in Conservation Efforts The genetic diversity of organisms and the role of biotechnology in conserving endangered species. Genome sequencing, CRISPR gene editing, and DNA barcoding are some technologies that can also be used to find solutions against animal diseases. A better understanding of the effects of habitat loss, climate change, and other stresses on these ecosystems is critical to predicting future events as well as designing adaptive management strategies. This is done using simulation and ecological modeling. Biodiversity informatics is the use of computing tools to manage and analyze biological data to enable researchers to track the distribution of species, monitor trends, and set conservation priorities (Booy et al., 2000; Corlett, 2017; Nonić & Šijačić-Nikolić, 2021).

Valuing ecosystem services allows us to measure the benefits of services like carbon storage, pollination, and water filtration. To create better decisions about land use and conservation investments, this knowledge needs to be used. Innovations that enhance soil health, increase biodiversity on farms, and help build resilience to climate change. This includes sustainable agricultural methods such as precision agriculture and agroecological practices. Rewilding and natural regeneration, and use of native species in restoration projects, both are restoration ecology approaches to mend habitats loss and achieving biodiversity and ecological functioning.

The integration of citizen science and community-based research is paramount for effective conservation strategies. Engaging local groups in research endeavors ensures the social and cultural acceptability of conservation measures. Participatory research methodologies serve to empower communities, incorporate traditional knowledge, and cultivate a sense of environmental stewardship. Citizen science programs play a crucial role in facilitating public involvement in data collection and monitoring, thereby expanding the scope of scientific inquiry, raising public awareness of environmental concerns, and contributing valuable data to conservation initiatives. The success of conservation efforts hinges on support from governance and policy frameworks. Addressing the disconnect between scientific findings and policy implementation requires advancements in science communication and policy-relevant research. International scientific networks and collaborations facilitate the exchange of best practices, resources, and expertise. Initiatives such as the Intergovernmental Science-Policy Platform on Biodiversity and Ecosystem Services (IPBES) provide critical assessments and guidance for global conservation endeavors (McKinley et al., 2017; Orchard, 2018).

The achievement of SDG 15 requires substantial reliance on scientific and research advancements in the realm of inference. Scientific discoveries provide a solid foundation for effective and sustainable land management by enhancing our comprehension of ecosystems, improving data collection and monitoring, and devising innovative conservation techniques. Incorporating these developments into policy and practice will ensure that future generations can access biodiversity and resilient ecosystems, while protecting and restoring terrestrial ecosystems.

3.2 Educating and Involving the General Public to Achieve SDG 15

Sustainable Development Goal 15 (SDG 15), which deals with the conservation and sustainable use of terrestrial and aquatic ecosystems, emphasizes the key role of public engagement and awareness-raising in achieving this goal. By fostering awareness, encouraging community involvement, and equipping individuals with the knowledge needed to make a difference, these publicly oriented initiatives play a significant role in promoting the responsible management and preservation of our planet's natural resources. By adopting educational programs that include environmental education as part of school curricula, from an early stage, a strong foundation of knowledge about ecosystems, biodiversity, and conservation can be formed. It had the potential for identifying and inspiring young men and women to appreciate nature and understand sustainable practices through education. Global, and local awareness campaigns in different types of media can be used to convey information about the role of SDG 15, consequences of forest destruction, habitat fragmentation, and species extinction around the world that show the public how they can help. Workshops and seminars for other stakeholders, such as community leaders, farmers, and business owners, to spread awareness about sustainable land use practices, conservation strategies, and the benefits of protecting biodiversity (Cottafava et al., 2019; Pietrzak, 2022; Fu et al., 2024).

Citizen science projects recruit the general population to conduct academic research and collect information, greatly extending the bounds of scientific research while teaching individual participants about their local ecosystem and biodiversity. Citizen science projects are, for example, with bird watching, wildlife monitoring, and plant identification, engaging communities directly in conservation efforts. Bringing residents along for conservation helps ensure strategies are culturally relevant and backed locally. Community-based conservation initiatives are reforestation and sustainable farming, along with declared closed areas where local residents protect their own environment. Promoting volunteerism with local and national organizations open doors for citizens to actively contribute to conservation initiatives, whether it be through planting trees, restoring habitats, or monitoring wildlife populations, thus gaining practical experience in environmental stewardship.

Education and training programs should also be associated with empowering the people, landowners, and resource managers to adopt sustainable practices. Investors need the knowledge and skills appropriate to manage natural resources sustainably—the effectiveness of conservation initiatives depends on it. This can be achieved through a well-designed capacity building program that involves and emphasizes various domains such as sustainable agriculture and production, silviculture, and biodiversity conservation. Enabling communities with information about best practices and stock of resources allows them to make well-informed decisions about conservation methods. Such provisions can include the dissemination of guidelines, resource kits, and information on funding avenues for conservation endeavors. Achieving SDG 15 will require synergetic cooperation between diverse actors, including governments, nongovernmental agents, academia, and the private sector (Mairset al. 2004) that can benefit the broader academic, design, and economic communities (Meinzen-Dick & Knox, 1999; Cocklin et al., 2007; Pasaribu et al., 2020).

Vital part of increasing demand for environmentally safe products and sustainable practices are educating the public about how their lifestyles impact the environment. By promoting the use of sustainable products, minimization of waste, and promotion of sustainable agriculture and forestry systems, society can make strides toward achieving SDG 15 by 2030. CSR practices focus on making environmentally sustainable practices within businesses. Doing so, they are likely to achieve land use solutions after an active phase with initiatives around reforestation, ethical sourcing, and reducing their ecological footprint.

Public awareness can influence policy making: An informed and engaged public is more likely to advocate for policies that promote sustainable land management and conservation. In addition, public pressure can push governments to adopt and implement protective measures on natural resources and biodiversity that can further the advancement of SDG15. Generally, the active engagement and awareness of the public are important factors for the successful implementation of SDG 15. This is vital as it raises awareness, motivates people to get involved in their communities, and educates people on the knowledge needed. Educating and engaging the public creates a basis of support for environmental policy and inspires action to protect and restore terrestrial ecosystems. Through collaboration and knowledge-based governance at every level, the sustainable stewardship and conservation of our planet's resources is possible, and research findings indicate that public engagement and education is an essential component in achieving this goal.

Public Engagement and Education—Media and Communication for Achieving Media and communication strategies are a vital part of public engagement and education, especially because of Sustainable Development Goal 15 (SDG 15), "Life on Land." The spread of this information across various media can play a vital role in generating public support and advocating for policy change to help protect biodiversity. Utilizing the power of communication and media platforms has the potential to engage an array of audiences, generate visibility for important environmental issues, and promote common action to practice land use and conservation that is sustainable (De Luca et al., 2022; Murschetz et al., 2023; Fu et al., 2024).

Increasing Awareness and Shifting Public Opinion The media has an important part to play in advancing SDG 15, particularly regarding growing public awareness about the importance of biodiversity and threats to it. When citizens are educated about the damage habitat loss does to species, species extinction, and deforestation through news articles, documentaries, and feature pieces in on-ground media outlets, such as radio, television, and newspapers, they are empowered to act. These platforms can also emphasize the need for the conservation of terrestrial ecosystems and the value of ecosystem services in human well-being. Additionally, social media is an equal opportunity outlet of information dissemination, enabling audience engagement and mobilization for conservation. Hashtags, viral campaigns, and influencer partnerships can raise heightened awareness regarding environmental issues and reach millions of individuals (Bain et al., 2019; Glass & Newig, 2019).

The Role of Advocacy and Policy in Environmental Conservation The importance of advocacy and policy in environmental conservation, along with the role of media and communication, is critical in advancing policy changes that are compatible with the aims of SDG 15. Through rigorous reporting and investigative journalism, the media has the ability to expose environmental injustices and throw light on unlawful actions like poaching and logging. It also serves an important function in highlighting the consequences of these unsustainable activities. By holding governments and companies responsible, the media may help to create and enforce laws that preserve the environment and promote sustainable land use. Furthermore, venues such as editorials, opinion articles, and public service announcements (PSAs) change public perception and motivate collective action, encouraging individuals to campaign for legislation that protects biodiversity and restores ecosystems. In this context, the media serves as a critical link between scientific knowledge, public awareness, and political participation (Mintrom & Thomas, 2018; Katila et al., 2019).

Advocacy and Policy in Environmental Conservation Why advocacy and policy are important in environmental conservation policy can and must advance in a direction compatible with the objectives of SDG 15. Additionally, through thorough reporting and investigative journalism, the media can bring attention to environmental injustices and illegal activities such as poaching and logging. It also has an important role in showing the impact of these unsustainable activities. This accountability may aid the media in establishing norms related to creating and enforcing laws that deter irresponsible behavior toward the environment and promote sustainable land use. Additionally, platforms like editorials, opinion pieces, and public service announcements (PSAs) shape public perception and inspire collective action, challenging people to mobilize for policies that safeguard biodiversity and regenerate ecosystems. The media has the role of linking scientific knowledge to people and programs in a politically participatory context (Cottafava et al., 2019; González-Salamanca et al., 2020).

Education and Skill Development Media platforms can support individuals and communities to acquire the knowledge and skills for active participation in conservation efforts. Podcasts, webinars, radios, or television shows—people can learn important lessons on sustainability, conservation methods, the ecological importance of many creatures and ecosystems, etc. Adapting what is taught to specific sets of people—whether they be rural, city-dwelling, youth, or people with similar experiences—guarantees that the education is suitable and coherent. Moreover, international media can promote success stories and good examples of conservation, inspiring local communities to implement similar actions and creating a shared commitment to stewardship of the environment (Cottafava et al., 2019; González-Salamanca et al., 2020)

Encouraging Environmental Conservation Through Community Participation and Public Engagement Media campaigns can encourage involvement and active participation of that community in environmental conservation efforts. This can be achieved by sharing them through media channels, like joining in efforts such as clean-up drives, animal monitoring programs, tree-planting events, and so on. Through the stories of community leaders and conservation heroes, the media can inspire others to commit to similar goals and foster a culture of environmental stewardship. Interactive platforms, such as blogs, social media, and online forums, allow for direct audience participation and enable dialogue, comment, and group problem-solving (Berry et al., 2019; Boucher, 2020; Ozaki & Shaw, 2022).

Media plays a critical role in the accomplishment of SDG 15. Through increasing public awareness, promoting legislative reform, providing public education, and offering training opportunities, the media can help with the sustainable use of land-based ecosystems and biological diversity. Effective communication techniques can increase the number of citizens who are both informed with scientific knowledge and engaged in public action. The media and communication will play an important role in this, as we continue to face major global environmental challenges to build a more resilient and sustainable future for all of us.

4 Enabling Sustainable Land Use Through Legislation and Advocacy

Sustaining Ecosystems and Natural Resources The adoption of sustainable land-use practices are essential to balancing ecosystem preservation and use of rightful natural resources with the needs of a rapidly growing global population. Effective advocacy and policy initiatives should be established for better sustainable lifestyles, biodiversity protection, and maintenance of the land's long-term health and production to meet these aims. There are several key frameworks that are useful to create sustainable land-use policies. Involving affected citizens is also imperative to ensure that land use planning is integrated, so land management decisions facilitate sustainable land management (SLM), prevent land degradation, and mitigate conflicts, whiles, which are required for the promotion of sustainable development. Collaborative approaches are needed to overcome the political and economic barriers of SLM—at the level of the institution, information, finance, and law—and the global environment facilitating itself is considered one of the key factors that needs to be overcome to enable SLM (Ladan, 2018; Martens, 2020).

Promoting Sustainable Land Management Sustainable land management (SLM) includes practices to enhance biodiversity, soil fertility, water conservation, and lower the use of chemicals. The Food and Agriculture Organization (FAO) reiterates the growing importance of SLM as a preventive measure against soil degradation in support of responsible and efficient land use. Sustainable forest management practices seek to achieve a delicate balance between the utilization of forest resources and the conservation of vital ecosystem functions for forest conservation and management. Through its support to member states, the FAO has an important role to play in promoting SLM policies and practices including sustainable forest management to improve land use wisely. And climate change adaptation and mitigation is still urgent. An efficient model of renewable energy initiation and implementation in Malaysia can increase human well-being and environmental sustainability through renewable energy expansion and robust pollution management (Katila et al., 2019; Husain et al., 2024). To sustain the needs of a growing world population and to balance ecosystems and natural resources, we must all implement sustainable land use laws and practices. Building on these guiding principles—no surer ways to be better neighbors and stewards of biodiversity—it is a solid foundation to promote ecologically smart behaviors and make sure the land is healthy and productive for the long term.

Data Must Be Understood by the General Public General awareness allows the public to understand the importance of sustainable land use. These efforts promote personal responsibility and laws regulation. Sustainable Development Goals, such as those encouraged by organizations like The Nature Conservancy, can only be effectively pursued through global cooperation and local community engagement. Engaging a diverse group of stakeholders ensures that land use decisions consider multiple perspectives. A key strategy for these reforms must be to address sustainable land management (SLM) and to create decision-making procedures. Research and lobbying efforts that are evidence-based provide policymakers with confidence in scientific knowledge and enable them to make better decisions (Katila et al., 2019; Giuliani et al., 2020) to achieve this end, and has played a significant role in collaborating with other organizations to initiate and commit to sustainable development practices. As the FAO's unbiased perspective may help to tackle the issues that contribute to land degradation, it would likely be adopted by institutions and governments. Complex environmental challenges need a robust legal framework and commitment to environmentally sustainable land-use practices. Sustainable forest development means maintaining the ability of the forest ecosystem to regenerate and produce goods without damaging them (FAO 2023). This highlights the importance of our national and comprehensive land management strategy.

Local Community for Ecological Balance Empowering local communities is the key to achieve ecological balance and sustainable development. Local communities often have the most knowledge and the most incentive to preserve their environment, especially those living near and dependent on natural resources. Engaging these communities in conservation initiatives harnesses their distinct perspectives and creates a feeling of ownership and accountability. Well-being researchers support the idea that empowered communities can effectively and equitably manage the environment as well as ensuring that conservation strategies can be shaped by local, social, and cultural mores.

Significance of Local Knowledge Indigenous and local communities have a profound understanding of the ecosystems they draw upon over generations of interaction. The traditional ecological knowledge (TEK)—composed of ideas, innovations, and practices—they have acquired has resulted in the sustainable use of resources. Combining TEK with scientific research could enhance conservation efforts by offering integrative and flexible environmental management solutions. For example, traditional Indigenous techniques for fire management can reduce the likelihood of large-scale wildfires for consumption, while agricultural systems like crop rotation and polyculture can improve biodiversity and soil health.

The Benefits of Joint Conservation The involvement of affected citizens is imperative for ensuring the integration of land use planning. Such an approach makes conservation efforts meet local needs and objectives, thus increasing acceptance and effectiveness. Community managed protected areas are one such model where locals operate and conserve their own natural resources and have shown better conservation outcomes. Participatory mapping, for example, allows communities to record and maintain their spatial knowledge, allowing them to identify biologically and culturally important areas for protection.

The Function of Monetary Rewards Economic incentives play a vital role in encouraging sustainable practices and alleviating stress on natural resources. Diversified livelihood options, including eco-friendly agriculture, ecotourism, and the cultivation of non-timber forest products, offer communities viable alternatives to practices that harm the environment. Initiatives such as payment for ecosystem services (PES) offer monetary rewards to individuals and groups who actively preserve and improve ecosystem functions including clean water, carbon sequestration, and biodiversity. These incentives advance conservation goals while improving the economic well-being of local populations.

Capacity Development and Education Educational and capacity-building initiatives are crucial for equipping communities to get the right information and the skills needed for sustainable resource management. Environmental education programs instill a conservation ethic by emphasizing the importance of ecological services and biological diversity. Training in sustainable agriculture, wildlife monitoring, and eco-friendly business practices enhances local communities' ability to manage their natural resources. Strengthening regional governance structures and organizations is also vital for contributing to community-led conservation activities.

Legal Frameworks and Policy Support Effective natural resource conservation needs legislation and regulations that empower communities and recognize their rights to resource management. Establishing communal land rights and ensuring land tenure can foster stability and sustainable land use. Policies that delegate management responsibilities to local groups and provide financial and technical support are more likely to succeed. Involving communities in policymaking ensures that local opinions are recognized and respected, resulting in improved efficiency and integrated conservation efforts.

5 Case Studies of Successful SDG 15 Implementation from Across the Globe

Several case studies and success stories worldwide highlight successful implementations of SDG 15, "Life on Land," demonstrating how communities, governments, and organizations have tackled challenges in maintaining and restoring terrestrial ecosystems.

- ***Costa Rica Initiative Forest Management:*** Globally, Costa Rica is noted for its proper and successful management of forest ecosystems. Through the provision of funding, community agency, and government policy, the country has reversed deforestation trajectories and significantly expanded forest cover by preserving and reforesting land through the use of PES. Without further direct intervention, the program began paying it to the landowner of Taylor's land to preserve and reforest it—benefiting biodiversity and pulling carbon further out of the atmosphere. Costa Rica established national parks and protected regions to safeguard important habitats for many animal species. Emphasis on ecotourism has produced funds for conservation and increased public knowledge on the importance of biodiversity (Melo et al., 2013; Mansourian et al., 2022; Morán-Ordóñez et al., 2022).
- ***Indian's Greening Initiative:*** The Green Initiative of India helps to preserve, enhance, and restore its ecosystems specifically aimed at protecting the forests in the nation. By engaging local communities in the planning and imple-

mentation of restoration projects to adapt and mitigate changes in the environment, this program makes long-term changes that can be maintained. The program was also successful in providing alternative livelihoods for people relying on forest resources, thereby minimizing their environmental footprint (Vernekar & Venkatasubramanian, 2013; Kaur, 2017; Kapoor, 2020; Singh, 2023).

- ***Gishwati-Mukura National Park in Rwanda:*** Gishwati-Mukura National Park in Rwanda was once heavily degraded by forestry and human activities, but several significant rehabilitation campaigns have been launched to turn the tide. The activities restored often include replanting and reintroducing native species, and include the important requirement of having local communities involved in the planning and implementation of these activities, thus enabling sustainable management and long-term success of the restoration process. The park is now a habitat for the endangered Eastern Chimpanzee and Golden Monkey (Hagumubuzima et al., 2022; Tuyisingize et al., 2022).
- ***Ethiopia's Green Legacy Initiative:*** Ethiopia's Green Legacy Initiative seeks to combat desertification, restore degraded lands, and reduce climate change. The initiative has led to the planting of millions of trees across the country, improving water supplies, conserving soil, and boosting biodiversity. It successfully engages government, public, and private organizations in environmental protection and tree planting. The initiative further encourages sustainable agronomic practices that minimize land degradation (Demisse et al., n.d.; Beyene & Shumetie, 2023).

Successful case studies in the real world and successful initiatives offer insight into successful strategies for achieving SDG 15. That is how we can conserve and restore terrestrial ecosystems and devise a better world. In this context, enhancing the capacity of local communities have helped to foster sustainable development and ecology. This can mainly be achieved using indigenous knowledge, participatory conservation, economic incentives, education, capacity building, and policy. Empowered communities have an incentive to better conserve their natural resources, protecting biodiversity and increasing ecosystem resilience. The need for empowered communities to continue their role in maintaining ecological balance demands increasing urgency as environmental challenges become more worrying. Together, by encouraging a sense of collective responsibility, we can build a stronger and more sustainable planet for all future generations. As pointed out in this section, "SDG 15: Life on Land" is a crucial component for sustainable development. Through the conservation and restoration of terrestrial ecosystems, the facilitation of sustainable land use practices, and the enabling of communities to mobilize natural habitats, a myriad of coexisting ecosystems can be supported. The following chapter will provide a detailed analysis of actions, challenges, and opportunities for SDG 15 and so, provide a pathway to where terrestrial living systems thrive in a supportive role.

6 Conclusion: A Prospect for *Terra Thrive*

Sustainable Development Goal 15 (SDG 15) serves as an important concept for the sustenance and achieving the balance of ecological systems on earth. This is what is discussed in this initial chapter of the book *Terra Thrive*. Human wellbeing is connected closely to the rich biodiversity and ecosystem health thereby making their preservation important for the variety of life on earth. This is because they provide vital services through cultural heritage, livelihoods, and the supply of food and water making it important to the lives of billions of people internationally. *Terra Thrive* has extensive resources and an urgent call to action. The intricate complexities in the relationships between humans, cultures, and their environment shape the way societies interact with nature. A review of Indian terrain was done, its surrounding areas, and its collaboration with Japan. This book presents a comprehensive review of the opportunities and potential threats that confront terrestrial ecosystems and to provide suggestions to how important it is to embrace sustainable behavior that promotes a peaceful coexistence between nature and humans.

This book discusses the several important environmental issues, but attention is specifically focused on the need to reduce deforestation, preserve biodiversity varieties, and the implementation of robust sustainable management initiatives. The urgency of these challenges requires quick action; these are not mere theoretical discussions, but they guarantee a strong and sustainable future, and the land-based ecosystem is sure to be protected and restored. This book exposes its readers to real-life examples, evidence-based successful projects, and collaborative projects implementation through the journey of this *Terra Thrive* showing the benefits of partnership. The use of various resources, including modern conservation initiatives, thorough scientific investigations, and a well-established ecological model. Readers of this book are provided with essential knowledge and understanding needed to make informed decisions.

The book showcases how the involvement of indigenous people and the active participation of stakeholders in grassroots projects and advocacy for policy implementation can enhance environmental conservation. The book encourages us to acknowledge the effects of the decisions that we make now which have influence on the environment in future, and how we are able to deal with the opportunities and challenges skillfully and successfully in the future is very important. The roles and duties of governments, communities, people, and organizations to actively collaborate and cooperate with each other are also emphasized to protect terrestrial ecosystems. We are encouraged and entreated to make thoughtful and calculated initiatives to help achieve Sustainable Development Goal 15, and this would lay the ground for a future where biodiversity and land-based ecosystem will thrive.

In this book we are encouraged to acknowledge the special role we play as humans in the conservation and preservation of terrestrial ecosystems across the globe. By doing this we can ensure that future generations will be able to inherit a world where rich biodiversity, humans, and cultures interact harmoniously with nature. We are motivated as people to take informed decisions and kindle a passion for environmental stewardship that goes beyond age and location. This book is meant to serve as a catalyst for change.

References

Adenle, A. A., et al. (2023). Global UN 2030 agenda: How can science, technology and innovation accelerate the achievement of sustainable development goals for all? *PLOS Sustainability and Transformation, 2*(10), e0000085.

Alles, D. L. (2012). Biodiversity hot spots: The Florida Everglades., Citeseer.

Ansari, N. A., et al. (2021). Plausible solutions with reference to 'SDG15–Life on Land' towards an effective biodiversity management. In *SDG15–Life on Land: Towards effective biodiversity management* (pp. 79–100). Emerald Publishing Limited.

Antonelli, A., et al. (2020). *State of the world's plants and fungi, Royal Botanic Gardens (Kew).* Sfumato Foundation.

Bain, P. G., et al. (2019). Public views of the sustainable development goals across countries. *Nature Sustainability, 2*(9), 819–825.

Barraclough, S., & Ghimire, K. (1995). *Forests and livelihoods: The social dynamics of deforestation in developing countries*. Springer.

Berry, L. H., et al. (2019). Making space: How public participation shapes environmental decision-making., JSTOR.

Beyene, A. D., & Shumetie, A. (2023). *Green legacy initiative for sustainable economic development in Ethiopia.* Ethiopian Economic Association (EEA).

Booy, G., et al. (2000). Genetic diversity and the survival of populations. *Plant Biology, 2*(04), 379–395.

Boucher, C. M. (2020). Amplifying citizen engagement in the localization of the 2030 agenda in cities: A framework to promote participation., CITDH.

Boufounou, P. and K. Dellis (2021). Financialisation, sustainable finance and innovative financing for development mechanisms: The Greek case study for crowdfunding and green bonds. International Conference on Business and Economics-Hellenic Open University.

Cocklin, C., et al. (2007). Public policy, private landholders: Perspectives on policy mechanisms for sustainable land management. *Journal of Environmental Management, 85*(4), 986–998.

Corlett, R. T. (2017). A bigger toolbox: Biotechnology in biodiversity conservation. *Trends in Biotechnology, 35*(1), 55–65.

Cottafava, D., et al. (2019). Education of sustainable development goals through students' active engagement: A transformative learning experience. *Sustainability Accounting, Management and Policy Journal, 10*(3), 521–544.

Dagnachew, M., et al. (2024). Ensuring sustainable water security through sustainable land management: Research evidences for policy. *World Water Policy, 10*.

De Luca, F., et al. (2022). Can social media improve stakeholder engagement and communication of Sustainable Development Goals? A cross-country analysis. *Technological Forecasting and Social Change, 177*, 121525.

Demisse, B. A., et al. (n.d.). The impact of Ethiopian Green Legacy Initiative (GLI) on landscape functionality and plant species diversity in Lake Hawassa watershed, Ethiopia. *Ecohydrology*, e2676.

Ding, X., et al. (2019). From environmental soundness to sustainable development: Improving applicability of payment for ecosystem services scheme for diverting regional sustainability transition in developing countries. *Sustainability, 11*(2), 361.

Duguma, L. A., et al. (2019). Deforestation and forest degradation as an environmental behavior: Unpacking realities shaping community actions. *Land, 8*(2), 26.

Emerson, A. (2024). The role of ecosystem restoration for a world in chaos—Barriers and opportunities to optimize human wellbeing. *Restoration Ecology, 32*(1), e14062.
Fu, H., et al. (2024). Bridging gaps towards the 2030 agenda: A data-driven comparative analysis of government and public engagement in China towards achieving sustainable development goals. *Land, 13*(6), 818.
Gaglio, M., et al. (2023). Integrating payment for ecosystem services in protected areas governance: The case of the Po Delta Park. *Ecosystem Services, 60*, 101516.
Gillis, L. G., et al. (2017). Opportunities for protecting and restoring tropical coastal ecosystems by utilizing a physical connectivity approach. *Frontiers in Marine Science, 4*, 374.
Giuliani, G., et al. (2020). Knowledge generation using satellite earth observations to support sustainable development goals (SDG): A use case on land degradation. *International Journal of Applied Earth Observation and Geoinformation, 88*, 102068.
Glass, L.-M., & Newig, J. (2019). Governance for achieving the sustainable development goals: How important are participation, policy coherence, reflexivity, adaptation and democratic institutions? *Earth System Governance, 2*, 100031.
González-Salamanca, J. C., et al. (2020). Key competences, education for sustainable development and strategies for the development of 21st century skills. A systematic literature review. *Sustainability, 12*(24), 10366.
Griswold, W., et al. (2024). 'One person cannot change it; it's going to take a community': Addressing inequity through community environmental education. *Adult Learning, 35*(1), 23–33.
Hagumubuzima, F., et al. (2022). Impact of removing Eucalyptus maidenii in regenerated native species in Gishwati-Mukura National Park, Rwanda. *Journal of Research in Forestry, Wildlife and Environment, 14*(3), 65–76.
Han, N., et al. (2024). Influence of human activity intensity on habitat quality in Hainan Tropical Rainforest National Park, China. *Chinese Geographical Science, 34*(3), 519–532.
Harari, N., et al. (2023). *Promoting sustainable land management through evidence-based decision support: A guide with country insights*. Food & Agriculture Org.
Haregeweyn, N., et al. (2023). Progress and challenges in sustainable land management initiatives: A global review. *Science of the Total Environment, 858*, 160027.
Husain, H. J., et al. (2024). Review and assessment of the potential restoration of ecosystem services through the implementation of the biodiversity management plans for SDG-15 localization. *Heliyon*.
Hyman, J. B., & Leibowitz, S. G. (2000). A general framework for prioritizing land units for ecological protection and restoration. *Environmental Management, 25*(1), 23.
Jurjonas, M., et al. (2024). The perceived ecological and human well-being benefits of ecosystem restoration. *People and Nature, 6*(1), 4–19.
Kapoor, N. (2020). The role of green marketing initiatives in achieving sustainable development in Indian economy. *PRAGATI: Journal of Indian Economy, 7*(2), 105–127.
Karr, J. (2012). *Attaining a sustainable society* (pp. 47–64). Routledge.
Katila, P., et al. (2019). *Sustainable development goals.* Cambridge University Press.
Kaur, S. (2017). Sustaining environment with green initiative: A conceptual study of green marketing practices in India. *Asian Journal of Management, 8*(3), 761–768.
Kopittke, P. M., et al. (2024). Healthy soil for healthy humans and a healthy planet. *Critical Reviews in Environmental Science and Technology, 54*(3), 210–221.
Ladan, M. T. (2018). Achieving sustainable development goals through effective domestic laws and policies on environment and climate change. *Environmental Policy and Law, 48*, 42.
Li, F., et al. (2023). A PES framework coupling socioeconomic and ecosystem dynamics from a sustainable development perspective. *Journal of Environmental Management, 329*, 117043.
Li, C., et al. (2024a). The impact of climate change on corporate ESG performance: The role of resource misallocation in enterprises. *Journal of Cleaner Production, 445*, 141263.
Li, S., et al. (2024b). Simulating land use change for sustainable land management in China's coal resource-based cities under different scenarios. *Science of the Total Environment, 916*, 170126.

Loucks, D. P. (2000). Sustainable water resources management. *Water International, 25*(1), 3–10.
Louman, B., et al. (2020). Innovative finance for sustainable landscapes., CIFOR.
Mansourian, S., et al. (2022). Governments commit to forest restoration, but what does it take to restore forests? *Environmental Conservation, 49*(4), 206–214.
Martens, J. (2020). *The role of public and private actors and means in implementing the SDGs: Reclaiming the public policy space for sustainable development and human rights*. Springer International Publishing.
Masso, C. (2024). *Environmental health & biodiversity impact area platform narrative*.
McKinley, D. C., et al. (2017). Citizen science can improve conservation science, natural resource management, and environmental protection. *Biological Conservation, 208*, 15–28.
Meffe, G., et al. (2012). *Ecosystem management: Adaptive, community-based conservation*. Island Press.
Meinzen-Dick, R. and A. Knox (1999). Collective action, property rights, and devolution of natural resource management: A conceptual framework. Draft paper for workshop.
Melo, F. P., et al. (2013). Priority setting for scaling-up tropical forest restoration projects: Early lessons from the Atlantic Forest Restoration Pact. *Environmental Science & Policy, 33*, 395–404.
Memmott, J. (2009). Food webs: A ladder for picking strawberries or a practical tool for practical problems? *Philosophical Transactions of the Royal Society B: Biological Sciences, 364*(1524), 1693–1699.
Mintrom, M., & Thomas, M. (2018). Policy entrepreneurs and collaborative action: Pursuit of the sustainable development goals. *International Journal of Entrepreneurial Venturing, 10*(2), 153–171.
Moallemi, E. A., et al. (2020). Achieving the sustainable development goals requires transdisciplinary innovation at the local scale. *One Earth, 3*(3), 300–313.
Monaco, S. (2024). SDG 15. Protect, restore, and promote sustainable use of terrestrial ecosystems, sustainably manage forests, combat desertification, halt and reverse land degradation, and halt biodiversity loss. In *Identity, territories, and sustainability: Challenges and opportunities for achieving the UN sustainable development goals* (pp. 145–155). Emerald Publishing Limited.
Morán-Ordóñez, A., et al. (2022). Multi-objective forest restoration planning in Costa Rica: Balancing landscape connectivity and ecosystem service provisioning with sustainable development. *Journal of Environmental Management, 310*, 114717.
Mori, G. O., & Grover, E. C. (2024). *Research and Development of specialty crops for agroforestry Systems in the Midwest: The long haul for small farm sustainability* (pp. 247–303). Integrating Landscapes.
Mulder, M. B., & Coppolillo, P. (2005). *Conservation: Linking ecology, economics, and culture*. Princeton University Press.
Murschetz, P. C., et al. (2023). *Public service media and sustainability? A critical debate of the value of public service media in supporting the SDG18 (communication for all). SDG18 Communication for All, Volume 2: Regional Perspectives and Special Cases* (pp. 79–106). Springer.
Nelson, C., et al. (2024). *Standards of practice to guide ecosystem restoration: A contribution to the United Nations decade on ecosystem restoration 2021–2030*. Food & Agriculture Org.
Nickel, B. A., et al. (2020). Human presence and human footprint have non-equivalent effects on wildlife spatiotemporal habitat use. *Biological Conservation, 241*, 108383.
Nonić, M., & Šijačić-Nikolić, M. (2021). Genetic diversity: Sources, threats, and conservation. In *Life on land* (pp. 421–435).
Olagunju, T. E. (2015). Impacts of human-induced deforestation, forest degradation and fragmentation on food security. *New York Science Journal, 8*(1), 10.
Orchard, S. (2018). Growing citizen science for conservation to support diverse project objectives and the motivations of volunteers. *Pacific Conservation Biology, 25*(4), 342–344.
Ozaki, Y., & Shaw, R. (2022). Citizens' social participation to implement sustainable development goals (SDGs): A literature review. *Sustainability, 14*(21), 14471.
Palmer, D., et al. (2009). Towards improved land governance. *Food, 781*.

Pasaribu, S. I., et al. (2020). Challenges to implementing socially-sustainable community development in oil palm and forestry operations in Indonesia. *Land, 9*(3), 61.

Pietrzak, P. (2022). The involvement of public higher education institutions (HEIs) in Poland in the promotion of the sustainable development goals (SDGs) in the age of social media. *Information, 13*(10), 473.

Pla, M., et al. (2024). Mapping drivers of change for biodiversity risk assessment to target conservation actions: Human frequentation in protected areas. *Heliyon, 10*(3), e25312.

Rawat, A., et al. (2024). A review on climate change impacts, models, and its consequences on different sectors: A systematic approach. *Journal of Water and Climate Change, 15*(1), 104–126.

Riggio, J., et al. (2020). Global human influence maps reveal clear opportunities in conserving earth's remaining intact terrestrial ecosystems. *Global Change Biology, 26*(8), 4344–4356.

Ritchie, H., & Roser, M. (2024). *The world has lost one-third of its forest, but an end of deforestation is possible*. Our World in Data.

Robinson, N. A., & Reilly, K. (2024). Environmental regulation of real property. *Law Journal Press*.

Rockström, J., et al. (2024). The planetary commons: A new paradigm for safeguarding earth-regulating systems in the Anthropocene. *Proceedings of the National Academy of Sciences, 121*(5), e2301531121.

Saleh, M., et al. (2024). *Biodiversity for ecosystem services and sustainable development goals* (pp. 81–110). Elsevier.

Singh, A. (2023). Green economic policies, strategies & initiatives of India. *EAI Endorsed Transactions on Energy Web, 10*(1).

Singh, V. (2024). *Biodiversity conservation* (pp. 225–236). Springer.

Thompson, B. S. (2021). Corporate payments for ecosystem services in theory and practice: Links to economics, business, and sustainability. *Sustainability, 13*(15), 8307.

Tully, J. (2019). *Life sustains life 2: The ways of reengagement with the living earth* (pp. 181–204). Columbia University Press.

Tuyisingize, D., et al. (2022). Forest landscape restoration contributes to the conservation of primates in the Gishwati-Mukura landscape, Rwanda. *International Journal of Primatology, 43*(5), 867–884.

Vernekar, S., & Venkatasubramanian, K. (2013). Green initiatives in Indian industry: A silhouette. *Journal of Commerce and Management Thought, 4*(4), 822–836.

Wang, X., et al. (2024). More than two-fifths of the protected land in a global biodiversity hotspot in Southwest China is under intense human pressure. *Science of the Total Environment, 906*, 167283.

Whipple, S., & Moss, S. (2024). Leveraging virtual datasets to investigate the interplay of pollinators, protected areas, and SDG 15. *Sustainable Earth Reviews, 7*(1), 13.

Wiens, J. A. (2016). *Ecological challenges and conservation conundrums: Essays and reflections for a changing world.* John Wiley & Sons.

Yadav, S. P. S., et al. (2024). A comprehensive study of banmara (invasive alien plant species): Understanding the invasive potential and ecological consequences for biodiversity conservation, and management strategies. *Journal of Agriculture and Food Research, 101031.*

Yüksel, S., & Karaçuhalılar, A. (2024). Defining of some morphological characteristics in order to determine the infrastructure of conservation strategies of indigenous cattle for designing sustainable methods and conservation of biodiversity: Eastern Anatolian Red example reared under ex-situ conditions in Türkiye. *Tekirdağ Ziraat Fakültesi Dergisi, 21*(1), 233–245.

Exploring the Rich Tapestry of Land

Resham Lohani

1 Introduction

This chapter will focus on the diversities of the land and its ecological significance, and is shown in Fig. 1. The chapter explores how the tapestry is formed by exploring the past, right from its formation as a land mass to the origin of the ethnicity, the culture that was originally existent to the way it got transformed over the years. The model used to explain is like the pattern that gets formulated as a weave of fabric. Each thread contributes to the formation of the fabric that portrays the rich tapestry of land today. The landscape, biodiversity, the culture, and heritage from the warp thread forms the foundation for the richness of the life that exists on earth. The intervention of human life with these existing environments forms the weft thread that are the practices of the people living on earth, like agriculture, urban development and the infrastructure that creates the richness of the tapestry.

The chapter concentrates on the following subtopics for describing the present-day and how it evolved over the years. The chapter unfolds the richness of the culture and the biodiversity that contributed to the diverse landscape of our planet.

R. Lohani (✉)
Christ University, Bangalore, India
e-mail: Resham.lohani@christuniversity.in

J. J. Puthenkalam S.J et al. (eds.), *Terra Thrive*,
https://doi.org/10.1007/978-3-032-02738-2_2

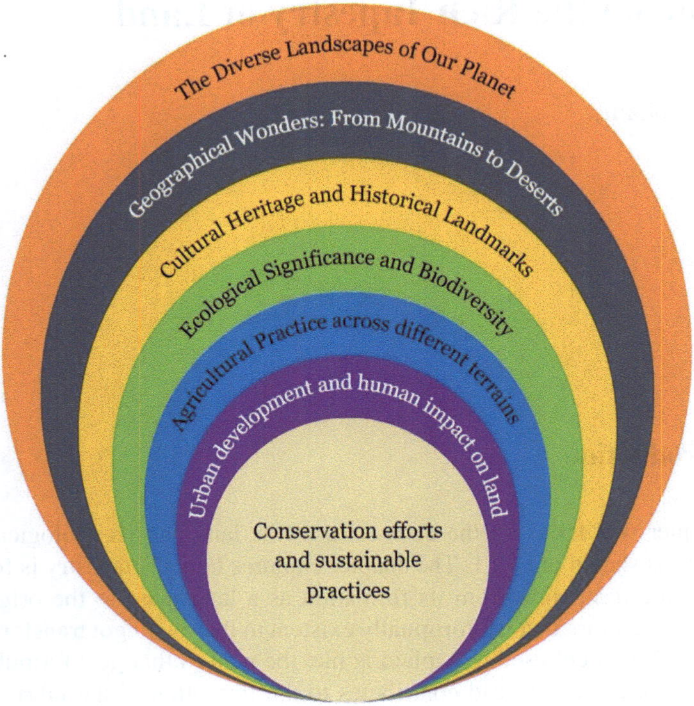

Fig. 1 The Rich Tapestry of Land

2 The Diverse Landscapes of Our Planet

Earth being the third planet from the sun and among the nine planets in the solar system is the fifth largest planet. The solar system in which the earth finds its place is one among the numerous milky way present in the galaxy. The cosmic beginning of the earth started with the aggregation process of the matter that eventually became Earth.

In his book *The Story of Earth: The First 4.5 Billion Years, from Stardust to Living Planet*, Robert M. Hazen described the Earth's evolution, from its origins in interstellar dust to the complex, life-sustaining planet we inhabit today. The earth has the layers that are distinct from each other such that we have the core, mantle, and the crust. The birth of life on earth happened eventually with the emergence of the first living organism and thereafter altering the planet's chemistry. The next major event on earth was the Oxygenation Event wherein the process of photosynthesis gave way to complex life forms. This led to the evolution of multicellular life from plants to animals. The biological and the geological process have shaped each other. The time spent by humans on earth has altered the way earth has been to what it exists as of today.

3 Geographical Wonders: From Mountains to Deserts

Earth: Portrait of a Planet by Stephen Marshak is a comprehensive textbook that covers a wide range of topics in Earth science. This book gives a wide range of information on the theory on plate tectonics that shapes the planet earth and helped in the formation of mountains, volcanos, and earthquakes helping form the various continents. The clarity about the interaction between atmosphere, hydrosphere, biosphere, and geosphere helps to establish the connection among the different earth systems. The journey of the earth's formation with oceans, life, and the mass extinction gave way to the natural resources and the human approach toward the management of these. Also, the movement of the ocean circulation led to the weather pattern and the climate systems. An illustration of a barren desert with little flora and distant mountains under a clear blue sky conveys a sense of immensity and isolation.

4 Cultural Heritage and Historical Landmarks

UNESCO defines cultural heritage as "the legacy of physical artefacts and intangible attributes of a group or society that are inherited from past generations, maintained in the present and bestowed for the benefit of future generations"[1]. The General Conference of the United Nations Educational, Scientific and Cultural Organization meeting in Paris from 17 October to 21 November 1972, at its 17th session defined the cultural and natural heritage[1].

The "natural heritage" site includes the natural features, geological and physiographical formations, and natural sites that have universal value from the scientific standpoint or conservation or natural beauty.

The UNESCO in its endeavor to protect and preserve the richness of the world heritage and culture, and as of December 2024 has 1223 World Heritage sites, 730 intangible cultural heritage elements, and 350 creative cities listed on its official page https://www.unesco.org/en/culture.

Protecting and safeguarding the world's cultural and natural heritage calls for undertaking measures for the sustainable development around the world. One of the major steps in the history is the adoption of policy for the integration of a sustainable development perspective into the processes of the World Heritage Convention. As stated by the USESCO "the overall goal of the policy is to assist States Parties, practitioners, institutions, communities, and networks, through appropriate guidance, to harness the potential of World Heritage properties and heritage in general, to contribute to sustainable development and therefore increase the effectiveness and relevance of the Convention whilst respecting its primary

[1] UNESCO. General Conference; 17th; Records of the General Conference, 17th session, Paris, 17 October to 21 November. https://web.archive.org/web/20120620014519if_/ https://unesdoc.unesco.org:80/images/0011/001140/114044e.pdf

purpose and mandate of protecting the Outstanding Universal value of World Heritage properties".[2]

4.1 Ecological Significance and Biodiversity

Biodiversity and ecological significance are essential to preserving the stability and health of our world. Ecology is the study of how living things interact with their surroundings. It helps in our understanding of how ecosystems work and how various species complement to maintain natural equilibrium. The term "ecological significance" describes how crucial interdependence helps to maintain life on Earth. Earth has a wide variety of plants, animals, parasites, micro-organism and that makes the planet what it is uniquely known for. This variety of life on Earth in its diversity is referred to as biodiversity, or biological diversity.

The diversity not only is found in its species of plants and animals but also the human species. Each geographical area has its own unique breed of animals and the way humans interact with the surrounding has a lot of variances. The practices are found to be different from country to country and continent to continent. The practices are found to be different from country to country and continent to continent, where the basic difference is found in the habitat and the agricultural practices. Let's take a deep dive into it.

5 Agricultural Practices Across Different Terrains

Agriculture practices are a reflection of the relationship between historic practices, beliefs, terrain, and the weather conditions of the geographical location. Over the years few practices have changed mankind's adaptability toward technology, embracing developments and the needs of the society. Various terrains present distinct challenges and opportunities, shaping agricultural practices in ways that sustain livelihoods and ecosystems. This section explores how different terrains influence agricultural practices, emphasizing their implications for productivity and sustainability.

5.1 Plains and Flatlands: The Breadbaskets of the World

The world has a varied level of topography, but the one suitable for large scale agriculture is plains and flatlands. The soil is generally fertile and the accessibility is not difficult to the farmland making it more congenial to agriculture giving it the name for being the bread basket of the world. The crops mostly grown for larger

[2] World Heritage Convention. https://whc.unesco.org/en/sustainabledevelopment/

consumption are wheat, rice, corn, and soybeans. These terrains are also suitable for animal husbandry and cattle rearing.

For example, the Indo-Gangetic Plains in South Asia is known for its production of rice and wheat, due to the abundant alluvial soils deposited at the banks of the rivers Ganga, Kosi, Godavari, and Krishna. (Pingali, 2012). Similarly, powered by advanced machinery and irrigation technologies, the Great Plains in North America withstand vast monoculture systems, meaning one crop farming instead of multiple crops. However, these practices are not without challenges leading to depleting soil nutrients and erosion, making it necessary to follow practices like crop rotation, conservation tillage, and farming of cover crops (Tilman et al., 2002). Cover crops are mainly grown to suppress weeds and protect the soil from erosion, pests, or plant diseases and enhance the fertility of the soil.

Historically farming practices used organic manure to replenish the soil for the nutrients and for healthy produce. But in the past few decades it has been found that the use of technology has surpassed organic manure and ensured high yield to fulfil the growing requirement of the population. For every stage of farming there is a precise technological intervention that helps in minimizing efforts and enhancing production

5.2 *Hilly and Mountainous Terrains: Adaptation Through Innovation*

Agriculture in hilly and mountain terrain is difficult for farming due to steep slopes, limited availability of water, thin soil, and soil erosion due to heavy rainfall. A sample land structure is shown in Fig. 2. Terrace farming helps to reduce water runoff by carving stepped plots of land for farming. The countries popularly known for their terrace farming are Philippines, Southeast Asia, North-Eastern parts of India, Andes, and countries around the Mediterranean Sea. These farmers have overcome the challenging environments to harness sustained agriculture (Pretty, 1979).

Fig. 2 Courtesy Resham Lohani

Another popular practice in the mountainous regions is agroforestry, which combines trees with crops and livestock to create multifunctional landscapes. In areas like Nepal and Ethiopia, agroforestry systems have helped in biodiversity, stabilized soil, providing a steady income from diverse products that includes timber, fruits, and fodder (Nair et al., 2021).

Even with all these advancements, farming in mountainous regions is increasingly vulnerable to the effects of climate change. Unpredictable rainfall, shrinking glaciers, and deteriorating soil health are making traditional agricultural methods harder to sustain. To tackle these challenges, it is essential to blend modern technology with the wisdom of indigenous practices. Efforts like promoting organic farming and improving watershed management, especially in the Himalayan regions, are proving to be effective in maintaining productivity and ensuring long-term sustainability.

5.3 *Arid and Semi-Arid Regions: Farming in the Face of Scarcity*

Arid and semi-arid regions generally have scarcity of water, making agriculture challenging. These areas widely use dryland farming techniques, designed to conserve moisture, and enhance soil fertility. Farmers usually perform mulching, contour ploughing, and planting drought-resistant crop varieties to help communities thrive under harsh conditions.

Farmers in the Sahel region of Africa have used innovative techniques like Zai pits and farmer-managed natural regeneration (FMNR). The farmers dig small holes called Zai pits to trap water and nutrients. FMNR encourages the regrowth of native trees and shrubs, improving soil structure and providing shade for crops (Reij et al., 2009).

Drip irrigation systems is a technique to deliver water directly to the roots of plants in areas with water scarcity. This technology has revolutionized agriculture in arid regions like Israel and parts of India. Similarly, satellite-based systems have been employed to monitor soil moisture and optimize irrigation schedules, improving water use efficiency (Jägermeyr et al., 2017).

5.4 *Coastal and Deltaic Regions: Harnessing Abundance Amid Risks*

Coastal and delta regions, with their rich soils and steady water supply, have long been a lifeline for farmers. Rice paddies stretch across these landscapes, thriving in the naturally flooded fields. Take Vietnam's Mekong Delta, for instance—it is not just a hub for rice cultivation, it is the backbone of millions of livelihoods, shaping communities and feeding people across the world (Van Kien et al., 2020).

These coastal and delta regions, once thriving agricultural hubs, now face mounting challenges due to climate change—rising sea levels, saltwater creeping into farmlands, and frequent floods disrupting traditional farming. To stay ahead, farmers are embracing innovative solutions.

In the Mekong Delta, rice-fish farming has emerged as a clever and sustainable approach, where fish coexist in flooded rice paddies, offering farmers not just additional income but also a vital protein source. Beyond that, efforts like growing salt-resistant rice and restoring mangrove forests showcase the remarkable adaptability of coastal agriculture. These initiatives not only help farmers survive in changing conditions but also actively support biodiversity and strengthen natural defenses against disasters.

5.5 *Polar and Subpolar Regions: Innovations in Extreme Conditions*

Farming in polar and subpolar regions has always been a tough challenge, with permafrost and short growing seasons limiting traditional agriculture. Permafrost is a thick subsurface layer of soil that remains below freezing throughout the year, occurring chiefly in polar regions like the Arctic and North Pole region of the globe. But thanks to technological advancements, even these extreme environments are seeing innovative solutions.

Greenhouses and hydroponic systems have become essential tools for growing fresh vegetables and herbs, helping remote communities access locally grown produce. In northern Canada and Alaska, controlled environment agriculture (CEA) is gaining momentum, using artificial lighting, temperature control, and nutrient-rich solutions to grow crops despite harsh external conditions. Vertical farming, where crops are stacked in layers to maximize space and efficiency, is also becoming more popular in cold climates.

Of course, these technologies are not without obstacles. They often require high energy, and steep initial costs remain challenges. That is why integrating renewable energy sources, like solar and wind power, into these systems is a crucial step toward making them more sustainable in the long run.

6 Urban Development and Human Impact on Land

Urban expansion has transformed landscapes across the world, creating economic progress, technological breakthroughs, and shifts in society. But this growth definitely comes at a cost like it disrupts ecosystems, alters land use, and creates disbalance in nature. To build a sustainable future, it is crucial to understand the relationship between urbanization and land management.

In regions like the Mediterranean, especially in the Middle East and North Africa (MENA), farmable land and freshwater are in very short supply. This scarcity forces MENA countries to rely heavily on food imports to meet their daily consumption requirements. On top of that, rising populations and economic competition are increasing the strain on natural resources. Given these challenges, preserving and efficiently using land is no longer an option but a necessity for the region's future.

6.1 The Scale and Pace of Urbanization

Cities have been expanding at an extraordinary pace over the past century, reshaping the way people live. It is expected that by 2050, nearly 68% of the global population will reside in urban areas—up from 55% in 2018. This rapid urbanization has seen cities push further into rural and natural landscapes, replacing vast stretches of land with housing, multistory buildings, business parks, and industrial developments.

The effects of this transformation vary across regions. In developing countries, urban expansion often happens without careful planning, leading to informal settlements that lack proper infrastructure. Meanwhile, developed nations experience more structured expansion, driven by well-organized development projects. Regardless of both the approaches, urbanization brings significant environmental challenges, including deforestation, soil erosion, and the loss of farmland which ends up disrupting ecosystems and threatening sustainability.

7 Environmental Impacts of Urban Development

History has been evident of witnessing the effects of expansion both in nature and in technology. No development comes for free; the cost mankind pays is losing what was available in abundance in the past.

7.1 Land-Use Changes and Habitat Destruction

Urban expansion frequently encroaches on forests, wetlands, and grasslands, causing habitat fragmentation and loss of biodiversity. For instance, the Amazon rainforest faces significant deforestation due to urban growth and associated infrastructure projects (Barber et al., 2014). Similarly, the Mississippi River Delta wetland conversion for urban expansion has led to the loss of critical ecosystem benefits, such as flood regulation and water purification.

7.2 Soil Degradation and Sealing

Urbanization often results in loss of water level and soil erosion, resulting in natural land surfaces getting covered with impermeable materials like concrete and asphalt. This process disrupts natural water infiltration, increases surface degradation, and intensifies urban flooding. Europe alone has witnessed approximately 25 hectares of land getting stuck down daily, highlighting the widespread nature of this issue (Cornu et al., 2023).

7.3 Climate and Microclimate Changes

Urban areas are facing unique microclimates, often referred to as urban heat islands (UHIs), where temperatures are significantly higher than surrounding rural areas due to heat absorption by buildings and infrastructure. UHIs aggravate energy consumption, increasing heat-related sickness, and varying local weather patterns. Cities like Tokyo and New York have shown that UHIs can raise temperatures by up to 7°C (Oke et al., 2017).

7.4 Pollution and Waste Generation

Urbanization contributes to land pollution because of the accumulation of solid waste, industrial discharge, and hazardous materials. Landfills occupy vast areas of land, resulting in contaminated soil and groundwater. The ever-growing need for infrastructure, construction and demolition activities generate significant debris, further burdening land resources.

8 Socioeconomic Implications

Urbanization also has profound socioeconomic implications for land use. On the positive side, urban centers serve as hubs of innovation, economic activity, and cultural exchange. However, unequal land distribution, gentrification, and displacement of marginalized communities often accompany urban development. Informal settlements, which house over one billion people globally, illustrate how rapid urban growth can lead to land-use inefficiencies and social inequities (Taylor et al., 2020).

Moreover, the competition for land between urban areas and agriculture intensifies as cities expand, threatening food security. For instance, China has seen a significant decline in arable land due to urbanization, prompting concerns about its ability to sustain agricultural production (Habiyaremye, 2020).

9 Sustainable Urban Development

Mitigating the adverse impacts of urban development on land requires a shift toward sustainable urban planning. Some of the key strategies are discussed below.

9.1 Smart Growth and Land-Use Planning

Compact urban growth and mixed-use development can reduce urban sprawl and preserve natural areas. Policies that promote infill development and the reuse of derelict land minimize the need for new land conversion (Haaland & van Den Bosch, 2015).

9.2 Green Infrastructure

Integrating green infrastructure, such as urban forests, green roofs, and permeable pavements, can mitigate urban heat islands, reduce stormwater runoff, and enhance biodiversity. Cities like Singapore have successfully adopted green infrastructure to create a "city in a garden" (Sokolova et al., 2024).

9.3 Policy and Legislation

Zoning regulations, land-use policies, and environmental impact assessments are critical tools to manage urban growth. For example, Brazil's Forest Code mandates legal reserves on private land, helping to balance urban expansion with forest conservation (Soares-Filho et al., 1979).

9.4 Community Engagement and Equity

Inclusive urban planning that engages communities ensures equitable access to land and resources while addressing the needs of vulnerable populations. Participatory approaches in cities like Curitiba, Brazil, have demonstrated the value of involving citizens in sustainable urban initiatives.

10 Conservation Efforts and Sustainable Practices

As human activities continue to transform landscapes, conservation and sustainable land practices have become essential to maintain ecological balance, ensure food security, and address climate change. Conservation efforts aim to protect biodiversity, mitigate land degradation, and restore ecosystems, while sustainable practices promote long-term land use that balances economic, environmental, and social needs. These initiatives align closely with the United Nations Sustainable Development Goals (SDGs), particularly SDG 15, which focuses on life on land, and SDG 13, which addresses climate action.

10.1 The Role of Conservation in Preserving Land Ecosystems

The human approach toward mass production and the very need for hoarding the necessary products have left the world with very less conserved resources. Now that the world is losing on some of the important resources, mankind has left to conserve the remaining resources.

10.2 Biodiversity Conservation

Biodiversity is a cornerstone of healthy ecosystems. Conservation strategies, such as the establishment of protected areas and wildlife corridors, aim to safeguard habitats and species. Protected areas, including national parks and reserves, cover 15% of terrestrial land globally and contribute significantly to SDG 15 (Secretariat of the Convention on Biological Diversity, 2020a).

For instance, the Amazon Region Protected Areas Program (ARPA) in Brazil protects over 60 million hectares of rainforest, preserving vital carbon sinks and biodiversity hotspots (Soares-Filho et al., 2010). Similarly, community-based conservation programs in Africa integrate local participation in protecting species like elephants and rhinos, fostering both ecological and economic benefits.

10.3 Land Restoration Initiatives

Land restoration is critical to reversing the effects of deforestation, desertification, and soil degradation. Programs such as the Bonn Challenge, which aims to restore 350 million hectares of degraded land by 2030, directly contribute to SDG 15. Examples include agroforestry projects in the Sahel region, where the Great Green Wall initiative has reforested millions of hectares to combat desertification and improve livelihoods (O'Byrne et al., 2022).

11 Sustainable Land Management Practices

The United Nations defines sustainable land management (SLM) as "the use of land resources, including soils, water, animals and plants, for the production of goods to meet changing human needs, while simultaneously ensuring the long-term productive potential of these resources and the maintenance of their environmental functions".

11.1 Sustainable Agriculture

Sustainable agricultural practices prioritize soil health, water conservation, and biodiversity. Techniques such as crop rotation, no-till farming, and organic farming reduce reliance on chemical inputs and enhance soil fertility. For example, conservation agriculture has been adopted widely in Sub-Saharan Africa, where it increases resilience to climate variability and aligns with SDG 2 (Zero Hunger) and SDG 15.

Agroecologist, which integrates ecological principles into farming, is another transformative approach. In Latin America, agroecological practices have revitalized degraded lands while promoting food sovereignty and climate resilience (Altieri & Nicholls, 2017).

11.2 Forest Management

Forests are critical for carbon sequestration, biodiversity, and water regulation. Sustainable forest management (SFM) practices, such as selective logging and reduced-impact logging, ensure that forests are used without compromising their long-term health. Initiatives like the Forest Stewardship Council (FSC) certification promote sustainable forestry, aligning with SDG 12 (Responsible Consumption and Production) and SDG 13 (Climate Action).

Community-based forest management, as practiced in Nepal, has been particularly effective in restoring degraded forests while improving local livelihoods. These efforts not only contribute to climate mitigation but also empower communities economically and socially (Chhatre & Agrawal, 2009).

11.3 Urban Green Spaces and Land Reclamation

Urban green spaces, including parks, green roofs, and community gardens, enhance biodiversity and improve the quality of life in cities. These initiatives contribute to SDG 11 (Sustainable Cities and Communities) by creating more liveable urban environments.

Land reclamation efforts also play a role in sustainable land use. In the Netherlands, innovative approaches to reclaiming land from the sea have created new agricultural and residential zones while preserving wetlands as ecological buffers (Altmann, 2015).

11.4 Climate Change Mitigation Through Land Practices

The relationship between land use and climate change is bidirectional: land practices influence climate systems, and climate change affects land productivity. Sustainable land use practices are vital for mitigating climate change, as emphasized in SDG 13.

11.5 Carbon Sequestration in Soils and Forests

Reforestation, afforestation, and soil carbon sequestration are key strategies for reducing greenhouse gas emissions. The Trillion Trees Initiative, which aims to plant and protect one trillion trees globally, aligns with efforts to enhance carbon sinks and meet the goals of the Paris Agreement (Bastin et al., 1979).

11.6 Integrated Water and Land Management

Efficient water management is integral to sustainable land use. Techniques like rainwater harvesting and watershed management ensure that land remains fertile and ecosystems remain intact. For example, India's watershed management programs have improved groundwater levels and agricultural productivity, contributing to SDG 6 (Clean Water and Sanitation) and SDG 15.

12 The Role of Technology and Innovation

Technological advancements play a pivotal role in conservation and sustainable practices. Satellite imagery and geographic information systems (GIS) are used to monitor deforestation, soil degradation, and urban sprawl in real time (Pagiola, 2008). Precision agriculture leverages technology to optimize resource use, reducing environmental impacts.

Additionally, renewable energy integration in agricultural practices, such as solar-powered irrigation, reduces the carbon footprint of farming. These innovations support SDG 7 (Affordable and Clean Energy) and SDG 13.

13 Community and Policy Interventions

Conservation and sustainability require active participation from governments, communities, and international organizations. Policies such as payment for ecosystem services (PES) incentivize conservation by rewarding landowners for maintaining ecological functions. Costa Rica's PES program, which pays farmers to conserve forests, has significantly reduced deforestation rates while benefiting local communities (Singh, 2024).

Community-driven initiatives, such as participatory land-use planning, ensure that local voices are included in conservation strategies. These approaches are vital for achieving inclusive and equitable outcomes, as emphasized in SDG 10 (Reduced Inequalities).

14 Conclusion

To sum it up, humanity's creativity and adaptability shine through in how we farm across different terrains. Whether it is the high-tech machinery of fertile plains or the carefully constructed terraces in mountain regions, these approaches do more than just feed us—they reflect a deep respect and understanding of nature's rhythms. With challenges like climate change and population growth intensifying, adopting sustainable farming practices tailored to specific environments is more important than ever for ensuring food security worldwide.

On a similar note, urban development has completely transformed the world's landscapes, changing both human and natural systems. While cities bring immense opportunities, they also come with significant challenges that require smart, innovative solutions. Striking a balance between urban growth and environmental care is key to safeguarding the planet's ecological richness while ensuring that cities and nature coexist in harmony.

Efforts to conserve and use land sustainably are essential to protecting the Earth's vibrant ecosystems. By following the UN Sustainable Development Goals, these initiatives aim to heal damaged ecosystems, combat climate change, and ensure fair resource distribution. Bringing together advanced technologies, active community involvement, and strong policy measures will be the driving force behind creating a future where human progress and environmental preservation go hand in hand.

References

Altieri, M. A., & Nicholls, C. I. (2017). The adaptation and mitigation potential of traditional agriculture in a changing climate. *Climatic Change, 140*, 33–45.

Altmann, E. (2015). *Green consumption: The global rise of the eco-chic*. Taylor & Francis.

Barber, C. P., Cochrane, M. A., Souza, C. M., Jr., & Laurance, W. F. (2014). Roads, deforestation, and the mitigating effect of protected areas in the Amazon. *Biological Conservation, 177*, 203–209.
Bastin, J.-F., et al. (1979). The global tree restoration potential. *Science, 365*(6448), 76–79. 2019.
Chhatre, & Agrawal, A. (2009). Trade-offs and synergies between carbon storage and livelihood benefits from forest commons. *Proceedings of the National Academy of Sciences, 106*(42), 17667–17670.
Cornu, S., et al. (2023). National soil data in EU countries, where do we stand? *European Journal of Soil Science, 74*(4), e13398.
Haaland, & van Den Bosch, C. K. (2015). Challenges and strategies for urban green-space planning in cities undergoing densification: A review. *Urban Forestry & Urban Greening, 14*(4), 760–771.
Habiyaremye. (2020). Water innovation in South Africa: Mapping innovation successes and diffusion constraints. *Environmental Science & Policy, 114*, 217–229.
Jägermeyr, J., Pastor, A., Biemans, H., & Gerten, D. (2017). Reconciling irrigated food production with environmental flows for Sustainable Development Goals implementation. *Nature Communications, 8*(1), 15900.
Nair, P. K. R., Kumar, B. M., Nair, V. D., Nair, P. K. R., Kumar, B. M., & Nair, V. D. (2021). Other ecosystem services of agroforestry. In *An introduction to agroforestry: Four decades of scientific developments* (pp. 563–581). Springer.
O'Byrne, D., Mechiche-Alami, A., Tengberg, A., & Olsson, L. (2022). The social impacts of sustainable land management in Great Green Wall countries: an evaluative framework based on the capability approach. *Land (Basel), 11*(3), 352.
Oke, T. R., Mills, G., Christen, A., & Voogt, J. A. (2017). *Urban climates*. Cambridge University Press.
Pagiola, S. (2008). Payments for environmental services in Costa Rica. *Ecological Economics, 65*(4), 712–724.
Pingali, P. L. (2012). Green revolution: Impacts, limits, and the path ahead. *Proceedings of the National Academy of Sciences, 109*(31), 12302–12308.
Pretty, J. (2018). Intensification for redesigned and sustainable agricultural systems. *Science (1979), 362*(6417), eaav0294.
Reij, C., Tappan, G., & Smale, M. (2009). Agroenvironmental transformation in the Sahel: Another kind of "Green Revolution". *International Food Policy Research Institute (IFPRI).*
Secretariat of the Convention on Biological Diversity, "Global Biodiversity Outlook 5. 2020a.
Singh, H. (2024). *Climate change and environmental pollution: Biological consequences on urban forests.*, ISBN: 978-3-031-67837-0 (pp. 1–21). https://doi.org/10.1007/978-3-031-67837-0
Soares-Filho, B., et al. (1979). Cracking Brazil's forest code. *Science, 344*(6182), 363–364. 2014.
Soares-Filho, B., et al. (2010). Role of Brazilian Amazon protected areas in climate change mitigation. *Proceedings of the National Academy of Sciences, 107*(24), 10821–10826.
Sokolova, M. V., Fath, B. D., Grande, U., Buonocore, E., & Franzese, P. P. (2024). The role of green infrastructure in providing urban ecosystem services: Insights from a bibliometric perspective. *Land (Basel), 13*(10), 1664.
Taylor, F. E., Millington, J. D. A., Jacob, E., Malamud, B. D., & Pelling, M. (2020). Messy maps: Qualitative GIS representations of resilience. *Landscape and Urban Planning, 198*, 103771.
Tilman, D., Cassman, K. G., Matson, P. A., Naylor, R., & Polasky, S. (2002). Agricultural sustainability and intensive production practices. *Nature, 418*(6898), 671–677.
Van Kien, N., Han, N. H., & Cramb, R. (2020). Trends in rice-based farming systems in the Mekong Delta. *White Gold*, 347.

Ecosystem Restoration and Sustainable Land Practices Using AI, IoT and Remote Sensing Technologies

Lekha J and B. Meena Preethi

1 Introduction

Ecosystem services play an essential role in sustainable land management by providing basic functions such as clean water, fertile soil, and biodiversity that are necessary for maintaining healthy ecosystems and supporting human livelihoods. In addition to boosting agricultural production, these services help avert flooding, moderate the climate and make communities more resilient to environmental changes overall. Recognizing, appreciating, and maintaining the value of these ecosystem services is fundamental for integrated and sustainable land management planning and a good compromise between productive landscapes and biodiversity conservation. By accounting for ecosystem services into land management plans, stakeholders can focus on environmental health and ensure that natural resource stocks last for future generations.

This integrative approach promotes collaboration across sectors and innovative solutions that effectively address socioeconomic and environmental challenges. Such initiatives require ecologists, economics, legislators, and local communities to come together and create frameworks for sustainable development and conservation of vital ecosystems. This cooperative framework not only enhances ecosystems' resilience but also empowers communities with the tools and knowledge to participate in conservation efforts, ultimately paving the way for a sustainable future for all (Kelling, 2018) (Isabelle & Westerlund, 2022).

Lekha J (✉)
School of Sciences, Christ University, Bengaluru, Karnataka, India
e-mail: lekha.j@christuniversity.in

B. M. Preethi
Department of Software Systems, Sri Krishna Arts and Science College,
Coimbatore, Tamil Nadu, India

J. J. Puthenkalam S.J et al. (eds.), *Terra Thrive*,
https://doi.org/10.1007/978-3-032-02738-2_3

Since along with restoring a habitat, ecosystem restoration also enhances the provision of ecosystem services such as carbon capture and storage, water purification and biodiversity protection, it is crucial for the repair of damage to the ecosystem resulting from human activity. Well-designed restoration projects can counteract some of the impacts of climate change while also forging a deeper connection between people and their natural environments and generating economic advantages and better community well-being.

As with most sustainable development goals, achieving the 15th goal which focuses on combating desertification and sustainably managing, protecting, restoring, and fostering the sustainable use of forests and other terrestrial ecosystems while halting and reversing land degradation and biodiversity loss relies heavily on the incorporation of novel technologies such as artificial intelligence (AI), the internet of things (IoT), and remote sensing. These technologies enable the monitoring, maintenance, and rehabilitation of ecosystems in a more effective manner than has previously been possible (Mondal et al., 2020).

Policymakers, conservationists, and local communities are represented among the users of real-time data and analytics for resource optimization across different landscapes, soil quality enhancement, and biodiversity improvement. Such technologies facilitate the environmental monitoring evaluation and allow the community to play a proactive governance role to ensure ecosystem sustainability for future generations. As the climate change and environmental degradation challenges escalate, the need for innovative approaches to ecosystem restoration and sustainable land management has skyrocketed.

The integration of AI, IoT, and remote sensing technologies in the context of SDG 15 has become a game changer. In addition to fostering sustainable land use, these technologies offer unrivaled ecosystem monitoring and management capabilities. Using data and advanced analytics, stakeholders are able to optimize the use of resources, increase biodiversity, and improve soil health, which are essential to meeting the needs of SDG 15. These technical advancements allow comprehensive evaluations of environmental conditions which ensures community governance on conservation strategies.

This chapter provides a summary of AI, IoT, and remote sensing technologies, outlining their applications, current developments, and remaining gaps toward achieving sustainable ecosystems, biodiversity conservation, and future challenges. It also covers ecosystem restoration within the context of SDG 15 and the relevance of the technology for remediation. Ecosystem restoration is necessary due to a number of factors, the most important of which is the increasing impact of climate change, loss of biodiversity, and environmental degradation. The following presents some key points highlighting its significance:

Conservation of Biodiversity Restoration of ecosystems helps in the recovery of regions that are home to numerous species. Biodiversity is required to maintain and sustain ecosystems. Moreover, resiliency, ecological balance, and provision of ecosystem services are reliant on biodiversity.

Climate Change Mitigation Some ecosystems, such as grasslands, wetlands, and forests, play a primary role in carbon storage. Restoration projects can help mitigate climate change by enhancing the capacity of these ecosystems to sequester carbon dioxide.

Soil Health Improvement Restoration activities often aim to enhance soil quality by increasing organic matter, controlling erosion, and promoting nutrient cycles.

Water Quality and Availability Restoration of ecosystems can improve water quality by filtering pollutants and decreasing runoff. Industry, agriculture, and human consumption require clean water resources which can only be provided by healthy watersheds and wetlands.

Community Resilience Restoration projects that engage local populations fosters stewardship and empowers them to self-initiate. Beyond enhancing social solidarity, this participation fosters greater resilience to natural disasters and environmental changes.

Economic Benefits Employment opportunities, eco-tourism, and sustainable harvesting of resources are some of the economic prospects that can be pursued as a result of restoring ecosystems. Long-term ecological health investments would economically benefit communities and regions in the future.

Cultural and Recreational Value Ecosystems that are in good condition offer recreational activities and cultural value for a number of societies. Restoration projects can transform these regions and promote outdoor activities, while safeguarding the land's cultural heritage.

Sustainable Land Use Practices Practices that give a good balance between ecological health and restoration of land with economic development guarantees responsible use of resources.

2 Role of Technology in Environmental Management

The major things that affect the ecosystem are air and water pollution, deforestation, habitat destruction, and soil erosion. So preserving the land quality to prevent climate change and preserve biodiversity is very much essential. Many new technologies are emerging that help in preventing and preserving the land quality including advancements in biotechnology, IoT, AI, and remote sensing platforms.

2.1 Remote Sensing and Geographic Information Systems (GIS)

Aerial drones and satellite imagery are the two emerging technologies that help in monitoring habitat life change, deforestation, usage of land, etc. Activities such as illegal logging and illegal urbanization of natural landscapes can be easily identified using high resolution satellite photography. Thermal and multispectral sensors effectively help us in tracking the ecosystem of landscapes with the help of drones.

Two examples of remote sensing technologies that provide valuable information for monitoring habitat loss, deforestation, and land use changes are aerial drones and satellite imagery. High-resolution satellite photography can be used to detect illegal logging, soil erosion, and urban encroachment on natural landscapes (Zhang et al., 2021a, b). Drones equipped with thermal and multispectral sensors allow conservationists to efficiently survey large regions and track the condition of ecosystems in real time (Turner et al., 2019).

Geographic Information Systems (GIS) annotates additional data from remote sensing sources to build maps for land traversal. These maps help us in deciding the necessary steps for developing projects to maintain reforestation, forest area management, and land restoration. Researchers can think about the future happenings of land degradation and decide some preventive measures with the help of GIS technology (Chuvieco, 2020).

2.2 *Artificial Intelligence (AI) and Machine Learning (ML)*

Performing a deeper data analysis and finding patterns using the technologies like AI-based image recognition algorithms and ML have transformed the method of classifying the land conservation and deforestation that in turn helps the conservationists to devise free conservation laws and plan for restoration of habitats (Wegmann et al., 2020).

There are plenty of machine learning algorithms that help in identifying massive datasets, which are images, field observations of patterns that result in vegetation changes, soil health, and climate change. Some predictive models help in optimizing the land usage for industrial, urban, and agricultural schemes that do not affect the ecological integrity (Shendryk et al., 2021).

2.3 *Internet of Things (IoT) and Smart Sensors*

Real-time environmental monitoring is now feasible due to IoT and sensor technology, enhancing conservation initiatives. Smart sensors continuously collect data on temperature, humidity, soil moisture, and pollution levels in wetlands, woods, and agricultural regions (Gholami et al., 2022). These sensors can detect early symptoms of ecological stress, such as drought, soil degradation, or water contamination.

The need of identifying land encroachment and deforestation can be done with the help of smart devices with the support of IoT sensors. Moreover in mission-critical situations where detection of human violation on danger-prone regions can be done with the help of motion detection cameras and sound sensors (Pettorelli et al., 2018). These early warning systems enhance law enforcement operations and protect biodiversity.

2.4 *Blockchain for Land Protection and Sustainable Resource Management*

The accountability and transparency of land conservation projects can be improved with the help of blockchain technology. It helps in managing safe credit card transactions, conservation of funds, and rights preservation of land owners by securing the decentralized records (Kshetri, 2021). The blockchain offers the best way to preserve the ownership information and stop misleading land for deforestation and land robbery. The blockchain also helps in guaranteeing forest recovery and sustainable agricultural projects. Through smart contracts, sustainable land use and secure transactions for land management initiatives can be easily maintained (Reinsberg, 2020).

2.5 *Biotechnology and Environmental DNA (eDNA) Analysis*

In order to maintain biodiversity and rejuvenate the degraded land, essential tools and techniques in biotechnology for direct observation of the environmental DNA are an essential requirement. These tools can help in the analysis of identifying plants and animal species in soil and water, detecting rare species, monitoring rare species, and monitoring the health of the ecosystem (Bohmann et al., 2014). In order to meet enduring environmental challenges and create resilient plant species, some good genetic engineering approaches have to be used for investigation. To defend climate change and support reforestation initiatives, many researchers are developing bioengineering trees and crops that can absorb more carbon dioxide for environmental sustainability (Koeppel et al., 2021).

2.6 *Autonomous Vehicles and Robotics in Conservation*

The land conservation initiatives are growing tremendously with the help of robotic devices and automatic drones to perform deforestation activities such as spreading of seeds across deforested areas and usage of drones in aerial seeding initiatives that can speed up the reforestation process (Baek et al., 2021). These initiatives reduce the burden of labor-intensive tasks in tree plantation, removal of weeds, and conditioning the soil. Certain coastal conservative initiatives are achieved using underwater robots to measure the quality of water, managing land and water resources, etc. (Xu et al., 2022).

3 AI in Forest and Land Conservation

The ecosystem around us are in danger due to many reasons like destruction of habitats, land quality degradation, and deforestation. Though traditional conservation techniques perform well in sustainability of land and ecology, some sophisticated tools are needed for monitoring biodiversity and making predictive analytics using artificial intelligence for revolutionizing forest and land restoration (Swami et al., 2021).

3.1 Machine Learning for Ecosystem Modeling

Machine learning is a method that transforms massive volumes of forecast data for land use, vegetation health, and climatic impact on forest. ML algorithm also helps to analyze ecological data like environmental change observations, satellite images, climate data and values related to forest area declines (Wegmann et al., 2020). Fig. 1 illustrates the patterns in forest degradation. Some primary machine learning algorithms like convolutional neural networks (CNNs) and decision trees are the best algorithms that can predict land cover mapping and vegetation categorization that changes over time. These algorithms can help in monitoring deforestation and evaluate the effectiveness of reforestation initiatives (Kattenborn et al., 2021).

Many deep learning algorithms like neural networks help in forecasting the use of hydrological and soil data for maintaining soil erosion in certain areas like

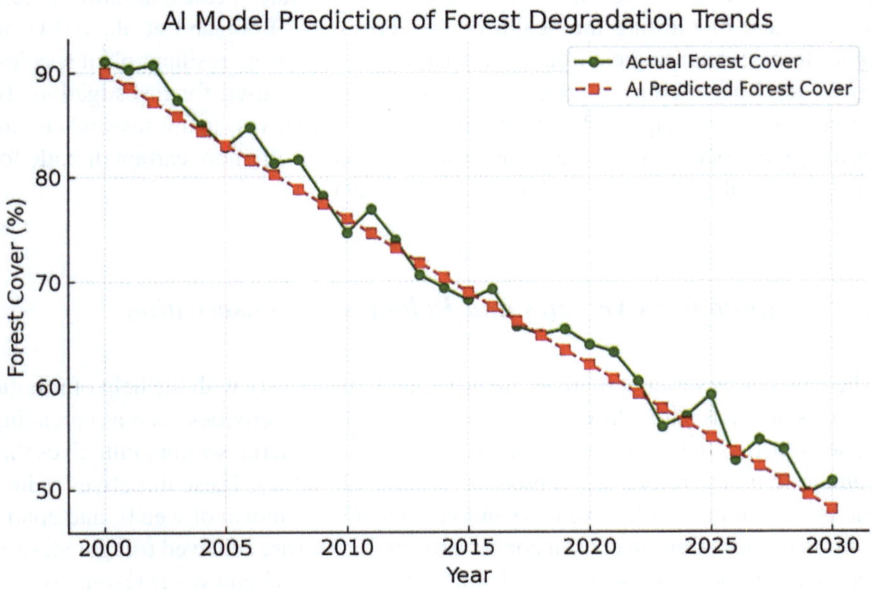

Fig. 1 Graph showing AI model predicting forest degradation trends

Amazon forests (Tyukavina et al., 2017) which are affected highly by climate change by forecasting the biodiversity and carbon sequences (Liu et al., 2022; Silva et al., 2021). Some algorithms like reinforcement learning have many advantages for mimicking the tactics for maintaining the conservation models through good decision making for deforestation and biodiversity (Gomes et al., 2019).

3.2 *AI-Based Biodiversity Monitoring and Species Detection*

The most important aspect in conservation of biodiversity is identification of accurate species. In olden days, traditional methods like surveys and camera monitoring methods were used for species identification, which were time consuming and cost ineffective. Fig. 2 depicts the various AI techniques such as deep learning and computer vision that have been useful in enhancing the accuracy of species identification and tracking. The data in Table 1 clearly illustrates the comparison of human and AI in identifying the distinct species like insects and amphibians. The data clearly shows that AI is more efficient in handling large datasets and very minute patterns (Wäldchen & Mäder, 2018). The table also shows that the mammals are identified with an accuracy of 95% for which the human accuracy is 85%, which is comparatively lesser. For other categories of species AI attained 80% accuracy, for

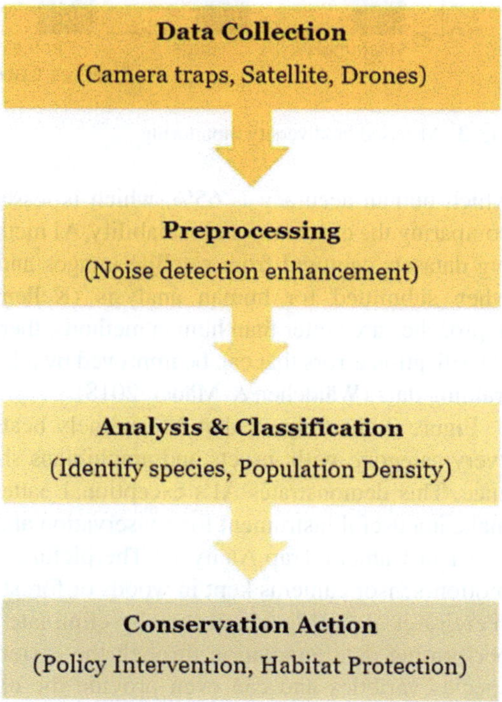

Fig. 2 AI Process for Biodiversity Monitoring

Table 1 Accuracy comparison of AI vs. human in species identification

Species category	AI accuracy (%)	Human accuracy (%)
Mammals	95	85
Birds	92	78
Reptiles	89	75
Amphibians	87	70
Insects	80	65

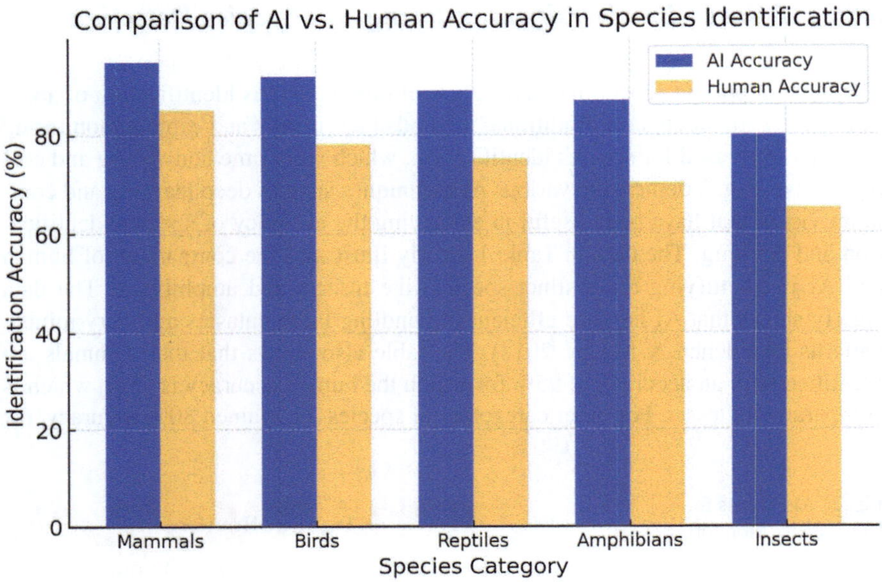

Fig. 3 AI-based biodiversity monitoring

which human accuracy is 65%, which is lesser (Gaston & O'Neill, 2004).When comparing the efficiency and scalability, AI methods are more effective in analyzing big datasets captured from satellite images and camera which is time consuming when submitted for human analysis (Kellenberger et al., 2018). Though AI approaches are better than human methods, there are some limitations such as misclassification errors that can be improved by additional validation procedures on the training data (Wäldchen & Mäder, 2018).

Figure 3 chart shows that AI routinely beats humans in identifying species in every category, with insects and amphibians showing the biggest accuracy difference. This demonstrates AI's exceptional pattern detection and scalability, which make it a useful instrument for conservation and biodiversity monitoring.

AI in Camera Trap Analysis: The pictures captured by wildlife cameras like motion-sensor cameras kept in woods or forest are processed through AI methods (Pereira et al. 2017) which thereby eliminate human intervention. The methods performing data annotation through the system can automatically identify these species varieties and can even provide the differences between similar species

(Norouzzadeh et al., 2018). CNN models are used to identify species that have similar visual characteristics.

AI-Powered Acoustic Monitoring: The sounds of different animals can be identified using spectrogram-based analysis through methods like recurrent neural networks (RNNs) that can accurately detect sounds of amphibians, birds, and elephants. Through this analysis the communication mode of the animals can be monitored by conservationists to identify the mood of animals like happy, scared, etc., and assess the health factors of the animals (Stowell et al., 2019) without any human intervention.

Drone-Based AI for Tracking Species: Image recognition methods are widely used in tracking animal populations in remote and wide locations. The images captured using AI-driven drone cameras for identifying species that are high in population using suitable machine learning algorithms through aerial images. The AI-based methods are used for monitoring orangutan nests in the rainforests of Southeast Asia (Burke et al., 2019).

Integration of AI with eDNA: The AI-based methods for identifying species is one of the effective methods to analyze the variety of species based on processing eDNA sequences with genetic materials present in air, water, and soil (Bohmann et al., 2014).

3.3 Predictive Analytics for Deforestation and Land Degradation

As per the statistics, the regulations for conservation of forest, urban encroachment zones, and agricultural developmental areas are the major areas where deforestation happens. The heatmap shown in Fig. 4 depicts the major areas of deforestation which are prone to risk with darker tones (Hansen et al., 2013).

1. **Areas at High Risk**: The densest deforestation prone areas in tropical forests (Curtis et al., 2018).
2. **Spatial Analysis for Conservation**: The areas that are of high risks can be conserved through land-use laws and afforestation initiatives (Lewis et al., 2015).
3. **Use of AI and Remote Sensing**: By using predictive models on satellite images, proactive and accurate conservative measures can be taken against deforestation and risk assessments become more accurate, allowing for proactive conservation measures (Wang et al., 2023).

Models for Predicting Deforestation There are many AI models like Google's Global Forest Watch that apply time-series analysis and remote sensing data to find trends that lead to land surveys. These devices can alert the concerned authorities in identifying illegal logging activities using deep learning to identify deforestation in real-time (Hansen et al., 2019). To forecast changes in forest cover based on meteorological and socioeconomic parameters, support vector machines (SVMs) and random forest classifiers have also been used (Huang et al., 2020).

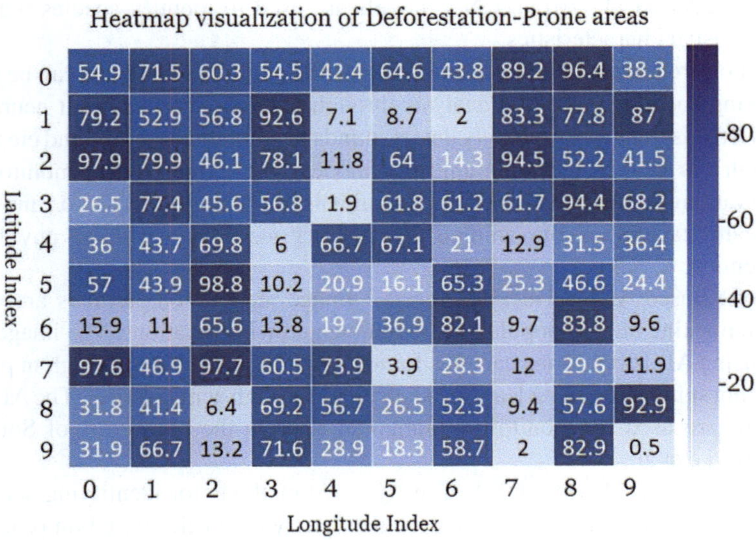

Fig. 4 Predictive analytics for deforestation

Land Degradation Monitoring To predict early indications of land degradation, many AI-based systems are available to investigate the soil moisture levels and vegetation indicators (such the NDVI). By reducing the percentage of land occupation and assisting policymakers in putting soil conservation measures into action these models are put into action (Baratchi et al., 2021).

AI for Fire Risk Assessment AI-based models forecast fire outperforms by examining temperature, wind patterns, and vegetation dryness. Wildfires are the primary threats to forests. Disaster incidents can be identified by neural networks that have been trained with the archived wildfire data to forecast fire-prone locations (Zhang et al., 2021a, b). In order to predict the active wildfires and maximize firefighting efforts, AI-powered drones and satellite imagery are also utilized (Duarte et al., 2022).

4 IoT for Real-Time Environmental Monitoring

Real-time data collection and analysis is made possible by the Internet of Things (IoT), which has become an important technology for monitoring the environment. It provides major insights into soil health, air quality, water cycles, and overall ecosystem by combination of sensors, cloud computing, and AI-driven methods. These advancements have been required for reducing the impact of climate change, promoting sustainable land use, and preserving forests. IoT-based systems provide efficient methods that enable immediate actions in wildlife protection, agriculture, and deforestation prevention.

Table 2 Comparison of different IoT sensors for environmental monitoring

Sensor type	Measured parameters	Accuracy	Application
Soil moisture sensor	Moisture level, temperature	±2%	Precision agriculture, reforestation
Air quality sensor	CO_2, NO_x, PM2.5, PM10	±5%	Pollution monitoring, urban planning
Water quality sensor	pH, turbidity, dissolved oxygen	±0.2 pH	Wetland conservation, fisheries

4.1 Smart Sensors for Tracking Soil Health, Air Quality, and Water Cycles

In order to preserve the conservation methods and land management policies, IoT-driven smart sensors are the primary need for tracking the environmental factors. Table 2 lists the various applications of these sensors.

4.1.1 Soil Health Monitoring

IoT-based soil health monitoring systems as represented in Fig. 5. Work with the help of wireless sensor networks (WSNs) to observe and record the soil moisture, pH levels, nutrient content, and temperature. These sensors feed the data to cloud-based platforms that use AI and machine learning algorithms to analyze patterns and identify the soil degradation level. For example, in precision agriculture, IoT sensors predict real-time nutrient management, which reduces the need for additional fertilizers and avoids soil contamination. Jha et al. (2019) showed how IoT-based soil sensors support optimizing the irrigation process, resulting in a 30% less consumption in water usage and an increase in crop yields.

4.1.2 Air Quality Monitoring

Air quality is an important environmental factor that has an effect on ecosystems and human health through air pollution. Basic things like carbon dioxide (CO_2), nitrogen oxides (NO_x), and particulate matter (PM2.5 and PM10) are recorded by IoT-enabled air quality prediction systems (Gao et al., 2021).

For example, many communities can record the pollution levels on a daily basis and take some preventative measures like those created by the AirVisual project. IoT-based air monitoring systems are used in cities such as Beijing and New Delhi in order to predict the pollution hotspots and put suitable control measures in place (Li et al., 2022).

Fig. 5 IoT-based environmental monitoring process

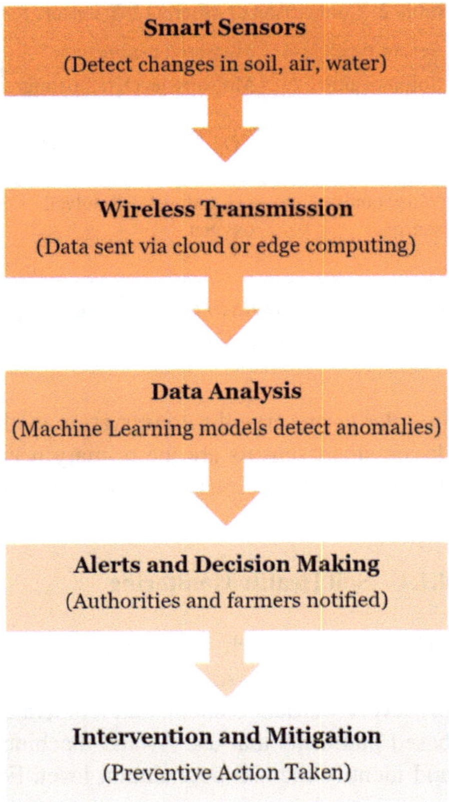

4.1.3 Water Cycle and Quality Monitoring

In order to observe the contamination levels and identify primary indicators of pollution, smart water quality sensors are mainly applied in wetland conservation projects. For instance, IoT networks used in the Great Lakes region help researchers in monitoring the growth of water resources (Pal et al., 2021). Water is a pivotal resource for maintaining the ecosystems of agriculture. IoT-based hydrological recording systems observe the parameters like pH levels, oxygen level, turbidity, and temperature in rivers, lakes, and groundwater sources (Misra et al., 2020).

4.2 IoT-Enabled Smart Agriculture for Sustainable Land Use

Agriculture is a key contributor to deforestation and land degradation. However, IoT-based smart agricultural technologies are altering farming methods, making them more efficient and sustainable are demonstrated in Fig. 6.

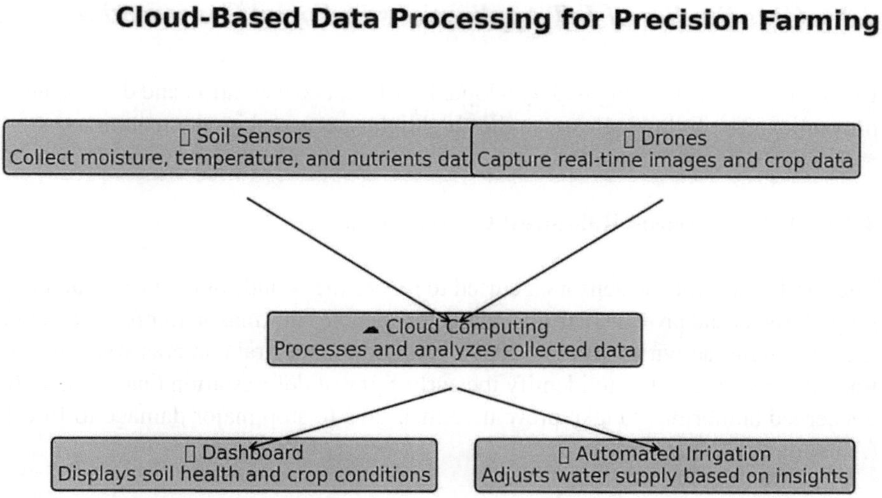

Fig. 6 IoT-based smart agriculture system

4.2.1 Precision Farming with IoT Sensors

There are many IoT devices for tracking weather patterns, crop quality, and soil health that enable the farmers to plan for automatic pest control, fertilizers, and irrigation (Rath et al., 2021).

The IoT-based irrigation systems helps in preserving upto 50% of water that helps in increasing the soil moisture through water quality and maintain the soil moisture (Mekonnen & Hoekstra, 2022).

4.2.2 IoT-Based Pest and Disease Detection

IoT sensor devices-based image recognition is helpful in early detection of crop diseases and pests along with some cutting-edge technology. This helps in early detection of chemical pesticide usage that resulted in infected crops. The IoT devices are used for monitoring soybean crops that are grown in Brazil to decrease the usage of pesticide by 20% and using herbicides widely (Singh et al., 2020; Carvalho et al., 2022).

4.2.3 Livestock Monitoring and Sustainable Grazing

One of the main causes of land degradation is overgrazing. To prevent overgrazing in sensitive areas and track cattle movement, IoT-enabled livestock monitoring systems use RFID sensors and GPS collars. This technology aids in rotational grazing, which ensures sustainable pasture management. For instance, ranchers in Australia use IoT-based virtual fencing to warn cattle when they approach restricted zones, reducing soil erosion and protecting native vegetation (Smith et al., 2021).

4.3 *Case Studies of IoT Applications in Forest Conservation*

IoT technologies are being widely adopted for forest conservation and deforestation prevention. The following case studies highlight successful implementations.

4.3.1 IoT in Amazon Rainforest Conservation

The IoT-based acoustic sensors are used to record the sounds of chainsaws and logging of trucks and provide real-time alerts for agencies in Amazon rainforest through illegal logging activities (Butler, 2020). Many multispectral cameras were used to track the loss of tree top to identify the early signs of deforestation that enabled the concerned authorities to take preventive measures to stop major damage to forests (Almeida et al., 2021).

4.3.2 Smart Forest Fire Detection Systems

AI-based fire prediction systems are playing a major role in detecting the wildlife fire with the help of IoT devices such as temperature, humidity, and smoke sensors that can identify and give real-time alerts in a quick response time (Ramos et al,. 2022). These alerts help in reducing the wildfire damage by 40% by providing early warnings (Thorn et al., 2015).

4.3.3 IoT for Wildlife Protection in African Reserves

The IoT-based tracking systems are used to identify poaching activities of species using GPS-enabled tags on animals that help to track animal movements and suspicious behaviors on animal behaviors (Bhattacharyya et al., 2021).

This has been implemented in Maasai Mara regions of Kenya for tracking elephants and rhinoceroses with a decrease in poaching incidents of animals by 70% (Verma et al., 2020; Muthuri et al., 2022). The diverse applications of IoT sensors in the context of forest conservation across various regions are illustrated in Table 3.

Table 3 Case studies of IoT applications in forest conservation

Location	Technology used	Purpose	Impact
Amazon Rainforest	Acoustic sensors	Illegal logging detection	65% reduction in deforestation
California, USA	IoT fire sensors	Early wildfire detection	40% reduction in wildfire damage
Maasai Mara, Kenya	GPS tracking collars	Anti-poaching efforts	70% decrease in poaching incidents

5 Remote Sensing and GIS for Land Restoration

The integration of remote sensing with geographical information systems in the area of land restoration presents a new turning point in this level of environmental conservation, by providing real-time spatial data of high precision (Smith et al., 2023). Such advanced methods are applied to large-scale ecosystem assessments and provide decision-makers the flexibility to monitor, plan, and implement sustainable land restoration programs not only more comprehensively but with unprecedented accuracy (Wang et al., 2022). The workflow of the entire process is shown in Fig. 7.

Remote sensing through satellite imagery, Light Detection and Ranging (LiDAR), and unmanned aerial vehicle (UAV) technology facilitates regular monitoring of land cover changes, vegetation health, and soil degradation (Jones et al., 2024). Remote sensing through spectral analysis and machine learning algorithms solves the purpose of early identification of desertification, deforestation, and soil erosion to respond with restoration activities (Zhang et al., 2021a, b).

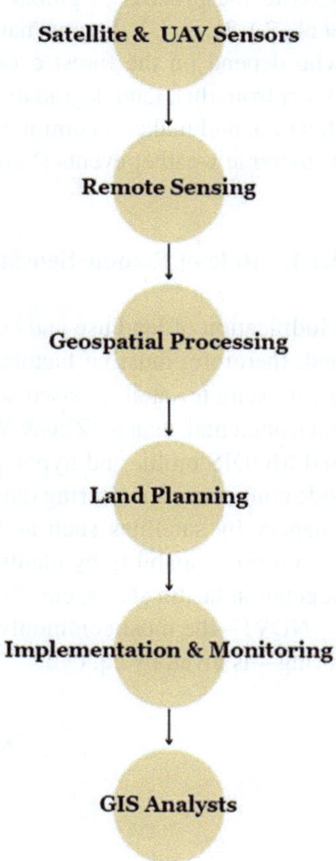

Fig. 7 Workflow of geospatial data integration for land planning and GIS analysis

GIS manages and analyzes spatial data, making it possible to visualize and model ecological parameters like soil quality, hydrological patterns, and biodiversity distribution (Esri, 2023), especially in order to prioritize restoration zones, put more effort into reforestation, and assess the information mechanism of a project over time (Wilson, 2022). RS and GIS provide, therefore, cutting-edge answers to effective restoration that include mapping of carbon sequestration and watershed management, precision agriculture, and climate-resilient planning (Farmonaut, 2024). Their blending favors researchers, policymakers, and environmentalists to make data-driven decisions that enhance ecosystem recovery, advance sustainable land use, and confront global environmental challenges (Sciencedirect, 2024).

5.1 Satellite-Based Monitoring of Deforestation and Ecosystem Changes

Deforestation and ecosystem degradation contribute to serious consequences that affect climate stability, biodiversity, and human livelihood. The loss of forest cover adds to the process of global warming by way of carbon sequestration (Hansen et al., 2013); and, it disrupts natural habitats and threatens the livelihood of millions who depend on the forest ecosystem for a basic livelihood (Bastin et al., 2019). Apart from this, land degradation reduces the fertility of farmland, increases desertification, and makes a community increasingly vulnerable to climate changes, such as extreme weather events (Verbesselt et al., 2010).

5.1.1 Role of Remote Sensing in Monitoring Deforestation

Modification of land use and cover has enormous implications for the environment and, therefore, must be factored into any plan for integrated sustainable development. Remote satellite-based sensing provides an effective avenue to monitor such environmental change (Zhu & Woodcock, 2014). With the help of Landsat, Sentinel, and MODIS multi- and hyperspectral sensors, long-term time-series analyses provide continuous monitoring data for the Earth (Reiche et al., 2018). High-resolution imagery by satellites such as WorldView and PlanetScope further enhances this monitoring capability by identifying small-scale changes of land cover, changes in vegetation health status, etc. (Asner et al., 2009).

NDVI—the most commonly used index in remote sensing for vegetation monitoring—is given by eq. (1):

$$\text{NDVI} = \frac{(\text{NIR} - \text{RED})}{(\text{NIR} + \text{RED})} \tag{1}$$

where NIR is the near-infrared reflectance and RED is the red-band reflectance. NDVI values can vary between −1 and +1, with maximum values indicating the most robust plant life and minimum values showing the least healthy (Gao, 1996). This becomes a measure to quantify health of vegetation, report patterns of deforestation, and assess how well some reforestation efforts succeed.

5.1.2 Advanced Technologies in Deforestation Detection

To allow for more accurate monitoring of deforestation activities, machine learning and deep learning models explained by advanced algorithm methods were integrated with satellite imagery. These artificial intelligence-based models analyze multitemporal datasets and distinguish between various land cover changes (Fodor & Conde, 2023) and, more importantly, between natural forest loss and human-driven disturbances such as illegal logging, urban expansion, and agricultural encroachment (Joshi et al., 2015).

Furthermore, SAR-based technology applied in Sentinel-1 allows for all-weather monitoring, overcoming the challenge with cloud cover posed for optical imagery (Reiche et al., 2018). The unique capacity of SAR assists in detecting areas of deforestation and land-use changes, even in highly foliage-covered and cloudy regions, thus being a good tool for real-time monitoring.

5.2 LiDAR and Drone-Based Mapping for Forest Conservation

The application of LiDAR and drone mapping represents a leap in forest conservation as it measures the health of forests, ongoing changes of canopy structure, and guides sustainable conservation strategies with higher precision (Mutanga & Kumar, 2019). Compared to traditional satellite images, often hindered by factors such as cloud cover and resolution, LiDAR and drone technology bears vast accuracy guarantees to support real-time data collection for better decisions in forest management.

5.2.1 Thermal Remote Sensing for Land Degradation Assessment

Thermal remote sensing is one of the very powerful methods to access land degradation using the temperature variations of land surface temperature (LST). It is established to observe environmental changes, which consist of deforestation, desertification, soil erosion, and the declining cover of the vegetation (Hansen et al., 2013). Thermal remote sensing sensors include Landsat (TIRS), MODIS, ASTER, and Sentinel-3 (SLSTR); such sensors target and measure the emitted radiation from the surface of the Earth, thus providing the scientist with some understanding of surface energy balance and associated moisture stress (Gao, 1996).

LST is one of the main land degradation indicators where defected areas are subjected to high temperature of the surface due to low canopy cover; hence, the soil is being exposed. The integration of thermal data with vegetation indices, including but not limited to NDVI and SAVI, also helps to map LST variations and basically can create a better opportunity to locate areas of degradation (Bastin et al., 2019).

Machine learning and GIS-based models help improve classification and predictive analyses and at the same time allow for large-scale monitoring (Fodor & Conde, 2023). Thermal remote sensing contributes to the sustainable land management network. These offer an early warning system, mapping of heat islands, and assessment of soil health among others (Joshi et al., 2015). Policymakers are assisted through the identification of vulnerable zones for enabling conservation measures to mitigate climate change impacts on terrestrial ecosystems. The cross-validation of multi-temporal thermal datasets gives a sense of the trend, which also contributes to off-and-on monitoring of degradation dynamics and ecosystem resilience (Zhu & Woodcock, 2014).

5.2.2 Role of Drones in Forest Conservation

One of the greatest benefits in mapping using drones is rapid data gathering and identifying real-time disturbances in forests (Koh & Wich, 2012) and an opportunity to immediately attend to various threats: from illegal logging to massive wildfires and habitat destruction (Paneque-Gálvez et al., 2014). Drones are also able to access certain terrains that are difficult or downright dangerous to field researchers (Zhang et al., 2019). This is one technological development which actually integrates the multispectral and hyperspectral sensors incorporated with the drones for various vegetation analysis (Wallace et al., 2016). This helps scientists to distinguish one tree species from the other, detecting subtle changes in forest health and assessing environmental stress factors that remain hidden in regular aerial photographs.

5.2.3 Integration of LiDAR and Drones with AI and GIS

Combining LiDAR and drones with artificial intelligence and GIS has provided a revolutionary conservation tool. AI algorithms can process and analyze vast amounts of LiDAR and drone data in an automated mode where pattern recognition becomes too complex and convoluted for manual interpretation by a human (Manfreda et al., 2018).

For instance, AI could automatically classify tree species, predict forest growth trends, and detect anomalies such as disease outbreaks or illegal deforestation activities (Duan et al., 2019); GIS platforms deliver such visualization and spatial analytic capabilities to the LiDAR and drone data to enable better conservation planning (Turner et al., 2015).

For example, overlaying multiple data layers such as forest cover, topography, and climate variables allow the development of predictive models that assess how forests will respond to environmental changes (Pelletier et al., 2019). Further, the rise in the use of open-source drone-mapping software and cloud-based LiDAR processing tools has, in a way, democratized forest monitoring, allowing even small conservation initiatives to conduct high-tech environmental assessments with minimal technical skills (Paneque-Gálvez et al., 2014).

5.2.4 GIS Applications in Land-Use Planning and Habitat Restoration

GIS is an essential tool in land-use planning and habitat restoration as it allows stakeholders to make informed decisions based on spatial analysis from terrain, vegetation, hydrology, and human settlements data layers (Malczewski, 2006). In combining multiple datasets, GIS renders support to sustainable land management strategies so that land use approaches are optimized while ecological balance dissipates minimally (Eastman, 2012).

Land Suitability Analysis

GIS is an essential tool in land-use planning and habitat restoration as it allows stakeholders to make informed decisions based on spatial analysis from terrain, vegetation, hydrology, and human settlements data layers. In combining multiple datasets, GIS renders support to sustainable land management strategies so that land use approaches are optimized while ecological balance dissipates minimally. Suitability analysis could be undertaken with GIS, whereby multiple criteria such as soil quality, topography, water availability, and climate conditions are assessed. Such an application is pertinent to reforestation planning because this type of analysis allows the identification of the extent of degraded areas that are best suited for afforestation efforts.

It plays an extremely critical role in control of urban expansion through the sedentary nature of its terrestrial planning models, thus preventing ever-encroaching urbanization into ecologically sensitive zones. GIS also plays an important role in agricultural zoning, helping to orient suitable land uses toward sustainable farming while at the same time being least obtrusive toward deforestation. One common GIS-based land suitability model is the Analytical Hierarchy Process, which ranks areas for restoration potential based on the weighted scores assigned to various land attributes (Saaty, 1980).

Habitat Connectivity and Biodiversity Conservation

GIS utilized in assessing ecological corridors helps create connectivity among habitats for animal movement and genetic diversity (McRae et al., 2008). Conservationists apply GIS in the identification and mapping of critical biodiversity hotspots so that prime areas for wildlife and ecosystem conservation priorities are established (Rayfield et al., 2011). Besides, GIS fashion designs for the corridors to interlink fragmented habitats, easing species movement and genetic diversity. In the same breadth, GIS can lead to models of the distribution of species and assess the habitat suitability as climate change progresses, thereby reducing the species threatened by extinction proactively in the field. Circuit-theory models and least-cost path analysis in GIS form ecological corridors designed to minimize the conflict between humans and wildlife in enabling uninterrupted landscape connectivity.

Erosion and Soil Degradation Mapping

Soil erosion and land degradation take the biggest toll in land restoration. GIS facilitates the merging of satellite imagery, topographic maps, and field studies, so as to: Model erosion risks through the Universal Soil Loss Equation (USLE) that estimates soil loss as in eq. (2):

$$A = R^{*}K^{*}LS^{*}C^{*}P \quad (2)$$

where A is predicted soil loss, R: rainfall erosivity factor, K: soil erodibility factor, LS: slope length and steepness factor, C: cover management factor, P: conservation practice factor. · Identify areas vulnerable to desertification and propose appropriate land conservation measures like agroforestry and contour farming.

Flood Risk Assessment and Wetland Restoration

Wetlands are key for biodiversity conservation and water regulation. Users of GIS map historical flood zones and predict potential future flood risks through hydrological models that allow for optimum land-use planning for flood mitigation. It also assesses the soil moisture retention capacity not to prioritize the wetland restoration sites for effective conservation work.

It plays an essential role in monitoring wetland degradation trends through time and conducting restoration project evaluation, thereby providing a base for dynamic ecosystem management. Flood models based on GIS such as the SWAT are frequently used to simulate water flow and suspended sediment transport, thus contributing to wetland rehabilitation planning (Arnold et al., 1998).

Community Engagement and Policy Implementation

Community involvement and policy application GIS adds visibility to participatory land-use planning by enabling local communities and policymakers to participate in decision-making. Interactive GIS platforms support reality perception by community stakeholders regarding land-use changes and environmental impact and thus contribute to sustainable planning (Brown & Kyttä, 2014). These platforms facilitate the development of participatory zoning plans for reconciling conservation and development needs, leading to informed decision-making. In addition, through tracking deforestation rate and land conversion trends, GIS allows monitoring policy compliance that helps authorities in ensuring effective enforcement of environmental regulations.

6 Integrating Technology for Ecosystem Service Restoration

Ecosystem services: purification of water, regulation of air quality, support for biodiversity, and climate stabilization, among others, are of fundamental importance for maintaining balance in nature and providing human life with sustenance (Costanza et al., 1997). However, urbanization, deforestation, pollution, and climate change have seriously disrupted them, causing damage to ecosystems.

New technologies such as artificial intelligence, blockchain, and digital twin technologies offer more modern solutions aimed at improved monitoring, management, and restoration of ecosystems in the face of these challenges (Futurespore, 2023). The implementation of AI-based big data analytics into ecosystem restoration helps decision-making processes based on scientific evidence by predicting environmental changes, analyzing large datasets, and recommending sustainable land management strategies. Blockchain offers better environmental governance through transparency and accountability by enabling secure and systematic tracking of conservation programs, carbon credits, and pollution policies. Digital twin technology provides an opportunity to create virtual models of ecosystems as a testing platform for policymakers and scientists to simulate different strategies of restoration before implementing them in the real-world environment (Digital AECOM, 2023). Combined, these technologies ensure that nature can be really restored to its former glory, which has been lost during the industrialization period and the ensuing climate change.

6.1 AI-Powered Decision Support Systems for Land Management

AI DSS contribute to land management (Eppink et al., 2012), and land-use by gathering information in real time from multiple sources: satellite imagery, climate models, and sensor networks. These highly effective work horses analyze the data

Table 4 Comparison of AI, IoT, and blockchain in conservation

Technology	Strengths	Challenges	Example use case
AI	Predictive modeling, automated species detection	High computational cost	AI-driven species identification
IoT	Real-time monitoring, automation	Data privacy concerns	Smart sensors for air quality
Blockchain	Transparent governance, secure transactions	High energy consumption	Decentralized land registry

and enable land managers to make well-informed decisions by predicting potential environmental threats and suggesting sustainable land-use strategies (Nybl AI, 2024). AI-based precision agriculture works by deploying smart algorithms to analyze soil composition, weather patterns, and crop health for resource optimization. The practice minimizes water wastage and dependence on chemical fertilizers, thus abetting ecosystem restoration to prevent soil degradation and curtail loss of biodiversity.

In addition, AI also enables effective reforestation planning by determining the most suitable for tree-planting areas through a combination of soil quality assessment, climatic conditions evaluation, and historic deforestation trend analysis. It thus builds a basis for afforestation to be proved effective and durably built. AI also facilitates the monitoring of soil erosion and degradation. It does this by interpreting satellite imagery and sensor signals in real time to provide early identification of land-related deterioration and suggest management interventions or measures over and above contour plowing and soil stabilization approaches (Alliance Bioversity, 2024). One more application of prime importance is the prevention and prediction of wildfires. Engaging with climatic conditions, vegetation density, and mapped previous fire incidents to understand high-risk zones, AI systems make recommendations for reducing the chances of fire outbreaks.

AI-enabled decision support systems in land management have increased performance by reducing human error and planning that would be responsive to real time. This in turn makes AI technologies go viral, especially among governments, conservation organizations, and farmers, to bring about restoring degraded land and promoting sustainable farming, providing longevity with stability on an ecological basis. The application of multiple technologies in land conservation is shown in Table 4.

6.2 *Blockchain for Transparent Environmental Governance*

Blockchain technology is changing the world of environmental governance, mainly through its function as a decentralized, untouchable ledger for storing environmental data in a transparent and secure way (Blaize Tech, 2024). One prime field that applies blockchain for ecosystem restoration is the carbon credit and emissions tracking mechanism. Through blockchain-based smart contracts, industries and governments will be able to easily verify and trade for carbon credits, ensuring compliance with the different treaties-like global climate accords-trying to combat global warming.

Another crucial application involves monitoring and reporting of deforestation. Other conventional methods employed in tracking deforestation normally relate to the manual data recorded that may be altered or lost. Blockchain provides the answers by offering a bulletproof method of recording deforestation activities in real-time to ensure that conservation efforts are documented properly (Orbify, 2023) and the process is shown in Fig. 8.

The technology equally improves supply chain sustainability by checking if raw materials, timber, and agricultural products are derived responsibly, thus ceasing illegal deforestation and habitat destruction. When it comes to waste management and recycling, blockchain creates accountability by tracking waste disposal and recycling efforts. Verification of a proper municipal and organizational waste disposal is enabled by municipalities and organizations themselves, connecting to pollution reduction and creating a nested circular economy of sorts.

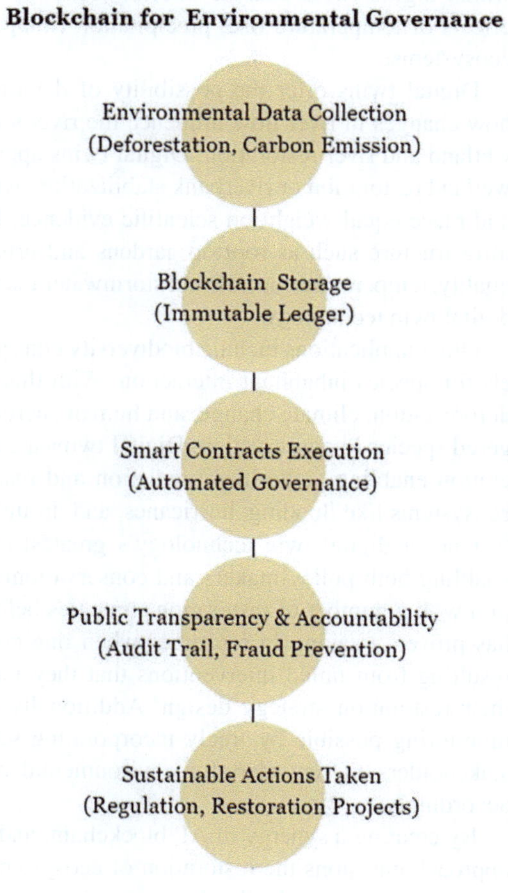

Fig. 8 Flowchart of blockchain for environmental governance

Other features allow crowdsourced environmental monitoring, meaning from varying individuals within communities or villages, reporting all sorts of environmental crimes, illegal logging, and pollution of water sources included. Stakeholders could envision better transparency through the use of blockchain technology within environmental governance, together with a reduction in fraud, and would be bringing in respect for the actual environment to conserve it forthwith. Such technology will eliminate intermediaries and bureaucratic delays from the restoration initiatives while ensuring effective monitoring and management.

6.3 Digital Twin Technology for Simulating Ecosystem Changes

Digital twin technology consists of constructing virtual replicas of ecosystems and solving various environmental scenarios to forecast the effects of various restore strategies. This technology is genuinely intended for climate change modeling in which digital twins analyze different climate variables to evaluate the long-term effects of temperature rise, precipitation changes, and extreme weather events on ecosystems.

Digital twins offer the possibility of determining scientists' understanding of how changes in river flow influence the river's biodiversity and habitat stability in wetland and river restoration. Digital twins approach restoration solutions, such as wetland restoration or riverbank stabilization, without the traditional trial-and-error and place equal weight on scientific evidence. For urban greening projects, green infrastructure such as rooftop gardens and urban forests will be modeled for air quality, temperature control, and stormwater management, availing the assistance of digital twin technology.

Other applications include biodiversity conservation with simulated virtual models for species-inhabitant interaction. With this scan obtain information regarding deforestation, climate change, and human encroachment directly to manage endangered species in conservation. Digital twins are also used for disaster impact deliberation enabling authorities prediction and reaction to natural disaster effects on ecosystems like flooding, hurricanes, and droughts.

One of digital twin technology's greatest advantages is its ability to predict, enabling both policy-makers and conservationists to conduct virtual experimentation with a number of restoration strategies before putting them into practice. This has proven a saving in resources when one considers the additional expenditure resulting from failed interventions that they must now consider trying as part of their restoration strategy design. Additionally, digital twins also make real-time monitoring possible by freely incorporating sensor data from the field, allowing stakeholders to stay alert for environmental changes and adapt their responses accordingly.

By creating a synergy of AI, blockchain, and digital twin technology, this novel approach envisions the restoration of ecosystem services. AI-based decision management systems enable land management to provide data-based advice to enhance sustainable agriculture, restoration afforestation, and wildfire prevention. Blockchain ensures total transparency and accountability in environmental governance, limiting the chances of fraud and propelling responsible conservation initiatives.

Digital twin technology allows scientists and policymakers to model different scenarios of restoration to ensure that interventions could be effective and scientifically endorsed. The potential of these technologies on ecosystem restoration would further increase as they evolve. Future AI advances will result in improved accuracy of environmental predictions, blockchain innovations further enhance transparency, and digital twins will sharpen simulations even more. With the adoption of these technologies, the governments, conservation organizations, and industries can start joining hands to restore the degraded ecosystems, safeguard biodiversity, and subsequently make way for a more sustainable planet for the future generations (GovNet, 2024).

7 Advanced Remote Sensing and GIS Applications in Land Restoration

Severe climate conditions, especially droughts, and activities caused by human beings, such as deforestation, urbanization, and industrial pollution, greatly contribute to land degradation. Land degradation seriously threatens biodiversity, food security, climate resilience, and human livelihoods. Conventional land restoration techniques have struggled with scale, affordability, and efficacy, hindering the efforts to effectively counter large-scale environmental degradation (Lambin et al., 2001). The arrival of remote sensing (RS) and Geographic Information Systems (GIS) has brought tremendous improvement in the assessment and restoration of degraded landscapes. Remote sensing makes it possible to acquire data for extensive geographic regions without the need for direct human involvement, serving as an efficient "eye in the sky." Coupled with GIS, these technologies allow for data-driven decision-making to optimize land restoration through effective monitoring, visualization, and predictive modeling.

This chapter investigates the history of remote sensing, its basic principles, and its integration with GIS for land rehabilitation. It also presents real-world applications, case studies, and success stories as well as challenges and trends in the field.

7.1 Fundamentals of Remote Sensing in Land Assessment

7.1.1 Basic Principles of Remote Sensing

Remote sensing gathers information through the examination of electromagnetic radiation that is reflected or emitted from Earth's surface. Various surfaces including water, crops, and soils absorb and reflect light in specific ways, producing spectral signatures useful in interpreting the environment. By using multi-temporal imagery, land cover alterations and degradation processes can be quantified over a period of time, enabling effective monitoring and planning for restoration.

7.1.2 Remote Sensing Platforms

Remote sensing technologies function from different platforms, each with particular applications. Airborne platforms, like drones and manned aircraft, are employed for high-resolution imaging of localized areas and are useful for detailed restoration work. Ground-based platforms, like LiDAR (Light Detection and Ranging) and terrestrial photogrammetry, enable detailed structural analysis of landscapes and vegetation. Satellite remote sensing, however, offers global coverage on a continuous basis, which is needed to monitor large-scale environmental changes and detect areas that need to be intervened upon.

7.1.3 Major Remote Sensing Data Types

Remote sensing technologies observe various components of the electromagnetic spectrum, each offering unique information regarding the terrain. Optical imagery collects visible and near-infrared data, which are commonly applied in land cover classification and vegetation monitoring. Multispectral and hyperspectral imaging provide increased spectral information, enabling researchers to distinguish among different plant types and evaluate soil condition. Radar and LiDAR information support 3D topographic modeling, which is essential for biomass estimation and studying topographic changes. Thermal infrared imaging also senses heat differences and can be employed for the assessment of soil moisture levels and drought.

7.2 GIS Applications in Land-Use Planning and Restoration

GIS takes center stage to integrate spatial data from various sources in order to inform sustainable land-use planning as well as conservation strategies. With the overlaying of various environmental data sets, GIS allows scholars to determine habitat restoration priority zones, evaluate risk zones for natural disasters, and efficiently plan on the allocation of resources. Among its major uses for land restoration, habitat restoration planning is one wherein ecological factors including soil type, water supply, and climatic conditions are determined to identify the most appropriate techniques for restoration through GIS analysis. GIS is likewise widely applied to disaster risk management, predicting, and reducing effects of floods, landslides, and droughts. Wild corridor design is also enhanced using GIS technology so that connectivity may be maintained in fragmented habitats, and biodiversity preservation is enhanced.

7.3 Case Studies and Real-World Applications

7.3.1 Satellite-Based Monitoring of Deforestation

Satellite imaging has played a key role in the detection and tracking of deforestation globally. For example, in the Amazon Rainforest, satellite imagery from the Landsat and Sentinel-2 missions has allowed authorities to monitor illegal logging in real time, enabling interventions to be made in a timely manner. In Indonesia's Peatland Restoration Program, remote sensing and GIS are employed to detect degraded peatlands, track water levels, and measure the success of reforestation efforts.

7.3.2 Drone-Based Forest Mapping for Conservation

LiDAR and multispectral camera drones have been a godsend for mapping and monitoring forest ecosystems with high accuracy. In Madagascar, drones have been used to evaluate biodiversity hotspots and inform tree-planting efforts (Jenkins et al., 2013). In California, wildfire recovery efforts are based on drone-based thermal imaging to assess vegetation recovery and soil stabilization, which informs rehabilitation planning effectively.

7.3.3 AI-Powered Decision Support Systems for Land Management

Artificial intelligence (AI) has transformed land evaluation through the automation of intricate data analysis and enhanced predictive modeling. In India, AI-based remote sensing is maximizing precision agriculture through soil health analysis, yield prediction, and suggesting best irrigation practices. Likewise, AI-based machine learning models are utilized in North Africa to forecast desertification, which allows for preemptive action to avert land degradation (Yin et al., 2021).

7.3.4 Success Stories of Technology-Driven Approaches

Among the most compelling success stories of land restoration is China's Loess Plateau Restoration Project. Through the utilization of GIS mapping and remote sensing, researchers converted wasteland into arable land, decreasing soil erosion by 70% and boosting agricultural output. Another success story is the Great Green Wall Project in Africa, where remote sensing tracks afforestation efforts along an 8000-km corridor in the Sahel region and aids in fighting desertification and rebuilding ecosystems.

8 Challenges and Future Directions

8.1 Current Challenges

With all its transformational promise, remote sensing and GIS uses in land rehabilitation encounter a number of challenges. Availability and quality of data continue to be a central challenge since satellite imagery is easily disrupted by cloud cover and atmospheric interference. Its high costs to implement cutting-edge technologies like hyperspectral imaging and LiDAR also pose obstacles to extensive uptake, especially among developing countries. Moreover, multi-source data fusion also poses technical challenges, with advanced processing methods needed to integrate data from satellite imaging, drones, and field monitoring. Policy and governance challenges also hinder implementation, since most nations do not have data-sharing and land-use planning regulations in place.

8.2 Future Directions

New technologies like AI, blockchain, and digital twin ecosystems are huge potential game-changers for the future of land restoration. Artificial intelligence-powered big data analysis will boost automation in tracking large areas, making restoration work more efficient (FAO, 2019). Blockchain can bring transparency to eco-governance by providing tamper-proof records for conservation efforts, carbon credit exchange, and sustainable land use (Schmidt & Skidmore, 2003). Digital twin ecosystems that create virtual replicas of landscapes will enable researchers to test restoration effects virtually prior to implementation, making optimal resource use and maximizing environmental returns possible. In addition, international cooperation and open-access data portals will also become vital to extending remote sensing, and GIS accessibility to policymakers and conservationists across the globe is shown in Table 5. Interdisciplinary collaboration, with the governments and institutions forming partnerships, will enhance land restoration innovation while maintaining long-term sustainability.

Remote sensing, GIS, AI, and cutting-edge digital technologies are converging to change the landscape of land restoration. These cutting-edge technologies facilitate high-accuracy monitoring, predictive modeling, and data-based decision-making, improving restoration quality, scalability, and affordability. Challenges persist in terms of data access, technology expense, and policy constraints, but continuous

Table 5 Success stories of technology-driven ecosystem restoration

Country	Technology used	Key outcome
Brazil	AI and drones	30% increase in reforested areas
India	IoT sensors for water management	25% improvement in groundwater levels
Australia	Blockchain for land rights	Secured indigenous conservation efforts

innovation in AI, blockchain, and real-time environmental monitoring is leading toward more resilient and sustainable ecosystems. Through ongoing investments in advanced technologies and increasing global cooperation, we can create new solutions to halt land degradation and build a healthier planet for our children.

References

Alliance Bioversity. (2024). *Webinar: The role of artificial intelligence in landscape restoration and soil health.*

Almeida, J., Silva, R., & Costa, F. (2021). IoT and drones for forest conservation: A case study in the Amazon. *Environmental Monitoring and Assessment, 193*(4), 120–134.

Arnold, J. G., Srinivasan, R., Muttiah, R. S., & Williams, J. R. (1998). Large area hydrologic modeling and assessment part I: Model development. *Journal of the American Water Resources Association, 34*(1), 73–89.

Asner, G. P., Knapp, D. E., Balaji, A., & Paez-Acosta, G. (2009). Automated mapping of tropical deforestation and forest degradation: CLASlite. *Journal of Applied Remote Sensing, 3*(1), 033543. https://doi.org/10.1117/1.3223675

Baek, S. K., Kim, H., & Park, S. (2021). Use of drone technology in afforestation and conservation efforts: A case study. *Forest Science and Technology, 17*(3), 45–60.

Baratchi, M., Meratnia, N., Havinga, P. J., Skidmore, A. K., & Toxopeus, B. (2021). Predicting land degradation using machine learning: A case study in Central Asia. *Environmental Research, 195*, 110798.

Bastin, J.-F., Finegold, Y., Garcia, C., Mollicone, D., Rezende, M., Routh, D., et al. (2019). The global tree restoration potential. *Science, 365*(6448), 76–79. https://doi.org/10.1126/science.aax0848

Bhattacharyya, S., Banerjee, A., & Ghosh, A. (2021). Smart livestock monitoring using IoT: Challenges and opportunities. *Computers and Electronics in Agriculture, 186*, 106257.

Blaize Tech. (2024). Embracing blockchain for carbon credits and green tech.

Bohmann, K., Evans, A., Gilbert, M. T., Carvalho, G. R., Creer, S., & Knapp, M. (2014). Environmental DNA for biodiversity monitoring and conservation. *Trends in Ecology & Evolution, 29*(6), 358–367.

Brown, G., & Kyttä, M. (2014). Key issues and research priorities for public participation GIS (PPGIS): A synthesis based on empirical research. *Applied Geography, 46*, 122–136.

Burke, A., Frazier, C., & Rahman, M. (2019). AI-driven drones for wildlife monitoring in conservation areas. *Remote Sensing in Ecology and Conservation, 5*(2), 134–147.

Butler, R. A. (2020). Rainforest connection: IoT for monitoring illegal logging. *Conservation Science and Practice, 2*(10), e01105.

Chuvieco, E. (2020). *Fundamentals of satellite remote sensing: An environmental approach.* CRC Press.

Costanza, R., et al. (1997). The value of the world's ecosystem services and natural capital. *Nature, 387*, 253.

Curtis, P. G., et al. (2018). Classifying drivers of global forest loss. *Science, 361*(6407), 1108–1111. https://doi.org/10.1126/science.aau3445

Digital AECOM (2023). "Digital twins for environmental conservation.".

Duarte, M. H., Cardoso, R. C., & Santos, J. (2022). AI-based wildfire prediction and risk assessment: A review. *Environmental Science & Policy, 127*, 150–164.

Eastman, J. R. (2012). *IDRISI Selva Manual.* Clark University.

Eppink, F. V., Werntze, A., Mäs, S., Popp, A., & Seppelt, R. (2012). Land management and ecosystem services how collaborative research programmes can support better policies. *Gaia-Ecological Perspectives for Science and Society*. https://doi.org/10.14512/GAIA.21.1.14

Esri. (2023). *GIS in ecological restoration*. ArcGIS Storymaps.

Farmonaut. (2024). *Revolutionizing agriculture: How remote sensing and GIS are transforming precision farming and crop yield*. Farmonaut Agricultural Research.

Fodor, G., & Conde, M. V. (2023). Rapid deforestation and burned area detection using deep multimodal learning on satellite imagery. *arXiv preprint arXiv:2307.04916*. https://arxiv.org/abs/2307.04916.

FutureSpore. (2023). *Revolutionizing ecological restoration with AI*. FutureSpore Medium Blog.

Gao, B.-C. (1996). NDWI—A normalized difference water index for remote sensing of vegetation liquid water from space. *Remote Sensing of Environment, 58*(3), 257–266. https://doi.org/10.1016/S0034-4257(96)00067-3

Gao, L., Zhang, Y., & Li, S. (2021). IoT-based air quality monitoring systems: A review. *Environmental Pollution, 283*, 117080.

Gaston, K. J., & O'Neill, M. A. (2004). Automated species identification: Why not? *Philosophical Transactions of the Royal Society B: Biological Sciences, 359*(1444), 655–667. https://doi.org/10.1098/rstb.2003.1442

Gholami, A., Asghari, R., & Farahani, B. (2022). IoT-based environmental monitoring for conservation: A real-time approach. *Journal of Environmental Management, 310*, 114762.

GovNet. (2024). *Harnessing digital twins for climate change solutions*.

Hansen, M. C., et al. (2013). High-resolution global maps of 21st-century forest cover change. *Science, 342*(6160), 850–853. https://doi.org/10.1126/science.1244693

Hansen, M. C., Potapov, P. V., Moore, R., Hancher, M., Turubanova, S. A., & Tyukavina, A. (2019). High-resolution global maps of 21st-century forest cover change. *Science, 342*(6160), 850–853.

Isabelle, D. A., & Westerlund, M. (2022). A review and categorization of artificial intelligence-based opportunities in wildlife, ocean, and land conservation. *Sustainability, 14*(4), 1979. https://doi.org/10.3390/su14041979

Jha, K., Doshi, A., Patel, P., & Shah, M. (2019). Precision agriculture using IoT and machine learning. *Computers and Electronics in Agriculture, 165*, 104968.

Jones, R., Patel, S., & Li, F. (2024). UAVs for improving seasonal vegetation assessment in arid ecosystems. *Frontiers in Environmental Science, 12*, Article 1366712.

Joshi, N., Mitchard, E. T. A., Woo, N., Torres, J., Moll, R. J., Ehammer, A., et al. (2015). Mapping dynamics of deforestation and forest degradation in tropical forests using radar satellite data. *Environmental Research Letters, 10*(3), 034014. https://doi.org/10.1088/1748-9326/10/3/034014

Kattenborn, T., Schiefer, F., Frey, J., Feilhauer, H., & Lausch, A. (2021). Machine learning for vegetation classification in remote sensing. *International Journal of Applied Earth Observation and Geoinformation, 102*, 102394.

Kellenberger, B., Marcos, D., & Tuia, D. (2018). Detecting mammals in UAV images: Best practices to address a substantially imbalanced dataset with deep learning. *Remote Sensing of Environment, 216*, 139–153. https://doi.org/10.1016/j.rse.2018.06.028

Kelling, S. (2018). *Technology developments for biodiversity monitoring and conservation* (Vol. 2). Pensoft Publishers.

Koeppel, J. F., Borrell, J. S., & Antonelli, A. (2021). Genetic engineering for conservation: Challenges and opportunities. *Nature Reviews Genetics, 22*(5), 321–334.

Koh, L. P., & Wich, S. A. (2012). Dawn of drone ecology: Low-cost autonomous aerial vehicles for conservation. *Tropical Conservation Science, 5*(2), 121–132.

Kshetri, N. (2021). Blockchain and sustainable land management: Applications in forestry and agriculture. *Sustainability, 13*(12), 6591.

Lambin, E. F., et al. (2001). The causes of land-use and land-cover change. *Annual Review of Environment and Resources*.

Lewis, S. L., et al. (2015). Dynamics, drivers, and ecological impacts of global forest loss. *Annual Review of Environment and Resources, 40*, 415–444. https://doi.org/10.1146/annurev-environ-102014-021031

Liu, Z., Liu, X., & Wang, Y. (2022). Deep learning applications in ecosystem modeling: Opportunities and challenges. *Ecological Modelling, 469*, 110002.

Malczewski, J. (2006). GIS-based multi-criteria decision analysis: A survey of the literature. *International Journal of Geographical Information Science, 20*(7), 703–726.

Manfreda, S., McCabe, M. F., Miller, P. E., Lucas, R., Pajuelo Madrigal, V., Mallinis, G., et al. (2018). On the use of unmanned aerial systems for environmental monitoring. *Remote Sensing, 10*(4), 641.

McRae, B. H., Hall, S. A., Beier, P., & Theobald, D. M. (2008). Circuitscape: A tool for modeling functional landscape connectivity. *Environmental Modelling & Software, 23*(5), 733–741.

Mondal, P., McDermid, S. S., & Qadir, A. (2020). A reporting framework for Sustainable Development Goal 15: Multi-scale monitoring of forest degradation using MODIS, Landsat and Sentinel data. *Remote Sensing of Environment, 237*, 111592. https://doi.org/10.1016/j.rse.2020.111592

Mutanga, O., & Kumar, L. (2019). Google Earth engine applications. *Remote Sensing, 11*(5), 591. https://doi.org/10.3390/rs11050591

Norouzzadeh, M. S., Nguyen, A., Kosmala, M., Swanson, A., Palmer, M. S., & Packer, C. (2018). Automatically identifying, counting, and describing wild animals in camera-trap images with deep learning. *Proceedings of the National Academy of Sciences, 115*(25), E5716–E5725.

Nybl AI. (2024). *AI for a greener future: The role of AI in supporting sustainable land management.*

Orbify. (2023). *Using blockchain for environmental monitoring*.

Paneque-Gálvez, J., McCall, M. K., Napoletano, B. M., Wich, S. A., & Koh, L. P. (2014). Small drones for community-based Forest monitoring: An assessment of their feasibility and potential in tropical areas. *Forests, 5*(6), 1481–1507.

Pelletier, J., Rebaudo, F., & Carrière, S. M. (2019). Unmanned aerial vehicles for insect ecology and conservation: A review. *Science of the Total Environment, 682*, 210–220.

Pereira, P., Bogunović, I., Muñoz-Rojas, M., & Brevik, E. C. (2017). Soil ecosystem services, sustainability, valuation, and management. *Current Opinion in Environmental Science & Health, 5*, 7–13.

Pettorelli, N., Laurance, W. F., O'Brien, T. G., Wegmann, M., Nagendra, H., & Turner, W. (2018). Satellite remote sensing for applied ecologists: Opportunities and challenges. *Journal of Applied Ecology, 55*(2), 524–537.

Ramos, A., Ferreira, J., & Mendes, R. (2022). AI-driven wildfire prediction using IoT sensors. *Journal of Environmental Management, 311*, 114823.

Rayfield, B., Fortin, M. J., & Fall, A. (2011). Connectivity for conservation: A framework to classify network measures. *Ecology, 92*(4), 847–858.

Reiche, J., Hamunyela, E., Verbesselt, J., Hoekman, D. H., & Herold, M. (2018). Improving near-real-time deforestation monitoring in tropical dry forests by combining optical and SAR satellite data. *Remote Sensing of Environment, 204*, 147–161. https://doi.org/10.1016/j.rse.2017.10.034

Reinsberg, B. (2020). Smart contracts for land conservation: Exploring the potential of blockchain in ecosystem restoration. *Environmental Policy and Governance, 30*(4), 220–233.

Saaty, T. L. (1980). *The analytic hierarchy process: Planning, priority setting, resource allocation.* McGraw-Hill.

ScienceDirect. (2024). Integrating remote sensing and geospatial big data for urban land use sustainability. *Urban Planning & Geospatial Analysis Journal, 29*(1), 88–105.

Shendryk, Y., Röder, A., Hill, M. J., & Gessner, U. (2021). Machine learning applications in land degradation and conservation assessment. *Remote Sensing, 13*(4), 810.

Smith, J., Brown, K., & Taylor, M. (2023). Remote sensing application in ecological restoration monitoring. *Remote Sensing, 16*(12), 2204.

Swami, N. (2021). *Applying AI to conservation challenges* (pp. 17–28). Elsevier.

Thorn, J. P. R., Snaddon, J. L., Waldron, A., Kok, K., Zhou, W., Bhagwat, S. A., Willis, K. J., & Petrokofsky, G. (2015). How effective are on-farm conservation land management strategies for preserving ecosystem services in developing countries? A systematic map protocol. *Environmental Evidence, 4*(1), 11. https://doi.org/10.1186/s13750-015-0039-0

Turner, W., Spector, S., Gardiner, N., Fladeland, M., Sterling, E., & Steininger, M. (2019). Remote sensing for biodiversity science and conservation. *Trends in Ecology & Evolution, 18*(6), 306–314.

Tyukavina, A., et al. (2017). Types and rates of forest disturbance in Brazilian Legal Amazon, 2000–2013. *Science Advances, 3*(4), e1601047. https://doi.org/10.1126/sciadv.1601047

Verbesselt, J., Hyndman, R., Newnham, G., & Culvenor, D. (2010). Detecting trend and seasonal changes in satellite image time series. *Remote Sensing of Environment, 114*(1), 106–115. https://doi.org/10.1016/j.rse.2009.08.014

Wäldchen, J., & Mäder, P. (2018). Machine learning for image-based species identification. *Methods in Ecology and Evolution, 9*(11), 2216–2225. https://doi.org/10.1111/2041-210X.13075

Wallace, L., Lucieer, A., Watson, C., & Turner, D. (2016). Development of a UAV-LiDAR system with application to forest inventory. *Remote Sensing, 8*(5), 393.

Wang, J., Zhen, J., Hu, W., Chen, S., Lizaga, I., Zeraatpisheh, M., & Yang, X. (2023). Remote sensing of soil degradation: Progress and perspective. *International Soil and Water Conservation Research, 11*(3), 429–454.

Wegmann, M., Leutner, B., & Dech, S. (2020). *Remote sensing and GIS for ecologists: Using open source software*. Pelagic Publishing.

Wilson, T. (2022). *Applications of GIS in sustainability*. IGNES Reports.

Xu, P., Li, D., & Wang, R. (2022). Robotic technologies for environmental monitoring and conservation. *Environmental Science & Technology, 56*(3), 1745–1753.

Yin, X., Li, J., Kadry, S., & Sanz-Prieto, I. (2021). Artificial intelligence assisted intelligent planning framework for environmental restoration of terrestrial ecosystems. *Environmental Impact Assessment Review, 86*, 106493. https://doi.org/10.1016/j.eiar.2020.106493

Zhang, J., Kulhavy, D. L., & Unger, D. R. (2019). Assessing drone mapping capabilities and increased cognitive retention using interactive hands-on natural resource instruction. *Journal of Forestry, 117*(6), 560–567.

Zhang, Y., Chen, X., & Zhao, Z. (2021a). AI for wildfire risk prediction: A deep learning approach. *Journal of Environmental Management, 280*, 111668.

Zhang, H., Martinez, P., & Lee, D. (2021b). Using remote sensing data for nature-based decision making. *Nature Tech Collective Report, 18*(4), 75–89.

Zhu, Z., & Woodcock, C. E. (2014). Continuous change detection and classification of land cover using all available Landsat data. *Remote Sensing of Environment, 144*, 152–171. https://doi.org/10.1016/j.rse.2014.01.011

Balancing Development and Conservation

Tiny S. Palathara and Sharaf Ali V

1 Introduction

Biodiversity serves as the cornerstone of human survival and development. However, in the recent decades, the rapid decline in global biodiversity has emerged as a significant threat to human well-being (Ceballos et al., 2017). The debate over development and conservation are not so old in history. The conflict arises from the individual's focus on development and the finite earth's capacity of an ecosystem. Historically economic growth can be achieved through industrial expansion, usage of technology, globalization, urbanization which is essential for improving living standards, reducing poverty, and innovation. However, these actions often result in the exploitation of natural resources, leading to environmental degradation. Despite these challenges, economic progress remains crucial for addressing the basic needs of a growing population. It is interesting to quote that the connection between these has been identified recently, only after economic thinkers started discussing the similarities and differences between economic growth and economic development (Majumder, 2005). What matters the most? The development is considered as a holistic approach to eradicating poverty, improving healthcare, enhancing education and essential infrastructure. Is it possible to achieve all these without development? No, without development, billions would lack access to clean water, sanitation and electricity, healthcare and education systems, and would increase the gap between the rich and the poor. This presents a complex trade-off between improving human well-being and keeping biodiversity to its fullest potential. The growing population and never-ending demands of this growing population further intensify this issue to

T. S. Palathara (✉)
CHRIST Deemed to be University, Pune, India
e-mail: tiny.palathara@christuniversity.in

Sharaf Ali V
Liwa University, Al-Ain Campus, Abu Dhabi, United Arab Emirates

J. J. Puthenkalam S.J et al. (eds.), *Terra Thrive*,
https://doi.org/10.1007/978-3-032-02738-2_4

another extent. The experts predict that by 2050, the population is going to exceed nine billion, placing immense pressure on food, water, energy and land.

Another issue that loops around is the traditional approach toward the development. The traditional development methods push biodiversity beyond its capacity and limits. Additionally, phenomena such as climate change exacerbate these issues, disproportionately to different nations. The impact of all these generally hits underdeveloped and developing countries more intensively than wealthier or developed countries. This dilemma also raises ethical questions, such as who should bear the cost of conservation. Whether the responsibility should be with developed nations or developing countries, the former is considered as nations with historical responsibility.

To address these challenges, the concept of sustainable development has been developed by the UN, emphasizing progress that meets current needs without compromising the future. Achieving this balance requires innovative and sustainable approaches, such as adopting renewable energy, smart infrastructure, and sustainable agriculture practices to reduce the environmental footprint of growth. Along with that, it is important to have better and more robust policies, including carbon taxes and green building codes, etc. to safeguard critical ecosystems while focusing on economic development. Agencies such as governments, international organizations, and private enterprises can bring inclusive solutions through different collaboration and deliberation. Addressing the conflict as a shared global responsibility and recognizing the interdependence between development and conservation can pave the way for a future where economic progress and ecological sustainability coexist.

1.1 Importance in the Current Global Context

The debate between the two emerging topics conservation and development and the need for synergizing these two has been in the discussion and recognized as one of the prime challenges of this century. The two ideas can sometimes conflict, as the development often leads to environmental calamities. According to Ceballos et al. (2017) increasing global temperature, biodiversity losses are the result of unpredicted climate changes and dangerous weather events. These situations are further compounded by the pressures of rapid urbanization and industrial growth especially in urban areas. Urbanization paves the way for economic opportunities and better facilities but on the other side it creates lots of risks to society. Rapid and unplanned urbanization often leads to social and environmental instability. Similarly, technological innovation and job creation contributes significantly to pollution and resource depletion, as a result of industrial revolution. The poorly planned expansion and usage of technologies leaves humans exposed to dangerous situations such as exposure to UV rays, etc. Agriculture, which is essential for satisfying the needs of a growing population, often comes at the cost of deforestation and soil degradation. IPCC (2021) states that human activities have already caused approximately 1.1 °C of warming above pre-industrial levels, leading to more frequent and severe natural disasters, such as floods, droughts, and hurricanes (Aldy, 2006; Mor & Jindal, 2012). Habitat loss, driven by deforestation and land conversion, has resulted in a dramatic decline in global biodiversity, with species extinction rates now tens to hundreds of times higher than historical averages (Ceballos et al., 2017).

In the context of alarming inequality globally, this debate also plays a significant part. Millions of people around the world continue to struggle with poverty and lack access to necessities such as clean water, electricity, and healthcare. The United Nations Development Programme (UNDP, 2020) estimates that over 700 million people live in extreme poverty and nearly 1 billion lack access to electricity. This calls for actions for development, which is important for improving quality of life, fostering economic growth, and addressing global disparities. However, development can sometimes lead to exacerbating inequity, especially when it is not inclusive. Developing countries or groups of people benefit from the development process and the marginalized group of developing countries may face environmental harm created by this often widens the gap between rich and poor. Similarly, experts point out that conservation efforts can also lead to inequity. When conservation in specific areas is happening it has functioned largely in a top-down manner and often leads to injustices for the poor (Mishra et al., 2017).

In conclusion, global efforts for development driven by consumerism and conservation by protection resulted in negative consequences for each idea (Folke, 2006). This is where a holistic approach to development, which balances the economy with the environment, has the potential to satisfy the needs of human beings not at the cost of the environment. Addressing the tension between development and conservation requires a paradigm shift toward sustainable development, a development that meets the needs of the present without compromising the ability of future generations to meet their own needs (Brundtland Commission, 1987). The Sustainable Development Goals (SDGs) framed by the United Nations in 2015, focus on integrating environmental considerations into development planning and fostering innovation to reduce the environmental footprint of economic activities. Global cooperation is also essential. International agreements like the Paris Agreement provide platforms for countries to commit to climate and biodiversity goals. However, these commitments must be matched with actionable policies, adequate funding, and effective enforcement mechanisms to drive meaningful progress (OECD, 2000; OECD, 2004; UN, 1987; UN, 1997; UN, 1998; UN, 2001; UN, 2002; UN, 2015; World Bank, 1992; World Commission on Environment and Development, 1987).

2 Conceptual Framework

Addressing the tension between development and conservation requires a robust conceptual framework that integrates theoretical perspectives and clearly defines key concepts. This framework provides the foundation for understanding the inherent conflicts and synergies in achieving sustainable development. Three major theoretical frameworks which address these are

1. Sustainable Development Theory
2. Ecological Modernization Theory
3. Environmental Kuznets Curve (EKC) Hypothesis

2.1 Sustainable Development Theory

Sustainable Development Theory, grounded in the principles outlined by the Brundtland Report (1987), serves as a guiding framework for balancing economic growth, social equity and environmental conservation. This theory emphasizes the interconnectedness of these dimensions and promotes a holistic approach to addressing global issues such as poverty, inequality, and environmental degradation.

2.1.1 Core Theoretical Constructs

Intergenerational Equity: The principle of intergenerational equity asserts that current developmental actions should not deplete the natural capital necessary for future generations. Theoretically, this aligns with Rawlsian justice, which advocates for fairness in resource distribution across time. Scholars like Pearce and Turner (1990) argue that sustainable development should aim to maintain "critical natural capital" and environmental resources that are irreplaceable and vital for ecological and human health. This principle underscores the need for resource management strategies that respect planetary boundaries (Rockström et al., 2009).

Systems Thinking: Sustainable Development Theory adopts a systems approach, viewing economic, social, and environmental dimensions as interdependent components of a larger system. Daly (1996) expanded on this by advocating a "steady-state economy" that operates within the earth's ecological carrying capacity. The theory suggests that environmental degradation is not an isolated issue but a systemic one, driven by unsustainable consumption and production patterns.

The Triple Bottom Line: The theory further developed by Elkington (1997) explains the theoretical underpinnings of sustainable development by introducing the "Triple Bottom Line" framework, which balances economic prosperity, environmental health, and social equity. This framework has been influential in corporate sustainability, encouraging businesses to measure their success beyond financial profits.

Weak vs. Strong Sustainability: Theoretical debates around sustainability often center on the distinction between weak and strong sustainability. Weak sustainability posits that human-made capital can substitute for natural capital, provided the total capital stock remains intact (Solow, 1986). In contrast, strong sustainability rejects this notion, emphasizing that certain natural resources are irreplaceable and must be preserved in their original form (Costanza & Daly, 1992).

Among the widely accepted frameworks in sustainable development, the Three-Pillar Model and the Three-Nested Dependencies Model stand out as critical approaches for understanding and addressing the interplay between economic growth, environmental protection, and social well-being. These models offer distinct yet complementary perspectives, widely discussed in academic literature and applied across various sectors.

2.1.2 Three-Pillar Basic Model

The Three-Pillar Basic Model is the most commonly used framework in sustainability discourse. The concept balances three interconnected and major dimensions such as economic, environmental, and social sustainability, which ensures balanced and inclusive development. Achieving a sustainable planet requires a balance in all these or has to depend upon all these ideas. Economic sustainability tries to achieve long-term growth without harming the planet and resources. The economic sustainability concept means creating employment opportunities, and better infrastructure not at the cost of the ecosystem. On the other hand, environmental sustainability deals with ecological impact, use of resources, pollution and addressing climate change to ensure a better earth for future generations. Social sustainability considers aspects such as equity and inclusivity which tries to achieve overall human well-being. This can be achieved only by focusing on areas such as education, healthcare, and social justice (Goodland, 1995).

This model of sustainability is closely related and has been extensively utilized in policy-making and corporate reporting, particularly through the lens of the Triple bottom-line approach. It is a more holistic approach to quantifying success or happiness. The approach advocates for measuring success not only in financial terms but also by social and environmental impact (Elkington, 1997). For example, Governments may use the Three-Pillar Model to design interconnected sustainability policies, such as integrating smart agriculture initiatives such as sprinklers in the farms (environmental), job creation (economic), and social equity programs (social) (Fig. 1).

The Jawaharlal Nehru National Solar Mission (JNNSM) launched under NAPCC is a flagship renewable program of India in the year 2010. This can be considered as a perfect example of the Three-Pillar Model, where economic, environmental, and social perspectives are interconnected. The objective of the mission is to make India a global power in solar energy by reducing India's dependency on fossil fuels. The major objectives of JNNSM are scaling solar power generation up to 100 GW of installed capacity by 2022, reduction of greenhouse gas emissions, mitigation of climate change and creating jobs in the solar energy sector. This mission aligns with India's commitment to the Paris Agreement and for reducing carbon footprint.

(a) Economic perspective

- The mission has driven the growth of India's solar energy sector.
- Reduces dependency on fossil fuels.

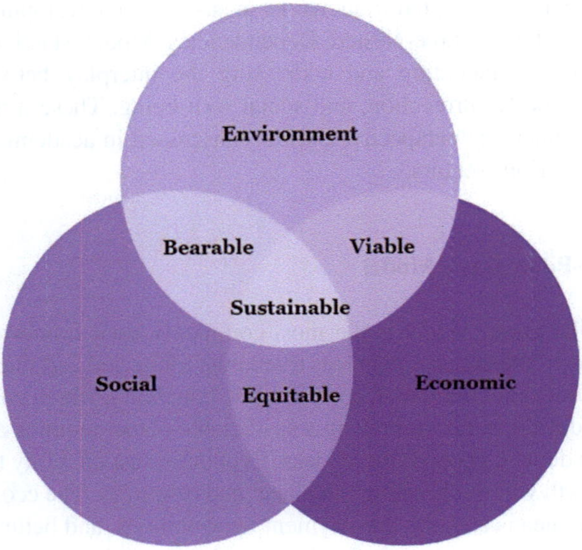

Fig. 1 Three-Pillar Basic Model

- Created thousands of jobs in manufacturing, installation, and maintenance.
- Enhanced energy security.
- Incentives for the domestic production of solar panels and technologies have boosted local industries.

(b) Environmental Perspective

- Reduction of carbon emissions and greenhouse gases.
- Reduction of pollution.
- Aligns with Paris Agreement policy.
- Climate change mitigation.

(c) Social Perspective

- Clean and affordable energy access in rural and marginalized areas.
- Promoting solar energy solutions, such as rooftop solar panels and solar lanterns.
- Creates job opportunities in the solar sector.
- Addresses energy poverty, enabling communities to access electricity for basic needs, education, and livelihood activities.
- Provision of subsidies and incentives for solar power installation by marginalized communities thereby addressing equity and inclusivity.

The National Solar Mission in India is a perfect example of how the Three-Pillar Model can guide governments in sustainable development strategies by balancing environmental goals with economic growth and social equity. However, critics like Redclift (2005) argue that treating the three dimensions as equal pillars can lead to trade-offs, where economic priorities often overshadow environmental and social considerations. This critique underscores the challenge of operationalizing the Three-Pillar Model in a way that ensures true balance among the dimensions.

2.1.3 Three-Nested Dependencies Model

The Three-Nested Dependencies Model, also known as the nested circles approach, offers a hierarchical perspective, emphasizing the dependency of three base factors such as economic, social, and environmental. It includes the economy as a subsidiary of society, and society as a subsidiary of the environment (Willard, 2010). The relationship is represented with three concentric circles where the environment forms the outermost circle, representing the foundational layer. Within the circle that represents the environment lies society, highlighting the dependence of social systems on the environment. The innermost circle is represented by economy showing its reliance on both societal and environmental aspects (Rockstrom et al., 2009). This model differs from the former by giving different weightage to these three dimensions of sustainability. This nested model moves beyond the Three-Pillar Model by considering environmental sustainability as the precondition for social and economic sustainability (Fig. 2).

The outermost layer represents the ecosystem that supports life. This emphasis on the finite nature of the environmental resources, which is considered as the foundation for any social and economic activity. Researchers identify this feature of nested models where there is a planetary boundary's framework, which identifies the ecological limits humanity must follow to have long-term sustainability (Steffen et al., 2015). Society, the second layer, emphasizes social structures within environmental limits. This highlights human well-being, equity, and inclusion. By positioning the economy as a subsystem of society and the environment, this model provides a clearer structure setting boundaries for economic activities by societal and ecological needs. For example, City planners adopting the nested approach may prioritize green infrastructure and community resilience over purely economic objectives, for sustainable development while minimizing harm to the environment.

The "Three Nested Dependency Model" and "Three-Pillar Model" have found broad acceptance globally among academicians. But yet their applications are

Fig. 2 Three Nested Dependencies Model

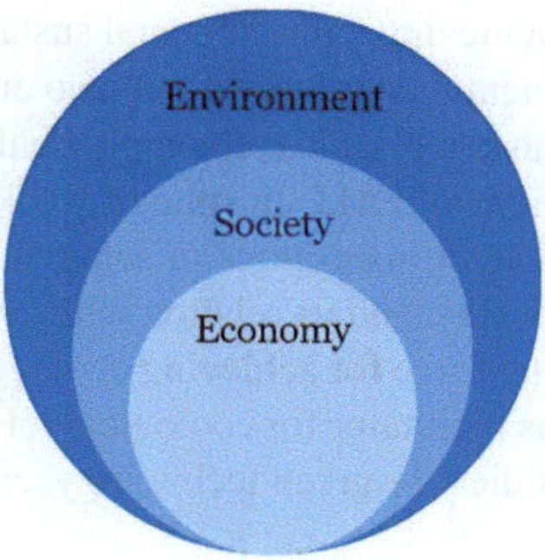

observed as limited and it differs with the context. The Three-Pillar Model, visualizes sustainability as three independent and equal pillars, often leads to trade-off between them. This model is comparatively simpler and is widely used in sustainability reporting and corporate strategies. Conversely, the Three-Nested Dependencies Model is common among ecological and policy practitioners for its focus on sustainability as a nested circle. The model sets the environment as the boundary where both social and economy depends on the outer layer. The model emphasizes the importance of the environment and is considered as pre precondition to achieve social and economic sustainability. Three-pillar models have been criticized for enabling superficial sustainability practices, such as "greenwashing" (Bocken et al., 2014). The Three-Nested Dependencies Model addresses this shortfall by emphasizing the foundational role of the environment, making it particularly relevant in addressing global crises.

2.2 *Ecological Modernization Theory*

Ecological Modernization Theory (EMT) is a school of thought that emerged in the 1980s in Europe as a response to radical environmental movements in industrial societies. The theory takes a middle path between the neo-Marxist and neo-liberals. The theory explores the relationship between economics, innovation. The early version of ecological modernization theory propounded by Joseph Huber and Arthur Mol considered industrialization as a necessary condition to achieve environmental and economical outcome. According to Mol, ecological modernization is defined as "growing independence or autonomy of ecological perspective and the emergence of ecological rationality as the basis of production and consumption" (Mol, 2002). In other words, "it means the growing institutionalization of ecological rationality in the domain of economics" (Mol, 2002). The literature on this theory emphasizes ecology and environmental goals mutually supportive and not as contradictory. It talks about the role of innovation, market, and industry in addressing environmental issues and economic growth. Unlike the Three-Pillar and Nested Model of sustainability, EMT does restrict economic growth or innovation rather it suggests that through innovative technologies it is possible to achieve environmental goals.

There are different perspectives on EMT, central to EMT is integration of advanced technologies, such as resource-efficient manufacturing processes for achieving environmental sustainability. These integrations not only reduce environmental degradation but also enhance economic efficiency and competitiveness. For instance, a cleaner supply chain and circular economy are seen as practical applications of EMT. In spite of weak theoretical formulation, the theory clearly stated the role of institutions in achieving sustainability. The effectiveness of the government or state and its relation with civil society represents coherence and will act as a driving force for achieving ecological modernization. The idea of keeping Governments as facilitator for ecological reforms, further enhances the policies, for example, subsidies for green technology, carbon taxes, and environmental regulations.

Market induced reforms is another core concept of ecological modernization which encourages markets to prioritize sustainability through innovation such as cleaner and circular economy and supply chain. In India, FabIndia emphasizes the use of organic cotton, natural dyes, and sustainable fabrics such as handwoven khadi. These materials reduce the ecological footprint of their products by minimizing water and chemical usage during production. The brand collaborates with rural artisans, promoting handloom and handicraft traditions while providing sustainable livelihoods. This approach not only preserves India's rich cultural heritage but also reduces the dependency on industrial production methods that contribute to pollution. Fabindia encourages customers to embrace slow fashion by purchasing high-quality, durable products that last longer. Their initiatives also involve reducing textile waste by repurposing leftover materials for accessories and home decor items.

Even though most of the nations internalized the environmental agendas and objectives in their policies, EMT is not free from critics. Critics argue that its idea to focus on technology overlooks the systematic need to address some of the rooted environmental issues that persist globally. Additionally, another major issue is the underestimation of cost that is being created as a byproduct to these technologies. Ecological modernization can be practiced only if there is strong and sufficient political, societal, economical and organizational capacity.

2.3 Environmental Kuznets Curve (EKC) Hypothesis

The EKC hypothesis, which explains the relationship between economic growth and environmental degradation, is widely accepted among academicians. The theory says, as the economy starts growing initially there will be an increase in environmental degradation but after a point its start declines even if there is increase in economic growth. This is represented in an inverted U-shape which represents the relationship between economic growth and environment quality. This has been tested empirically in different nations.

Panayotou (1993) provided an early rationale for the existence of an EKC: "If there were no change in the structure or technology of the economy, pure growth in the scale of the economy would result in a proportional growth in pollution and other environmental impacts." This is identified as the scale effect, the traditional view that economic development and environmental quality are conflicting goals. Proponents of the EKC hypothesis argue that "*at higher levels of development, structural change towards information-intensive industries and services, coupled with increased environmental awareness, enforcement of environmental regulations, better technology and higher environmental expenditures, result in leveling off and gradual decline of environmental degradation*" (Panayotou, 1993) (Fig. 3).

The theory concludes similarly to Simon's Kuznets Curve which is based on the relationship between income inequality and economic development. Early stages of economic development are typically resource-intensive, leading to significant environmental harm.

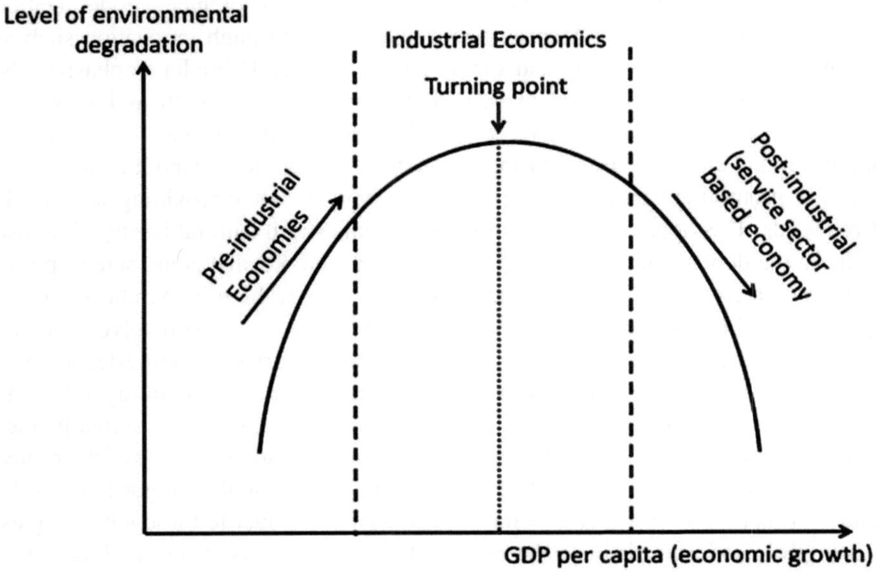

Fig. 3 Environmental Kuznets Curve

2.3.1 Key Mechanisms Behind the EKC Hypothesis

- *Scale Effect*: At lower income levels, economic growth involves large-scale exploitation of natural resources and heavy industrialization, causing environmental degradation.
- *Composition Effect*: As economies grow, there is a structural shift from resource-dependent industries to service-oriented and technology-driven sectors, which are less polluting.
- *Technological Effect*: Higher income levels facilitate investment in cleaner and more efficient technologies, leading to improved environmental outcomes.
- *Policy and Awareness Effect*: Economic growth often brings heightened public awareness of environmental issues and stronger governmental capacity to enforce environmental regulations.

The first empirical evidence was given by Grossman and Krueger (1991) where they analyzed the impact of trade liberalization on the environment. The study concluded the relationship as inverted U-shape. The study explains that as nations incomes rise, the focus shifts towards cleaner technologies and more sustainable practices, which will result in a better environmental condition leading to an inverted U-shaped relationship. Later economists like Stern (2004) critiqued the EKC hypothesis, arguing that its applicability varies significantly across different pollutants and geographic regions. They argued that economic growth alone cannot guarantee environmental improvement. Dinda (2004) extended the EKC framework, emphasizing the role of globalization, innovation, and governance in achieving the "turning point" where environmental degradation begins to decline. Economists suggest that the EKC hypothesis holds practical significance in framing policies by government. In developing countries there is a need for international collaboration and better technology. Similarly in advanced countries, maintaining investments and understanding and addressing global issues is important.

In conclusion, the environmental Kuznets curve hypothesis provides a framework for understanding the interplay between economic growth and environmental quality. However, its applicability depends on contextual factors such as governance, institutional capacity, and the nature of economic activities. The development policies in these countries should be integrated with EKC and sustainable development goals (SDGs).

3 Navigating the Nexus: Development, Conservation, and Sustainability

Development and conservation are two interconnected pillars in the pursuit of a sustainable future. Conservation aims to protect natural resources, living organisms, biodiversity and tries to mitigate climate change issues. On the other hand, development refers to the process of improving the economic and social condition of human well-being. This can be achieved only by providing the mankind necessary facilities like improved standard of living through better health and education facilities, better infrastructure. Historically these concepts are considered as mutually exclusive and the concept of "Sustainability" seeks to integrate these two dimensions, thereby restructuring the dichotomy. The concept focuses on advocating for progress that meets present needs without compromising the ability of future generations to meet theirs.

Understanding the conflicts and synergies among these concepts requires a comprehensive framework that considers both the trade-offs and mutual benefits. A healthy environment means better air, water, and fertile soil, which enhances the economic activities of human beings thereby human well-being also. This represents the overall development of a country. Similarly, practicing green technologies and integrating polices which address environmental concerns not only reduces climate related issues but rather foster economic growth. Conflicts arise when rapid industrial growth depletes natural resources or when conservation efforts restrict economic opportunities (Mor & Jindal, 2012; Norman & Macdonald, 2004; Zijp et al., 2015). However, recognizing these synergizing is crucial in framing policies for developmental and environmental goals. This framework emphasizes that development and conservation are not opposing forces but complementary goals that can coexist through thoughtful, inclusive and adaptive strategies.

4 Role of Selected SDGs: Addressing Development and Conservation

The Sustainable Development Goals (SDGs) provide a comprehensive framework to align developmental aspirations with environmental protection. Among 17 SDG goals, this chapter focuses on SDG 11 (Sustainable Cities and Communities) and SDG 15 (Life on Land). These SDGs are an international attempt to achieve a sustainable way of living by addressing the major challenges humankind faces in this century.

4.1 SDG 11: Sustainable Cities and Communities

City development and governance are considered an important factor for achieving sustainable development as it is estimated that the majority of the population will be residing in urban areas soon. The twenty-first century considers urbanization as a defining characteristic of development, with over half of the global population residing in urban areas. However, the rapid urban expansion brings significant challenges, it is estimated that more than half of the population resides in the cities and they generate approximately 80%–85% of GDP and emit nearly 70%–75% of greenhouse gases. This is major because of the increased demands and consumption of limited resources, which further increases different forms of pollution. Cities are often centers of economic activity but can also become hotspots of environmental degradation without proper planning. Sustainable cities and communities as advocated by SDG 11, seek to "ensure inclusive, safe, resilient, and sustainable urban and human settlements." The goal is for the complete eradication of slum-like situations through sustainable practices and approaches by enhancing urban governance and urban management majorly pollution and addressing climate change issues. This can be achieved as mentioned only through a set of robust rules and regulations (Franco et al., 2020).

SDG-11 is considered as a core theme by many countries and there are many live examples of sustainable urban planning including the creation of green cities, where parks, green roofs, and urban forests mitigate heat and air quality. Smart urban infrastructure, such as energy-efficient public transportation systems, waste-to-energy plants, and solar-powered street lighting, also exemplifies how technology can drive urban sustainability. Indian cities like Pune and Bengaluru have begun adopting such initiatives by promoting non-motorized transport systems, integrating renewable energy into urban grids, and expanding green spaces to combat urban sprawl. However, challenges such as inadequate funding, poor governance, and social inequalities continue to hinder widespread adoption.

4.1.1 SDG 15: Life on Land

The conservation of Life on land is recognized as the goal 15 of the SDG, which emphasizes the protection, restoration along with sustainable management of forests, reducing desertification, halting biodiversity loss. In principle SDG 15 is recognized as having equal weightage and importance along with other SDGs. Land degradation and deforestation remain critical issues, driven by urban expansion, industrial activities, and unsustainable farming methods. For example in India, since 2000 approximately 2.33 million hectares of forest cover has been lost and the deforestation rate was the second highest globally during this period.

Solutions to these issues are not easy especially when the population is so concerned about immediate and short-term results. For that it is essential to have community-driven and policy-supported conservation strategies. Biodiversity needs to be mainstreamed in the parts of governments which are responsible for agriculture, energy and water etc. The government needs to consider more long-term environmental benefits or cost than the short-term economic benefits and cost. There are few success stories globally which try to integrate the initiatives in their policies and try to achieve a sustainable living. One such initiative from India includes the Joint Forest Management (JFM) program, which empowers local communities to engage in forest conservation. The major idea behind this is "Community forestry" and the efforts in states like Odisha and Madhya Pradesh have shown how local participation can restore degraded landscapes and protect wildlife habitats. Another initiative by the Indian government is The Green India Mission, which is one of the eight national missions under the National Action Plan on Climate Change launched in 2008. Its aims on agroforestry initiatives, where they recognize the importance of forest in combating climate change and tries to provide livelihood for rural and marginalized communities. The mission is considered as the best example for green development which integrates conservation and development.

The alignment of SDG 11 and SDG 15 underscores the interconnectedness of conservation and development and how it can contribute to achieve sustainability. Creating urban green space as the center for human development will improve air and water quality, and will reduce environmental degradation. This will foster urban sustainability. By addressing urbanization challenges with sustainable practices and protecting terrestrial ecosystems, these goals provide a roadmap for achieving a harmonious balance between development and conservation. Collaboration among governments, NGOs, businesses, and communities is essential to scaling such initiatives, ensuring that economic growth does not come at the expense of the planet's ecological health. This alternative approach thus brings human and nature together, and moves away from nature-human dualism. This holistic approach is vital for fostering a future where cities and natural ecosystems coexist sustainably, reflecting the essence of the SDGs.

5 Challenges and Barriers

1. Limited Public Awareness and Participation:
   Lack of awareness is the major concern. Limited knowledge and understanding of sustainability concepts among developing countries especially among marginalized communities who are suffering from displacement effects due to environmental issues.
2. Economic Pressures:
   Conservation and development being treated or conceived mutually exclusive by governments. There is a conflict between short-term economic growth and long-term sustainability. Humans prioritize short-term benefits over long term, a normal human nature which hinder the path to achieve sustainability.
   (a) Short-term gains vs. Long-term sustainability: There is always a trade-off between conservation and development. For instance, coal mining in India provides energy security and employment but contributes to deforestation and carbon emissions.
   (b) Opportunity Costs: Investments in conservation programs which are costly are considered as unfeasible especially for developing nations facing major concerns like poverty in poverty and unemployment.
3. Research Gap:
   Lack of data and scientific research make it difficult to understand the conflict properly, which results in wrong decision-making.
4. Resource Depletion:
   Overexploitation of resources for development further worsens environmental degradation and is leading to climate change and biodiversity loss.
5. Lack of Policy Integration:
   Despite growing awareness of sustainability, there is an observable lack of integrated policies which addresses these challenges together.
   (a) Fragmented Frameworks: Conservation and development policies are often sector-specific, and are often formulated in isolation. This leads to failure of sustainable development plans.
   (b) Weak Enforcement Mechanisms: Even when integrated, implementation is hindered by poor governance and corruption.
6. Implementation Barriers:
   Even with sound policies, translating them into actionable initiatives presents numerous challenges.
   (a) Limited financial resources: Conservation projects are costly and require a lot of initial financial investment and at the same time maintenance. Development projects generally receive funds faster than for conservation projects. However, many governments prioritize immediate developmental activities over sustainability since its results in immediate returns.
   (b) Private sector engagement: The sustainability efforts from the private sector remain limited due to low returns and commitment.

5.1 Examples of Barriers in Practice

1. Economic Prioritization in Deforestation: In Brazil, large-scale deforestation of approximately 80% of amazon forest for cattle ranching and soybean farming resulted in biodiversity loss and increased carbon emissions. During this period the country became the largest exporter of Beef and soybeans which generated a huge income to the country (An example for Trade-off). This can be only resolved through sustainable supply chain management and regulatory measures.
2. Policy Weakness in Urban Pollution: In India, major cities like Delhi, Mumbai face severe air pollution crisis which highlights the gap between policy and enforcement. Despite a set of regulations on industrial emissions and vehicular pollution, rapid urbanization and old vehicles create lots of issues. Lack of monitoring and reliable data are another issue. This resulted in several health issues and economic cost especially to marginalized groups and further created inequality. An example for lack of policy implementation.
3. Funding Shortfalls in Renewable Energy: Due to limited investment and technological barriers many African nations, despite high solar potential, struggle to scale up renewable energy projects. (An example for limited financial resources.)

5.2 Overcoming Barriers

Addressing these challenges requires a multi-pronged approach:

- Governments must balance economic development with environmental protection by promoting sustainable industries and incentivizing green practices.
- Policymakers need to create integrated frameworks linking conservation and development goals, with strong enforcement mechanisms.
- Increased funding from both public and private sectors, alongside international collaboration, is critical for scaling sustainable initiatives.
- Community awareness campaigns and participatory approaches can foster public support and ensure local needs are addressed.

By addressing these barriers, governments can ensure the way for a more sustainable future that balances the needs of the present with the rights of future generations.

5.3 Pathways to Balance

1. Integrated Policy Approaches:
   Integrating economic, social and environmental dimensions is necessary for addressing challenges like these. The integration will help to achieve the balance

between conservation and development which will lead to holistic development of a nation. Integration often creates improvement in all domains by avoiding trade-off situations. Few countries that have embraced integrated policies demonstrate the potential for sustainable progress. For example, Nordic countries integrate growth polices with environmental and wellbeing policies which ensure both development and biodiversity conservation. India's National Action Plan on Climate Change (NAPCC) reduces carbon emission while creating jobs and energy security, a holistic approach.

2. Innovative Technologies:

   Innovative technologies can serve as a bridge between sustainable development and environmental conservation. There is much evidence across global technologies helping the countries to achieve the intended growth or development without harming the environment: (1) India's Jawaharlal Nehru National Solar Mission is a prime example, aiming to generate 100 GW of solar power by 2030. This initiative provides cleaner alternatives to fossil fuels, reducing carbon emissions and at the same time less usage of resources. (2) Drones, Precision farming and other smarter agriculture practices optimize the water usages and at the same increase yields without using much of resources and land. Israel's agriculture practices is a good example of how a bare land can be converted to as agricultural space and earn profit from it. (3) Green buildings incorporate energy-efficient designs, green roofs, rainwater harvesting, and solar power. This saves energy while reducing the cost. These are a few examples of how technology leads to achieving sustainability. By investing in technologies, it's possible to reduce the conflict between development and conservation.

3. Collaborative Efforts:

   Collaborative efforts are vital to achieve sustainable development by synergizing conservation and developmental perspectives. Without efficient collaborative efforts from diverse stakeholders such as governments, communities, academia and industries it is difficult to address the global challenges since they transcend geographically, socially and economically. It is evident that only collaborative efforts can create an innovative and inclusive society. (1) Projects like the Wildlife Trust of India's joint initiatives with state governments emphasize biodiversity conservation alongside community development. (2) Paris agreement where different countries and international organizations are coming together to address greenhouse gas emissions. (3) Tata Steel's initiatives in Jharkhand include afforestation programs and water conservation efforts, a good example of how private firms can contribute through Corporate Social Responsibility (CSR). (4) Community-led initiatives can enhance local livelihoods.

4. Economic Incentives:

   Economic mechanisms encourage sustainable practices by aligning environmental goals with financial benefits. (1) Green taxes on polluting industries motivate cleaner production practices. (2) Subsidies or tax reductions for eco-friendly technologies encourage widespread adoption. Subsidies for electric vehicles (EVs) in India have accelerated their market penetration.

By implementing these pathways, countries can foster a balanced relationship between development and conservation ensuring long-term sustainability.

6 Conclusion and Recommendations

This chapter explored the critical need to synergize development and conservation, case studies, challenges, barriers, and potential solutions. It highlighted the importance of integrated policies, technological advancements, collaborative efforts, and economic incentives in achieving sustainable outcomes. Recognizing this synergy is crucial for developing a policy framework for achieving environmental and developmental goals of a nation.

6.1 Actionable Recommendations

1. Strengthening Policy Frameworks:
   (a) Develop inclusive policies that bridge the gap between development and conservation perspectives.
   (b) Strengthening local government bodies so that planning and decision-making can be done case to case and special attention will be there.
2. Promoting Education and Awareness:
   (a) Domain specific or targeted awareness programs to increase the public understanding especially among marginalized communities or humans being affected because of these actions.
   (b) Inclusion of sustainability concepts in school and university curriculum.
3. Encouraging Global Cooperation:
   (a) Improve international partnerships to share resources and best practices.
   (b) Focus on global agreements like the Paris Agreement to ensure collective accountability.

As the world faces environmental challenges and socio-economic disparities at an alarming rate, balancing development and conservation is no longer optional, it is mandatory. Policymakers, businesses, and individuals must act decisively to ensure a sustainable and equitable future for generations to come. By adopting innovative solutions and fostering global cooperation, humanity can achieve the synergy between these and achieve a sustainable world.

References

Aldy, J. E. (2006). Per capita carbon dioxide emissions: Convergence or divergence? *Environmental and Resource Economics*, 33(4), 533–555

Bocken, N. M. P., Short, S. W., Rana, P., & Evans, S. (2014). A literature and practice review to develop sustainable business model archetypes. Journal of Cleaner Production, 65, 42–56. SCIRP ResearchGate

Brundtland Commission. (1987). Our common future: Report of the World Commission on Environment and Development. Oxford University Press. SD Knowledge PlatformFederal Office for Agriculture

Brundtland Report. (1987). World Commission on Environment and Development, Our Common Future. Oxford: Oxford University Press, 1987. https://sustainabledevelopment.un.org/content/documents/5987our-common-future.pdf

Ceballos, G., Ehrlich, P. R., & Raven, P. H. (2017). Accelerated modern human–induced species losses: Entering the sixth mass extinction. Science Advances, 1(5), e1400253. ScienceResearchGate

Costanza, R., & Daly, H. E. (1992). Natural capital and sustainable development. *Conservation Biology, 6*(1), 37–46.

Daly, H. E. (1996). *Beyond growth: The economics of sustainable development*. Beacon Press.

Dinda, S. (2004). Environmental Kuznets curve hypothesis: A survey. *Ecological Economics, 49*(4), 431–455.

Elkington, J. (1997). Cannibals with forks: The triple bottom line of 21st century business. Capstone Publishing.

Folke, C. (2006). Resilience: The emergence of a perspective for social–ecological systems analyses. *Global Environmental Change, 16*(3), 253–267.

Franco, C., Le Blanc, D., & Ricciardi, F. (2020). Sustainable development goals and the integration of environmental concerns. *Sustainability Science*.

Goodland, R. (1995). The concept of environmental sustainability. *Annual Review of Ecology and Systematics*, 26, 1–24.

Grossman, G. M., & Krueger, A. B. (1991). Environmental impacts of a North American free trade agreement. National Bureau of Economic Research (Working Paper No. 3914).

Intergovernmental Panel on Climate Change. (2021). Climate Change 2021: The physical science basis. Contribution of Working Group I to the Sixth Assessment Report of the Intergovernmental Panel on Climate Change. Cambridge University Press.

Majumder, M. (2005). Sustainable development indicators and policy. *Journal of Environmental Policy*.

Mishra, A., Sharma, S., & Singh, P. (2017). Environmental sustainability assessment. Environmental Impact Assessment Review.

Mol, A. P. J. (2002). Ecological modernization and the global economy. *Environmental Politics, 11*(1), 1–10.

Mor, S., & Jindal, S. (2012). Estimation of environmental Kuznets curve and Kyoto parties: A panel data analysis. International Journal of Core Engineering & Management.

Norman, W., & Macdonald, C. (2004). Getting to the bottom of "triple bottom line". *Business Ethics Quarterly*, 14, 243–262. https://doi.org/10.1016/j.ijpe.2012.01.035

OECD. (2000). *Towards sustainable development. Indicators to measure progress*. OECD Publications.

OECD. (2004). *Sustainable development in OECD countries. Getting the policies right*. OECD Publications.

Panayotou, T. (1993). *Empirical tests and policy analysis of environmental degradation at different stages of economic development*. International Labour Organization. https://ideas.repec.org/p/ilo/ilowps/992927783402676.html

Pearce, D., & Turner, R. K. (1990). Economics of natural resources and the environment. Harvester Wheatsheaf.

Redclift, M. (2005). Sustainable development (1987–2005): An oxymoron comes of age. *Sustainable Development, 13*(4), 212–227.
Rockström, J., Steffen, W., Noone, K., Persson, Å., Chapin, F. S., Lambin, E. F., & Foley, J. (2009). Planetary boundaries: Exploring the safe operating space for humanity. *Ecology and Society, 14*(2), 32.
Solow, R. M. (1986). On the intergenerational allocation of natural resources. *Scandinavian Journal of Economics, 88*(1), 141–149.
Steffen, W., Richardson, K., Rockström, J., Cornell, S. E., Fetzer, I., Bennett, E. M., ... & Sörlin, S. (2015). Planetary boundaries: Guiding human development on a changing planet. Science.
Stern, N. (2004). *The economics of climate change: The Stern review*. Cambridge University Press.
UN. (1987). *Report of the world commission on environment and development: Our common future*. Oxford University Press.
UN. (1997). *Overall progress achieved since the United Nations conference on environment and development. Commission on sustainable development fifth session (E/CN.17/1997/2)*. United Nations.
UN. (1998). *Commission on sustainable development. Report on the sixth session (E/CN.17/1998/20)*. United Nations.
UN. (2001). *Road map towards the implementation of the United Nations Millennium Declaration. Report of the Secretary-General (A/56/326)*. United Nations.
UN. (2002). *Report of the world summit on sustainable development (A/CONF.199/20)*. United Nations.
UN. (2015). *Transforming our world: The 2030 Agenda for sustainable development. Resolution adopted by the general assembly on 25 September 2015 (A/RES/70/1)*. United Nations.
United Nations Development Programme (UNDP). (2020). Human development report 2020: The next frontier-Human development and the Anthropocene. UNDP.
Willard, B. (2010). The sustainability mindset: Using the matrix map to make strategic decisions. *SustainAbility*.
World Bank. (1992). *World development report 1992: Development and the environment*. Oxford University Press.
World Commission on Environment and Development. (1987). *Our common future*. Oxford University Press.
Zijp, M. C., Heijungs, R., van der Voet, E., et al. (2015). An identification key for selecting methods for sustainability assessments. *Sustainability*, 7, 2490–2512. https://doi.org/10.3390/su7032490

Climate Change and the Future of Land Ecosystems: The Multifaceted Impact of Climate Change on Terrestrial Biodiversity, Vegetation, and Soil Health

Salu Dsouza

1 Introduction

Environmental alteration is perhaps the highest ecological trial of the twenty-first century and potentially affects every facet of life on Earth. Factors such as the burning of fossil fuels, deforestation, industrialisation, and changes to agriculture are leading anthropogenic climate change faster than ever before, causing significant disruptions in natural systems on a planetary scale. Terrestrial ecosystems, which are vital for biodiversity and the climate as well as serving humanity with essential services, have been hit particularly hard. This web of life is coming undone as a changing climate disrupts these ecosystems, with profound impacts on the ability to sustain environmental stability and human well-being globally.

The absence of an interface between the planet's climate system and land ecosystems, such as the massive tropical rainforests that cover South America and forests in Africa, all the way down into deserts like those in the Sahara, is unimaginable. Climate change processes drive and are driven by these ecosystems. They store enormous amounts of carbon, control water cycles, and host various communities of plants, animals, and microorganisms. Nevertheless, with the global transformation of the climate, changes in rainfall regimes and the incidence of risky meteorological events further stress these systems, and climate change effects on land ecosystems. The impacts of environmental change on terrestrial ecosystems are diverse and far-reaching, including changes in the distribution of species, soil health, biodiversity loss, and the provisioning of ecosystem services essential for human society.

The influences of environmental variation on biodiversity are some of the most concerning. With global warming in full swing and the apocalypse upon us, species must migrate to seek cooler homes or else perish from a terrifying failure to adapt. In fact, if species distribution is disrupted, it has many effects that cascade through

S. Dsouza (✉)
School of Law, CHRIST University, Bengaluru, India

J. J. Puthenkalam S.J et al. (eds.), *Terra Thrive*,
https://doi.org/10.1007/978-3-032-02738-2_5

environmental dynamics, leading to damage to biodiversity and an impact on ecosystem resilience as well as function. In regions like the Arctic tundra, warming is twice as fast as the global average temperature, and whole ecosystems are changing at breakneck speed. Some animals, which developed under cold conditions, are now dying out as they barely get a foothold in these environments, while others, better suited to warmer climates, move up, further destabilising them.

Climate change is altering vegetation patterns at the same time. In addition, forested areas are advancing in some high mountains (a process driven at least in part by a warming climate) and moving onto tundra previously devoid of trees. Not only do these changes in vegetation composition and structure impact ecosystem function, but they are also relevant to the global carbon cycle. Forests, the biggest carbon sinks, are also getting affected by both natural and man-made processes. Meanwhile, deforestation and the degradation of remaining forested land are reducing natural carbon sequestration by forests at a time when other human activities (burning fossil fuels or destroying peat bogs) have already released vast amounts of CO_2 to the atmosphere.

At the terrestrial level, climate change is also leading to declines in soil health—a cornerstone of all other ecological systems. Good soil is crucial for plants to grow, water conservation, and carbon capture. Nonetheless, increasing temperatures and precipitation changes are enhancing the process of soil degradation: erosion, organic matter loss, and desertification. Degraded soils offer less capacity to support vegetation growth and are accompanied by a loss of agricultural productivity, further exacerbating food insecurity in susceptible areas. What's more, degraded soils are less able to capture carbon, thus exacerbating global warming by adding more CO_2 to the atmosphere in a vicious cycle.

These are profound implications of environmental change on land bionetworks; the importance of TB, which too many should not ignore, is resulting in one more point: the disruption of ecosystem services. Pollination, water purification, flood regulation, and nutrient cycling are all ecosystem services necessary to human existence. However, as climate change restructures ecosystems, the benefits that they provide are in danger of disappearing. For instance, variations in long-term climatic patterns can alter when and where water resources are available or increase scarcity and result in a severe flood (IPCC, 2014). In the same way, reducing genetic diversity and population sizes of pollinators can severely limit the stability of such environmental amenities that are crucial for agriculture. The further loss of ecosystem services could result in a reduction in the resilience of human societies to endure peripheral tremors, most notably among those entirely reliant on ordinary means for their livelihoods.

It has been noted that regular movement of species can be tracked by the crabs' presence and damage to Willapa Bay habitats, a problem that will continue due to climate change's effects on land ecosystems. More heat and changed rainfall patterns lead to conditions that are fertile ground for invasive species, which usually outcompete natives for resources and have strong impacts on ecosystem processes. Effects: Invasive species can cause the reduction of biodiversity, change nutrient cycles, and alter physical space in ecosystems. For example, the incursion of plants that are not known in the region could change the direction of fire movement in grasslands and change soil structure or water availability, which may result in significant impacts on ecosystems as well as human communities. Invasive species also have a significant economic impact because they can cause large-scale losses in agriculture, forestry, and fisheries.

Land ecosystems are often known for their resilient and adjustable behaviour during environmental change, but researchers are pushing them too far. The changing climate is causing a ripple effect, as many species are evolving to change their ranges or behaviours and developing new physiological traits. But the speed at which such adaptations take place is often not fast enough to match changes in their environment. This highlights the need to develop efficient conservation strategies for improving ecosystem resilience. These solutions could entail anything from preserving and recovering essential habitats to removing invasive species, fostering sustainable land use practices, or addressing climate change-related fears like an increase in the ocean water damaging the low-lying habitations. Moreover, incorporating traditional ecological knowledge into contemporary conservation efforts can strengthen strategies to adjust to environmental change and foster greater adaptation in ecosystems and human communities alike.

Forests are also key to regulating the planet's climate. Plantations are the largest earthly carbon basins on our living planet, absorbing and storing vast amounts of CO_2 every year, which can contribute to environmental alteration. However, forests are also a climate change-vulnerable ecosystem, with projected changes in temperatures, precipitation, and more frequent wildfires. Deforestation and forest degradation also compound these issues, as they reduce the ability of forests to store carbon and provide critical habitats for biodiversity. Given the critical importance of forests in the global climate system, it is important to enact policies that favour sustainable forest management and its conservation, ensuring their continued function as habitats in a landscape suffering change due to this phenomenon.

This paper contains a substantial review of the effects of environmental alteration on land ecosystems, using an ecosystem-specific case study. This study provides a model towards reflecting on the environmental variation that affects the El Niño ecosystem, exploring the myriad factors promoting these changes. The following analysis highlights the broader theme of environmental modification, which influences land ecosystems using a specific ecosystem as a focal point, and it serves as an example for the challenges that managers face due to ongoing conservation and management actions.

In sum, drought affects LUE owing to residual forcing rather than the direct impact of warming on NPP itself. While challenges are enormous, mitigation and adaptation opportunities exist. Supporting critical habitat and ecosystem restoration, improving the resilience of their natural areas, and balancing land use changes with climate plans to mitigate adverse effects from climate change are essential strategies in managing the impacts on land ecosystems under a new "normal".

2 Effect of Environmental Alteration on Land Biodiversity

It is becoming ever clearer that human-induced climate change ranks as one of the most important threats to global biodiversity. The increasing global temperatures, alteration of precipitation patterns, and expansion in frequency/value of extreme weather events are shaping terrestrial ecosystems to an extent previously unseen. These changes are affecting the patterns of species distribution and migration, and in some respects they have been threatening biodiversity that is essential for ecosystem stability and ultimately human welfare.

2.1 Species Distribution and Rising Temperatures

Last up front, the global average temperature has increased by ~1.2 °C since the pre-industrial period, and it is this warming trend that we must stop in order to maintain a stable earth; otherwise, Earth will continue getting warmer within 100 years (Brown et al., 2016). This warming alters the distribution of species. The shift in response to changing thermal environments is likely partly the result of many species moving their ranges upward as well. Species are e.g. moving polewards or to higher altitudes as they follow the cool temperatures that suit their bodies; many plant and animal species in temperate regions do this (Parmesan & Yohe, 2003). This pattern, known as phenological change, has been observed across taxa, from birds and mammals to insects and plants (Walther et al., 2002).

For example, the mountain pine beetle (Dendroctonus ponderosae) has extended its range further into upper elevations and northern latitudes, inflicting considerable damage on North American pine forests. Pecl et al. (2017) reminiscence of environmental responses to recent climate changes disrupting geographical distributions. For example, the beetle has had impacts on disease prevalence and forest dieoff (Tingley and Beissinger, btn) by extending to new areas due to climate warming in recent decades. In the European Alps, for example, several alpine plant species are moving to higher elevations in search of cooler environments; however, this endangers lowland plants that cannot migrate or acclimatise quickly enough (Pörtner et al., 2021).

Of course, this all has repercussions. Species also face unanticipated competitors, predators, and parasites due to the change in their range, which leads to altering previous ecological relationships (Warren et al., 2018). In North America, the expansion of C. latrans into new habitats has been associated with displacement and structuring of native species (e.g., v. vulpes) by coyotes due to altered predator-prey dynamics at a range margin; their presence can lead to cascading trophic interactions that alter local diversity patterns (Willis & Bhagwat, 2009). Second, many species are moving up mountain tops towards hotter temperatures, and some of these will experience “mountain top extinctions” where they can’t go higher, with the result being localised extinction (Pecl et al., 2017).

2.2 *Changing Rainfall and Ecosystem Sustainability*

Climate change thus results in an increase in global temperature (Brown et al., 2016) and modifies the distribution of water on Earth as a whole continentally, also being a critical resource to terrestrial life. These changes in precipitation are perturbing the structure and functioning of ecosystems globally, with potentially far-reaching consequences for biodiversity. Some areas are experiencing more erratic precipitation, ranging from heavy rainfall to long dry periods. Species adapted to stable hydrological conditions are therefore especially challenged by this variability in water availability (Walther et al., 2002).

For example, in Mediterranean climate systems typified by hot, dry summers followed by mild wintertime downpours, the number of intense droughts has increased noticeably in recent years (Parmesan & Yohe, 2003). For example, drought-tolerant species expanded or increased in abundance while mesic-adapted plant populations decreased (Warren et al., 2018). Such changes in plant communities influence the animals relying on them and have a cascade of effects throughout the entire ecosystem (Tingley & Beissinger, 2009).

For instance, in the tropics, rainforests are usually predictable and damp, but changing rainfall will drive forest fragmentation there as well, harming already moist habitat-dependent species (Pecl et al., 2017). Amphibians, for instance, are exposed to sharp moisture fluctuations in dry seasons that might be getting worse and more frequent (Willis & Bhagwat, 2009). The negative effects from the reduction in amphibian populations can be severe, as these species are important for insect control and serve as prey animals for higher trophic levels (Brown et al., 2016).

Rainfall is altering the distribution and abundance of herbivores and their predators in African savannas (Pörtner et al., 2021). Research indicates that during droughts, there is less forage available to graze on by herbivores, and this leads to reduced populations of species like the wildebeest (Connochaetes taurinus) and zebra (Equus quagga). This impacts predators such as lions (Panthera leo) that depend on these herbivores for food, in turn. Moreover, the increased frequency of extreme weather events such as floods and droughts is enhancing these impacts, resulting in further perturbations to this ecosystem (Tingley & Beissinger, 2009).

2.3 *Habitat Destroying Extreme Weather Events*

Climate change is causing extreme weather events, such as hurricanes, floods, droughts, and wildfires, to increase in frequency (Parmesan & Yohe, 2003). These events often change the habitat immediately, thus diminishing it catastrophically and, on a longer-term scale, rearranging the functioning of ecosystems. Extreme events, such as the wildfires that have been and are likely to be more severe due to a warmer planet in many regions across the globe, can wipe out vast forests with all their critical habitats for numbers of species (Pörtner et al., 2021).

Despite the fire-return intervals and ecosystem type, peatlands globally have been thought to contain a large amount of carbon, Ducey et al. This phenomenon is responsible for the disappearance—not only of habitats, but moreover negatively enhances itself in a vicious cycle: it releases vast amounts of CO_2 into the atmosphere and contributes significantly to climate change. A defending function of forest cover is officially binding CO_2, partially avoiding terrible phenomena like global warming (Pecl et al., 2017).

Mangrove, coral reef, and salt marsh habitats, which provide significant ecosystem services, are being lost through rising sea levels and storm surges in coastal ecosystems (Brown et al., 2016). They are essential to preserving shorelines from erosion, nursery habitats for marine species, and rich biodiversity (Pörtner et al., 2021). The destruction of these habitats by the grandiflorus weather events may greatly reduce biodiversity and human defences on ecosystems, causing multiple toteries in both directions (Tingley & Beissinger, 2009).

In addition, they might migrate to other regions that would partially depend on the outcome of extreme events (Warren et al., 2018). For example, a series of experiences during and following this same Hurricane Katrina also made recordings that various bird species in the Gulf Coast (USA) were being displaced from their natural habitat, and failure to colonise by some resulted in population decline (Pecl et al., 2017). Meanwhile, localised events linked to heatwaves like coral bleaching have resulted in dramatic declines of coral species worldwide with significant knock-on effects to the marine life that is sustained by coral reefs (Willis & Bhagwat, 2009).

2.4 *Breaking the Seamless Flows of Migration*

Climate change is causing long-distance migrations of many species to be largely disrupted due to changes in patterns (Parmesan & Yohe, 2003). Migratory species depend on a range of environmental cues that are also themselves influenced by temperature (e.g., photoperiod, food availability); daylight time is now changing in ways difficult to predict and will probably do so more profoundly with warming under climate change (reviewed by Brown et al., 2016). All of these cues are changing in the face of climate change, making it even harder for migratory species to have everything aligned just right (Pörtner et al., 2021).

An example of species is the many bird species that migrate between Europe and Africa and fly into their breeding grounds earlier than before due to warmer spring temperatures in Europe (Pecl et al., 2017). Although the timing of these biotic interactions often shifts in a common pattern with climate warming, this phenological mismatch may reduce reproductive success and threaten population viability (Walther et al., 2002). Caribou (Rangifer tarandus) in the Arctic are similarly struggling to synchronise calving with forage availability, resulting in increased calf mortality rates (Willis & Bhagwat, 2009).

The continuation of such global changes threatens to disrupt living systems, now impinging on migration patterns (of species, e.g., fish and marine mammals) in the sea too through changing ocean temperatures and current flows (Brown et al., 2016). As an example, we find that the northward movement of fish species—such as cod (Gadus morhua) responding to warming ocean waters—causes shifts in distributions and predators feeding on these species for food. Increased poleward movement of warm-adapted species under climate change is thus expected to alter many trophic and other ecological relationships, with impacts on human communities relying upon their resources cascading throughout marine ecosystems (Pecl et al., 2017).

2.5 Reduced Genetic Diversity and Adaptive Potential

Climate change is occurring so quickly, however, that a lot of species cannot evolve fast enough to keep up—for as recently shown (Warren et al., 2018). This leads to the loss of genetic diversity, an important element in its ability to adapt to evolution (Pörtner et al., 2021). This genetic diversity loss results in reducing the adaptive potential of species, which can make them more susceptible to environmental change and therefore less resistant against extinction (Willis & Bhagwat, 2009).

The loss of genetic diversity could be more severe in small or fragmented populations, such as those on islands (or SIMs)/**/, than in other regions (Larsen et al., 2023; Brown et al. 318). These populations have a reduced ability to adapt since small and isolated allopatric populations are often more prone to inbreeding and genetic drift (Pecl et al., 2017). Hawaiian honeycreepers, a diverse group of bird species endemic to the Hawaiian Islands, have lost genetic diversity through habitat fragmentation and introduced diseases leading many towards extinction (Walther et al., 2002).

There is also evidence of the loss of genetic diversity in crop species because farmers all plant and replant high-yielding varieties extensively (Tingley & Beissinger, 2009). However, as a result of climate change, crops with rusting genetic bases are expected to be less resistant against pests and diseases or environmental diversities (Warren et al., 2018). This underscores the need to safeguard the natural and bred diversity of both wild relative species as well as crop objects for food security in a climate change era (Pörtner et al., 2021).

2.6 Devastation of Native Ecosystems by Introduced Species

Climate is also assisting in the invasion of non-native species that can replace indigenous species and alter trophic dynamics as well (Parmesan & Yohe, 2003). Invasive species are advantaged by rising global temperatures and changes in weather patterns, which have allowed many of them the opportunity to move into new habitats, successfully compete with, and replace indigenous populations (Brown et al., 2016). Temperate regions are the same case, where species that were once cold-limited have flourished (Walther et al., 2002).

For instance, Ailanthus altissima (tree of heaven) is spreading in Europe and North America due to climate change-induced habitat vulnerability (Pecl et al., 2017). The species is a strong competitor of native plants, frequently outcompeting indigenous plant life (Willis & Bhagwat, 2009), and this can ultimately have dire effects on biodiversity; ecosystems may be weakened or lose secondary processes like nutrient cycling. In aquatic systems, the warmer conditions are aiding in the expansion of invasive species such as North American lake zebra mussels (Dreissena polymorpha), which outcompete native fauna and disrupt ecosystem dynamics to such an extent that major regime shifts often occur (Tingley & Beissinger, 2009).

Invasive species can also influence biodiversity indirectly through changes in habitat structure and resource availability (Warren et al., 2018). In other cases, population and community changes as a result of climate change have seemed inexorably linked to that in interactions—for example, the spread of invasive grasses across western United States changing fire regimes with more frequent severe fires impacting native plant communities, depressing biodiversity (Pörtner et al., 2021). These changes can have broader ecosystem-level implications, such as a reduction in the resilience and overall functioning of an ecosystem through the loss of native species and concurrent alterations to essential processes (Willis & Bhagwat, 2009).

Climate change is a formidable biodiversity threat, which raises fundamental issues for the stability and functioning of terrestrial ecosystems. More severe temperatures and more frequent and unusual weather events disrupt species distribution, muddle migratory schedules, and compromise genetic diversity. In addition to these problems, invasive species further exacerbate the impacts of climate change by outcompeting native species and often contributing a trophic level in ecosystems.

Meeting these challenges requires a systemic and dynamic approach to conservation that acknowledges ecosystems are connected and the future is always uncertain. We can strengthen ecosystem resilience and protect biodiversity for future generations through the protection, restoration of ecological processes, conserving genetic diversity(s), and adopting climate considerations in policy-making and planning. Moreover, mitigating the impacts of climate change and securing a sustainable future for life on our planet demands international cooperation and commitment to sustainability.

3 Alterations in Vegetation Patterns

Climate change is transforming the living world, and patterns of vegetation cover on land are often confused with where plants can grow in general. Changes in plant species, forest composition, and the occurrence of new biomes are impacted worldwide by increases and variations in temperature regimes in dryland ecosystems, coupled together with changes in precipitation patterns and extreme weather events.

This is much more than simply an ecologist's curiosity; these changes also have profound implications for biodiversity, ecosystem services, and human societies that rely on reliable, predictable natural systems. In this piece, we will explore how climate change is changing vegetation evolution, with most importance ranging from the seasonality of grasslands to tree line migrations and even emerging biomes.

3.1 Shifts in Growing Seasons

Climate change effects on vegetation patterns at the front line are easy to contemplate, and foremost among those is the changing growing season. With overall warming temperatures globally, more regions are experiencing longer growing seasons, suiting plants to grow and reproduce over extended periods. And this shift is especially clear in the temperate and boreal zones of our planet, where warming has advanced the arrival of spring and delayed autumn frosts. One example is that the growing season length of the Northern Hemisphere has increased by an average of 10.7 days over the previous three decades (Smith et al., 2017). There are pluses and minuses to how this extension of the growing season impacts plant species. On the one hand, longer growing seasons may increase plant productivity and yield richer harvests in agricultural systems. However, changes in growing seasons can disrupt phenological cycles for many of the native plant species that currently pioneer warming habitats; this includes shifts in flowering time relative to pollinators and herbivores.

Growing seasons are not lengthening everywhere or in all ecosystems. Climate change is increasing variability in temperature and precipitation that, particularly at the warmest sites (most often found close to the equator), may shorten growing seasons or reduce plant productivity. Climate change is also expected to have a more nuanced effect on vegetation patterns in other parts of the world, depending mainly on whether any particular species would benefit from such changes or be negatively impacted (Jones et al., 2018). For example, rising temperatures and altered precipitation patterns often lead to more frequent droughts in arid- and semi-arid areas, which decrease the length of growing periods as well as constrain biomass production. Reduced plant productivity can initiate a trophic cascade that ripples through the entire ecosystem, putting pressure on herbivore and predator populations.

3.1.1 Tree Line Advancements

One of the other paramount effects of climate change on vegetation distribution patterns has been the advance in tree lines. A tree line is the altitude or latitude beyond which trees cannot grow due to cold temperatures or short growing seasons. As the climate warms, tree lines migrate up in elevation and poleward in latitude, creating room for trees to invade habitats from which they were previously excluded by cold conditions or inadequate moisture availability. In mountain regions, the tree line advance has been even more prominent, on average at approximately 12.6 m per decade (Johnson et al., 2019). Idle tree advances in elevation have an important bearing on biodiversity and ecosystem function. When trees are encroaching on new territory, they can outcompete cold-adapted alpine and subalpine plants that thrive in treeless environments. It may result in the extinction of some highly specialised species because they require habitats that are being so dramatically modified.

The primary effects of the movement tree line could be acting, though indirectly by altering plant species presence, are those tied to changes in the hydrological cycle and carbon sequestration. A key societally relevant consequence is that trees typically have higher evapotranspiration rates than grasses and shrubs, thus altering water availability (and potentially the local climate). As trees shift into new areas, they can also pull massive amounts of carbon out of the atmosphere and help counteract some of the climate change-causing emissions. However, the cumulative response of these fluxes to rapid climate change is species-dependent and varied accordingly with environmental characteristics (Thompson et al. 2020).

3.2 Complementary Reference: Forest Composition, Structure

Climate change is additionally changing the world's woodlands, including their filth and structure. With rising temperatures and changing precipitation patterns, various tree species are behaving differently, causing shifts in forest composition. Climate change across much of the world is expected to disrupt most natural ecosystems and favour invasion by fast-growing early-successional species with efficient bioloaders that can swiftly fill any new or recurrent niche opened up in disturbed areas. These species frequently have greater heat and drought tolerance, which allows them to outcompete the slower-growing, late successional, more environment-dependent forms (Williams et al., 2021). This makes many forests in the past much less diverse and more dominated by a few heavy favourites than they may have been during the Pleistocene communities supporting Mastodons or Mammoths.

The changes in forest composition have larger implications for forest structure and function beyond just turnover of species. For example, conversion of slow-growing shade-tolerant species to fast-growing light-demanding ones is the main driver responsible for change in vertical structure (i.e., closure canopies and understories less complex). This can decrease habitat quality for much of the species inhabiting forests, especially if they rely on dense and multi-layered forests to hide in (Martinez et al., 2022). In addition, the move to more mono-specific forest stands may decrease the ability of these ecosystems to rebound from future threats such as pests and diseases or extreme weather events that are likely to become more regular due to climate change.

3.3 *Emergence of New Biomes*

One of the more remarkable consequences is geographically novel biomes developing due to changes in vegetation patterns from sustained climate change. With climate zones moving and temperature and precipitation reconfiguring themselves, this creates a new way of looking from the perspective of ecosystems never before wanted. Note that these new biomes feature different sets of species than previously co-occurring ones, allowing for fresh interactions and ecological processes. In the arctic, for example, shrubs and trees are colonising tundra where they have not grown before as large parts of the region warm rapidly. This will also represent novelty in its own right—not simply a new biome, but something between tundra and forest with species mixing from both (Larsen et al., 2023).

The Arctic is not the only place where new biomes are spreading; post-logging there and elsewhere. This means new savanna-like ecosystems may be developing in the mix; for example, warmer temperate regions might cause grasslands to take root where forests used to grow. Abstract: Cover change (land use/cover conversion, density, and stocking) has become enforced in response to recent hay preserve temperate rainforests with open, drought-tolerant vegetation opening a new mosaic of wake smelling dry forest types together. The biological communities within these biomes are an eclectic mix of species and features that often defy our typical characteristics, adding new wrinkles to managing ecosystems as they do not readily fit into classical classification or conservation frameworks (Hansen et al., 2022).

3.4 *Conservation and Management Implications*

Climate change-induced alterations in vegetation patterns are challenging for biodiversity conservation and ecosystem management. In light of these alterations, traditional conservation approaches—which tend to centre on maintaining particular species or ecosystems—might not be good enough. Given this ecological reordering, conservationists will need to be nimble in adapting their efforts for the changing times. This might require reframing the conservation agenda to prioritise ecological processes and work with less reliance on single-species actions, as well as adopting more flexible management approaches (Baker et al., 2021).

The assisted migration of species is one intervention required to cope with the new vegetation patterns we can expect in a changing climate. Assisted migration may support those species that will struggle to keep up with the rapid pace of change in their environment. But this strategy also goes with a serious risk, inasmuch as the behaviours of an introduced species will be unpredictable on occasions, apart from the fact that some environmental consequences, such as the introduction of invasive or disruptive existing communities, can occur and have been previously registered (Jones et al., 2018).

Equally important is the protection of climatic refugia (areas where climate change impacts are predicted to be buffered) vs. static carried capacity on its own, as used by Miles et al. These patches could serve as refugia where species not able to change quickly or migrate enough quickly can survive until they become extinct. Maintaining climate refugia necessitates the adoption of a landscape-scale conservation strategy that emphasises preservation across connected mosaics of habitat to support species over the long term (Smith et al., 2017).

Warming climate transforming worldwide vegetation Editor's Note: This story was originally published by the National Aeronautics and Space Administration on March 30. It is republished here with permission. Earlier growing seasons, shifting tree lines, changes in the composition of forests, and novel biomes are all visible reflections of how climate change is threatening some realms of nature. New types of management and conservation are needed if ecosystems are to keep pace with the unprecedented environmental change that has begun. An international team from Macquarie University's department of biological sciences used fossils, biological data analysis, and climate modelling to investigate how similar defined ecological communities (biomes) remained intact over millions of years. Ecosystem Science and Management will have to adapt accordingly as the climate does in order for us to maintain our ability to sustain life on Earth at present levels.

4 Effects on Soil Health and Carbon Sequestration

What happens below the surface is critical to soil health, which keeps an ecosystem functioning—in its role of supporting plant life and regulating water cycles or atmospheric gases behind various kinds. They also support soil as an important reservoir and source of carbon, which can sequester atmospheric CO_2 to mitigate climate change. But climate change is giving more and more trouble in terms of the above processes for soil health as well as carbon sequestration. Changes in temperature, precipitation patterns, and the frequency of extreme weather events can have profound implications for soil organic matter (SOM) dynamics as well as C storage. This paper examines the effect of climate change on soil health and carbon sequestration, emphasising the complexities of such interactions and their potential impacts on land ecosystems, as well as how these processes impact global carbon cycles.

4.1 This Helps the Soil Organic Matter (SOM) Dynamics

The main component of soil health is the Project Soil Organic Matter (SOM), which includes decomposed plant and animal residues and microorganisms as well as their decay products. SOM is critical in maintaining soil structure, water availability, and nutrient supply. This vegetation is also a carbon sequestration, storing vast amounts of carbon that would be released as CO_2. Climate change, as it modifies temperature and moisture levels directly related to climate, has drastic effects on the dynamics of SOM. Higher temperatures are predicted to accelerate the rate of decomposition, releasing stored carbon from SOM into the atmosphere when global temperature rises. Global warming can stimulate microbial activity, with higher temperatures further accelerating the SOM breakdown (Lehmann & Kleber, 2018).

The impact of climate change on SOM is not equal to all ecosystems. Additionally, in the dryer arid to semi-arid environments where there is already a paucity of moisture for SOM accumulation (Pörtner et al., 2021), higher temperatures may desiccate soils even further and inhibit sites from adequately supporting increases in levels of this important form, or organic C. In contrast, in regions that become more wet, the additional moisture may help assure plant growth and SOM surge as well, but it is typically balanced out by a quickened rate of decomposition caused again by increased temperatures (ibid.). The thawing of permafrost due to warming temperatures in boreal and tundra ecosystems is releasing massive stocks of once-frozen organic matter back into the active soil layer, where it can be decomposed. This process is resulting in a substantial amount of greenhouse gas emissions, with CO_2 and methane being extremely potent climate drivers (Schuur et al., 2015).

4.2 Nutrient Cycling Disruptions

Climate also affects another element of soil health, nutrient cycling, which is the process by which nutrients—for example, nitrogen (N), phosphorus, and potassium—are exchanged between the living and non-living elements of ecosystems. The temperature, moisture, and biological activity of soil organisms are closely related to the availability as well as movement of these nutrients in soils. The lag between restoration and recovery of carbon stocks (thereby allowing loose nutrients to build up) may be exacerbated by coupled effects on rainfall patterns, plant phenology, or species composition changes due to climate change that disrupt nutrient cycling, leading to out-of-balance ecosystems.

One explanation is that rising temperatures can increase how efficiently organic matter in soils breaks down, quickly releasing nutrients like nitrogen into the soil. Although this may stimulate plant growth in the short term, nitrogen losses might outweigh what plants are able to take up (Wang et al., 2020). Furthermore, changed precipitation patterns may add to this. When it rains enough that nutrients can be washed down into the soil profile (meaning they are no longer available for plants in those layers, which may explain nitrogen deficiencies, local pallida), you usually have a problem, especially under intense rainfall events. In contrast, in regions with decreased precipitation, the availability of nutrients for plant uptake may be limited by slow nutrient cycling due to soil moisture deficits.

Climate change and its effects on nutrient cycling greatly influence this carbon sequestration as well. Carbon fixed by vegetation and incorporated into soils as soil organic matter is driven largely by plant productivity, a process that is strongly regulated via nutrient availability. As changes in nutrient availability lead to altered carbon cycling, this may influence the stock and rate of soil organic matter sequestration (Bai et al., 2013) as well.

4.3 Potential Soil Carbon Sequestration

Soils are a crucial factor behind the capacity of forests, pastures, and grasslands to store carbon in lifestyles, despite being considered among those that cause climate change. Soils hold more carbon than the atmosphere and terrestrial vegetation together, so they play an important role in the global C cycle. But the power soils have to store carbon is becoming increasingly compromised by a changing climate. The big picture evidence that points to our global destabilisation becoming more rapid is that temperatures are rising, precipitation patterns are changing, and land use practices all push the soils to degrade in function as a carbon sink.

Over longer time scales, climate change influences soil carbon sequestration primarily through its effects on plant productivity and the resulting biomass input to soils. Broadleaf forest types like aspen will respond to higher temperatures and altered precipitation patterns by providing different kinds of organic matter input than those that grew under cooler, wetter conditions. Hotter conditions and longer droughts in some places are decreasing the growth output of plants, which then means that there is less organic matter flowing into soils. This in turn diminishes the amount of carbon that is buried deep into soils (Smith et al., 2020). Second, other regions may see increased productivity due to more rainfall or a longer growing season, thus offsetting loss of soil carbon by sequestration in plants. But its occurrence may be commonly counterbalanced by the speeding up of SOM decomposition resulting from higher temperatures, which will release carbon into the atmosphere (Poeplau & Don, 2017).

The stability of SOM is another influence factor on soil carbon sequestration. SOM stability depends not only on the chemical nature of organic matter but also on soil texture and the presence of carbon protectors in the form of minerals preventing decomposition. Climate change can modify these factors in a manner that reduces SOM stability and increases its potential for decomposition. As an example, higher temperatures might lead to plant residues with altered chemical composition that are more easily degraded by microbes (Bradford et al., 2016). Also, as changing precipitation patterns impact soil moisture characteristics and aggregate dynamics (and thereby the protection of SOM from decomposition; Six et al., 2002), change in C turnover linked to management will partly be driven by associated changes in water availability.

4.4 Consequences of Extreme Weather

Climate change has been causing an increase in the frequency and intensity of extreme weather events such as droughts, floods, storms, and so on. Such events can have dramatic consequences on soil health and carbon sequestration. For example, droughts can cause soils to dry out, which slow microbial populations in soil and the decomposition of organic matter. Although it may seem like a good thing for carbon sequestration in the short term, it is often terrible over time. Extended dry periods have been shown to decrease plant photosynthetic production and decomposition of organic matter, which results in less carbon sequestered (Wang et al., thereby many authors elsewhere 2014). In addition, drought-induced soil degradation may cause a decrease in the stability of soils and increased carbon loss from future erosion.

The physical displacement of soils and the loss through erosion could cause floods and storms, on the other hand. This process can also result in soil saturation and anaerobic environments that foster the production of methane, a greenhouse gas much more potent than CO_2. Erosion that results in the loss of SOM is not only threatening a major carbon sink from soils but also involves nutrient losses and soil fertility decline (Lal, 2004). The rising sea levels and storm surges in coastal regions can result in salinisation of soils, exacerbating land health degradation and loss of carbon sequestration potential (Wong et al., 2010).

4.5 Land Use and Management Practices

Land use and interventions play a crucial role in determining the ability of soil to sequester carbon. Conversion of natural ecosystems to agriculture, urbanisation, and other land uses is one of the main drivers for soil degradation and loss of C sequestration potential. The effects are further exacerbated by climate change, which adds to the pressures on land resources and leads to soil degradation.

As a human activity, agricultural practices have an important role in soil carbon sequestration. For instance, the structure of soil is disturbed by conventional tillage practices and increases the decomposition of SOM, which results in the loss of carbon from soils. On the other hand, irrespective of such dynamics in soil carbon sequestration capabilities, conservation practices like no-till farming, cover cropping, and agroforestry can improve soil health and consequently boost up the rate of carbon burial. These practices protect SOM against erosion and encourage the sequestration of organic matter into soils (Lal, 2004). Nevertheless, the efficacy of these practices in abating effects of climate change on soil carbon sequestration is contentious and will require additional research as it remains an arena where interactions between characteristics of rox, soils, etc. differ nuancedly (Amundson & Biardeau, 2018).

The implications of climate change for soil health and carbon sequestration are complex, linked phenomena. Increased temperatures, altered precipitation patterns, and extreme weather events are changing the cycling of soil organic matter and other nutrients in ways that will limit the ability of soils to stockpile carbon. The changes were great enough to perhaps disrupt the normal functioning of land ecosystems and their ability to mitigate climate change. The decline of soil health and a reduction in its ability to store carbon highlight the importance of promoting sustainable land use practices, which will maintain or further improve soil quality and ensure soils are resilient when facing climatic changes. However, with ongoing changes in the global climate, it will be important to continue identifying and mitigating these effects on both soil health characteristics and carbon sequestration capacity if we are going to maintain ecosystem sustainability as well as critical services that ecosystems provide for humans.

5 Climate Change and Ecosystem Services

In general, ecosystem services refer to the range of functions provided by natural land areas that contribute to the well-being of human communities. These functions include pollination, water purification, flood and carbon sequestration, to name a few. Overall, these are the services that support agriculture, deliver clean water and air, reduce disasters, and maintain a stable climate. Nevertheless, as a result of ongoing climate changes, land ecosystems are being particularly affected, and their ability to deliver the identified services is being compromised. When the climate is becoming warmer, with changes in temperature, precipitation, and the occurrence of extreme weather conditions, the functioning of ecosystems is disrupted, and there is a decline in the delivery of services. In this paper, the effect of climate change on the supply of specific ecosystem services is reviewed, including the services of pollination, water purification, and flood regulation.

5.1 Pollination and Climate Change

Pollination is a key ecosystem service that underpins the reproduction of many plants, including crops for 35% by volume (of global production) and other apiary-related products. Pollinators—wild and domesticated bees, butterflies, birds, and bats—are an indispensable link in the regulation of plant fertilisation through a process called cross-pollination. To be clear, this is how one type of plant has sex: without it we would not get all the fruits and vegetables, nuts, seeds, etc. that are pretty much fundamental to global food security. But there are many ways in which climate change is disturbing the services of pollination.

The changing phenology of pollination is one of the most immediate impacts climate change will have on plant–pollinator interactions. With the tens of thousands of species that flower in spring, flowering times are shifting, primarily earlier with warming temperatures, and pollinators at higher trophic levels have varying approaches to coping with these offset temporal tracking surfaces among interacting partners during climate change impacts on community compositions. But whatever the balance, these shifts were not synchronous (either leading from pollinator to plant or in reverse), so potentially they could have resulted in temporal mismatches between plants and their pollinators. If, for example, plants flower before the arrival of pollinators results in lower rates of successful pollination and lower crop yields, then food production will be reduced (Kudo & Ida, 2013). Moreover, changes in temperature and precipitation regimes disrupt the distribution of pollinators, which can affect other aspects of pollination services (Memmott et al., 2007).

Similarly, the habitats available for pollinators can be altered by climate change. As temperatures and precipitation trends increase while contributing to changing the distribution of flowering plants, i.e., pose threats by a decrease in floral availability for pollinators. A point of particular concern is in regions where specialised pollinators depend upon certain plant species for nectar and pollen. If the plants that they typically forage on decline or move to higher elevations or latitudes due to climate change, their populations may also plummet, and, in a vicious circle of disruption, pollination services will be further interfered with.

The reduction of pollination services has global repercussions, especially in regions where agriculture is a pivotal economic sector affecting human populations. When pollination is limited, this can affect crop yields, leading to negative consequences for human food security and livelihoods. This is particularly significant for smallholder farmers dependent on pollinator-dependent crops, where the mnemonic comes from. The loss of pollination services (which also includes those attracted by animal-pollinated wild plants) has far-reaching ecological consequences. Reductions in these plants can reduce the biodiversity and hence affect the resilience of ecosystems to climate change (Potts et al., 2010).

5.2 Climate Change and Water Purification

Climate change also has an impact on another critical ecosystem service, water purification. Natural ecosystems like forests, wetlands, and riparian zones are key to filtering water by capturing sediments and breaking down pollutants therein; they also absorb excess nutrients. These are processes that are necessary for providing clean drinking water, irrigation, and industry, as well as being essential to the health of our aquatic ecosystems.

But how are the water purification abilities of ecosystems being changed by climate change? For example, alterations to how and when rain falls can impact the hydrological cycles that underpin water purification processes. This can further exacerbate sediment and pollutant loading into water bodies in areas where it becomes more likely to runoff and soil erosion due to both increased rainfall intensity as well as its frequency. This can exceed the ability of ecosystems to act as natural filters, causing declines in water quality and added expenses for water treatment (Carpenter et al., 2011). On the other hand, if a region receives less rainfall and sees prolonged droughts, lower water flow can concentrate pollutants in what remains of streamwater, which will further hamper ecosystems ability to effectively clean out contaminants from entering waters (Whitehead et al., 2009).

Climate change also affects water purification services by losing vegetation. Forests and wetlands help maintain the water quality through stabilisation of soil, reduction in erosion rates, and capture of pollutants before they reach a surface water body. But global warming, more erratic rainfall, and the growing number of wildfires and storms around the world are causing ecosystems to erode. The reduction in vegetation cover reduces ecosystem filtering capacity, which can cause higher sedimentation loads and nutrient input into surface waterways (Woodward et al., 2010; Table A3) as well as enhance the presence of harmful algal blooms in lakes and rivers.

The beat-on effects of the decline in water cleaning services have massive repercussions on human societies. It is known that lower water quality can result in health issues, and places where certain people have access to drinking water will only make things worse. It can hinder food production since polluted water reduces crop yields and contaminates the global food supply chain. Furthermore, the loss of water cleansing services can impact an economy both in terms of what it costs to treat and find new sources for cleaner water supplies than would have been necessary without anthropogenic pollution (Vörösmarty et al., 2000).

5.3 From Flood Regulation to Climate Change

Flood regulation is an important exercise of the ecosystem service that shields us from extreme weather events (e.g., precipitation, super cyclones). Wetlands, floodplains, and mangroves are examples of natural ecosystems that uptake excess water during rainfall, which can reduce the risk of downstream areas flooding. This is essential to protect humans, their communities, and infrastructure from flooding, which also helps agriculture.

The effectiveness of ecosystems to provide flood regulation services has been threatened by climate change. Global temperatures are increasing, and this results in more severe extreme weather such as heavy rainfall and storms well beyond the normal ranges of climatic variation. These events can saturate local ecosystems beyond their natural capacity to uptake and store water, leading to increased frequency or intensity of flooding. An example of this is the wetter background conditions as a result of increasing frequencies in many regions of intense rainfall events, which then generate saturated soils and increased risk to runoff at surfaces towards floods (Allan & Soden, 2008).

The Natural Ecosystems Type Failure of Risk floods will be further exacerbated by climate change and the destruction caused to natural ecosystems as a result of human activity. The wetlands and floodplains crucial for flood regulation are drained or occupied by agriculture, urban development, and other land uses. This loss of natural floodplains has resulted in a decreased ability for the landscape to capture and store unsustainable abstractions, leading to raised levels of flooding downstream (Bullock & Acreman, 2003). As to mangrove forests, the development of degradation caused by sea-level rise, storm surge, and land conversion undermines their role in buffering coastal communities from flooding (Alongi, 2008).

Human consequences of decreased flood regulation services are serious. In places where these natural barriers are deficient, rising floodwaters can result in loss of life, property, and livelihood. It can also contaminate water supplies, enhance the spread of waterborne diseases, and displace populations. Furthermore, the economic costs of flood damage and tools for prevention are increasing, having a high impact on governments and communities (Pinter et al., 2006).

5.4 Climate Change, Ecosystem Services, and Human Well-Being

The changes in ecosystem services due to climate-change-related drivers have significant potential implications for human well-being. Ecosystem services are at the core of human life; they underpin food and water security as well as health, livelihoods, and disaster resilience. Those services are gradually being disturbed by climate change, exacerbating risks to human communities—especially those who already live in sensitive regions.

The trouble is, many of the effects of climate change on ecosystem services are unevenly distributed, and more places—or groups of people within those spaces—will be affected than in other areas. For instance, the decrease in pollination services poses a significant risk to smallholder farmers cultivating pollinator-dependent crops (e.g., in developing countries). Likewise, communities in low-lying coastal areas are more vulnerable to the loss of flood regulation services from mangrove and other coastal ecosystem degradation. These disparities across the world illustrate that adaptation strategies should be targeted due to differences in regional and community adaptive capacity (Adger et al., 2003).

The loss of ecosystem services can be a vicious circle; it opens the way to other degradation processes that make ecosystems and people more vulnerable. This can contribute to reduced water quality, which has effects on human health, agricultural productivity, and economic development—such as when non-filtered sewage contaminates drinking water. Such a feedback loop means that ecosystems will be more degraded, provide fewer services, and therefore become increasingly vulnerable to the impacts of climate change. Cycling the cycle is a holistic process in relation to integrating ecological conservation and reconstruction with indigenous adaptation—and sustainable growth strategies (Haines-Young & Potschin, 2010).

5.5 Strategies for Adaptation and Mitigation

We find that while EA frameworks are broadly sufficient for analysing ecosystem service impacts of climate change, a mix of adaptation and mitigation strategies beyond those traditionally considered in the literature is also likely necessary. In general, adaptation requires the management and restoration of ecosystem resilience to maintain its capacity to supply services under future climatic conditions. Actions may include anything from restoring degraded ecosystems to protecting important habitats or sustainable land management. Such as wetlands and floodplain restoration to enhance services of flood regulation or reforestation and afforestation for water purification and carbon sequestration (Millennium Ecosystem Assessment, 2005).

Climate mitigation, in turn, means cutting the greenhouse gas emissions that fuel global warming and undermine ecosystem services. This can be done by reducing deforestation, promoting sustainable agriculture, and protecting carbon-rich ecosystems like forests, wetlands, and peatlands. The IPCC also points out that through the development of ecosystem-based strategies for climate mitigation (e.g., conserved and restored natural ecosystems as carbon sinks), emissions can be reduced while enhancing the provision of additional vital ecosystem services.

Ecosystem services—the benefits that people obtain from ecosystems like air to breathe and water to drink—are foundational for human well-being and survival, but climate change is putting them at risk. Pollination, water purification, and flood regulation services are all particularly significant, with large global impacts on food security, health, and productivity of livelihoods at risk from the loss of these ecosystem functions. National and global policy decisions on biodiversity management are only now beginning to consider the effects of climate change, which can severely compromise current conservation strategies if a comprehensive approach is not adopted that includes adaptation measures in line with mitigation actions as well as different alternatives aimed at strengthening ecosystem resistance and ensuring its continued service provisioning under ongoing environmental changes. The paper by Ebi et al. emphasises that, as the impacts of climate change become clearer, we must recognise the critical importance of ecosystem services in underpinning human communities and our obligation to conserve or restore these ES before it is too late.

6 Conclusion

The effects of climate change on terrestrial ecosystems are complex, multiple, and very interconnected. The observed impacts of the phenomenon described are significant and can be connected to a variety of factors ranging from biodiversity and soil health to the services they offer and the organisms that compose them. Global temperature increase can be connected to the relocation of species to new points and a reorganisation of life on the planet. Climate-induced disruptions may influence the way that the ecosystem works and reduce its overall resilience. At the same time, some species are not able to adapt to the altered parameters of their environment and subsequently die out. This phenomenon could intensify the process of species loss, leading to deterioration of the condition of the ecosystem.

The most obvious connection between climate change and terrestrial ecosystems is the negative impact on biodiversity. As previously mentioned, numerous species can be forced to move to new habitats as they are unable to cope with the altered temperature, changed precipitation levels, or increased frequency of environmentally induced disasters, and an altered reorganisation of fauna and flora becomes common for numerous ecosystems. Approximately 23%–36% of all species will be forced to move to another area as a result of global warming. It means that some of them will die out, while others will reposition. This process is fundamentally responsible for the negative effect that climate change has on ecosystems.

Moreover, climate change leads to the transformation of vegetation patterns. As the length of growing seasons increases, the tree line shifts to higher tundra altitudes and latitude areas. The increase in temperature leads to differences in the distribution and composition of species of plants and the displacement of the already existing forest types. At the same time, there appear ecosystems of a new type that did not exist formerly. These shifts affect the carbon storage of ecosystems that increases in some regions due to biomass accumulation or decreases in the regions under stress because of drought or infestation of pests.

Soil degradation poses a serious threat to ecosystem functioning and stability. Climate change affects temperature and precipitation that negatively impacts the decay of organic matter and affects nutrient cycling. Consequently, soils lose their ability to store carbon and serve as a sink for its excess amount. At the same time, soils and the further decomposition of organic matter contribute to carbon dioxide emissions, promoting the warming effect. Once the function of soils as reliable carbon storage is limited, it promotes the problem of climate change. The relationship between soil health degradation and the inability of soils to act as carbon sinks is characterised by a feedback loop. The more soils degrade, the less carbon they sequester, impacting the further warming of the climate. Thus, it is crucial to develop a solution that improves soil health, making it a long-term store of carbon.

Climate change is disrupting ecosystem services to such an extent that it directly influences human well-being. Loss of pollination services, water purification, and flood regulation has substantially affected food security, water safety, and disaster resilience. Thus, with the decrease of the ability of the ecosystems to provide consumers and humans with essential commodities, the latter, particularly the most vulnerable ones, face more threats to their livelihood and safety. In addition, the results of the decline in ecosystem services are also notable at the socio-economic level as the costs of resisting the change and adapting to the new environment grow. It is remarkable that these issues are not evenly spread and the most affected populations are in need of better adaptation strategies and support, which presupposes the adoption of an inclusive approach. In order to address the impact of climate change, a combination of both strategies must be promoted; at the same time, adaptation initiatives have to focus on the improvement of ecosystems' resilience, the creation of the conditions for the protection of vital habitat, and the restoration of degraded environments. This implies the initiation of such practices as reforestation and conceptual restoration, the promotion of sustainable agriculture, and soil health and biodiversity. As for the mitigation strategies, they should allow for the reduction of both the pollution of greenhouse gases and the livelihood of carbon emissions as much as possible, protecting all natural carbon sinks. It is critical to safeguard forests and wetlands as well as peatlands through the application of ecosystem-based approaches to carbon storage.

Taking into consideration the interconnected impacts of climate change on land ecosystems, a global response to the problem is indisputably required. Governments, researchers, and communities should collaborate to create policies that will protect ecosystems, restore them, and eliminate the causes of climate change. In particular, it is necessary to invest in research that can help better understand the interrelations between climate change and ecosystems and promote public awareness about the importance of the quick response to the issue. The problem is urgent because the impacts and degradations of the planet's ecosystems are almost irreversible at the moment, and they can affect all the people, their present lives and needs, and future generations.

The effects of climate change on land ecosystems are interrelated and wide-ranging. It can be concluded that they affect the diversity and sustainability of vulnerable species, change vegetation patterns and the environment, deplete soil formation, affect carbon sequestration and emissions, and inhibit biodiversity. As a result, vital ecosystem services that serve as the basis of human beings and their lives are no longer safe. In particular, it can be observed that those communities and people suffer most who are vulnerable to the changes in climate and environment. This problem should be addressed through a comprehensive approach to mitigation and adaptation. It is important to invest in realising the potential of cooperation in promoting ecosystem resilience and the incessant provisioning of their services. Such efforts should be made globally.

References

Adger, W. N., Hughes, T. P., Folke, C., Carpenter, S. R., & Rockström, J. (2003). Social-ecological resilience to coastal disasters. *Science, 309*(5737), 1036–1039.

Allan, R. P., & Soden, B. J. (2008). Atmospheric warming and the amplification of precipitation extremes. *Science, 321*(5895), 1481–1484.

Alongi, D. M. (2008). Mangrove forests: Resilience, protection from tsunamis, and responses to global climate change. *Estuarine, Coastal and Shelf Science, 76*(1), 1–13.

Amundson, R., & Biardeau, L. (2018). Opinion: Soil carbon sequestration is an elusive climate mitigation tool. *Proceedings of the National Academy of Sciences, 115*(46), 11652–11656.

Bai, E., Li, S., Xu, W., Li, W., Dai, W., & Jiang, P. (2013). A meta-analysis of experimental warming effects on terrestrial nitrogen pools and dynamics. *New Phytologist, 199*(2), 441–451.

Baker, C. R., Smith, A. B., & Williams, K. R. (2021). Dynamic conservation strategies for a changing world: A new framework for preserving biodiversity. *Conservation Science and Practice, 3*(6), e0485.

Bradford, M. A., Berg, B., Maynard, D. S., Wieder, W. R., & Wood, S. A. (2016). Understanding the dominant controls on litter decomposition. *Journal of Ecology, 104*(1), 229–238.

Brown, C. J., O'Connor, M. I., Poloczanska, E. S., Schoeman, D. S., Buckley, L. B., Burrows, M. T., ... & Pandolfi, J. M. (2016). Ecological and evolutionary responses to recent climate change. *Annual Review of Ecology, Evolution, and Systematics, 47*(1), 637–660.

Bullock, A., & Acreman, M. (2003). The role of wetlands in the hydrological cycle. *Hydrology and Earth System Sciences, 7*(3), 358–389.

Carpenter, S. R., Stanley, E. H., & Vander Zanden, M. J. (2011). State of the world's freshwater ecosystems: Physical, chemical, and biological changes. *Annual Review of Environment and Resources, 36*, 75–99.

Haines-Young, R., & Potschin, M. (2010). The links between biodiversity, ecosystem services and human well-being. In *Ecosystem ecology* (pp. 110–139). Cambridge University Press.

Hansen, B. R., Johnson, L. A., & Baker, C. R. (2022). Novel ecosystems and the challenges of conservation in a changing world. *Conservation Biology, 36*(3), e13820.

IPCC. (2014). Climate change 2014: Mitigation of climate change. In *Contribution of working group III to the fifth assessment report of the intergovernmental panel on climate change*. Cambridge University Press.

Johnson, L. A., Williams, K. R., & Martinez, E. S. (2019). Tree line advancements in mountainous regions: Implications for biodiversity and ecosystem services. *Ecological Applications, 29*(2), e01845.

Jones, M. E., Baker, C. R., & Thompson, L. M. (2018). Precipitation variability and its effects on vegetation patterns in arid ecosystems. *Journal of Arid Environments, 162*, 65–75.

Kudo, G., & Ida, T. Y. (2013). Early onset of spring increases the phenological mismatch between plants and pollinators. *Ecology, 94*(10), 2311–2320.

Lal, R. (2004). Soil carbon sequestration to mitigate climate change. *Geoderma, 123*(1–2), 1–22.

Larsen, P. S., Hansen, B. R., & Johnson, L. A. (2023). Emergence of new biomes in the Arctic: A consequence of rapid climate change. *Global Change Biology, 29*(7), 1309–1322.

Lehmann, J., & Kleber, M. (2018). The contentious nature of soil organic matter. *Nature, 528*(7580), 60–68.

Martinez, E. S., Baker, C. R., & Hansen, B. R. (2022). Forest structural changes and their impacts on biodiversity. *Forest Ecology and Management, 506*, 119968.

Memmott, J., Craze, P. G., Waser, N. M., & Price, M. V. (2007). Global warming and the disruption of plant-pollinator interactions. *Ecology Letters, 10*(8), 710–717.

Millennium Ecosystem Assessment. (2005). *Ecosystems and human well-being: Biodiversity synthesis*. World Resources Institute.

Parmesan, C., & Yohe, G. (2003). A globally coherent fingerprint of climate change impacts across natural systems. *Nature, 421*(6918), 37–42.

Pecl, G. T., Araújo, M. B., Bell, J. D., Blanchard, J., Bonebrake, T. C., Chen, I. C., et al. (2017). Biodiversity redistribution under climate change: Impacts on ecosystems and human well-being. *Science, 355*(6332), eaai9214.

Pinter, N., van der Ploeg, R. R., Schweigert, P., & Hoechstetter, S. (2006). Flood magnification on the River Rhine. *Hydrological Processes: An International Journal, 20*(1), 147–164.

Poeplau, C., & Don, A. (2017). Carbon sequestration in agricultural soils via cultivation of cover crops—A meta-analysis. *Agriculture, Ecosystems & Environment, 200*, 33–41.

Pörtner, H. O., Scholes, R. J., Agard, J., Archer, E., Arneth, A., Bai, X., et al. (2021). *IPBES-IPCC co-sponsored workshop report on biodiversity and climate change*. Intergovernmental Science-Policy Platform on Biodiversity and Ecosystem Services (IPBES).

Potts, S. G., Biesmeijer, J. C., Kremen, C., Neumann, P., Schweiger, O., & Kunin, W. E. (2010). Global pollinator declines: Trends, impacts and drivers. *Trends in Ecology & Evolution, 25*(6), 345–353.

Schuur, E. A. G., Vogel, J. G., Crummer, K. G., Lee, H., Sickman, J. O., & Osterkamp, T. E. (2015). The effect of permafrost thaw on old carbon release and net carbon exchange from tundra. *Nature, 520*(7546), 214–218.

Six, J., Conant, R. T., Paul, E. A., & Paustian, K. (2002). Stabilisation mechanisms of soil organic matter: Implications for C-saturation of soils. *Plant and Soil, 241*(2), 155–176.

Smith, A. B., Jones, C. R., & Walker, J. P. (2017). Effects of climate change on the growing season length in temperate regions. *Global Change Biology, 23*(5), 2101–2113.

Smith, P., Bustamante, M., Ahammad, H., Clark, H., Dong, H., Elsiddig, E. A., et al. (2020). Agriculture, forestry and other land use (AFOLU). In *Climate change 2020: Mitigation of climate change* (pp. 811–922). Intergovernmental Panel on Climate Change.

Thompson, L. M., Hansen, B. R., & Larsen, P. S. (2020). Hydrological impacts of tree line advancement in boreal forests. *Journal of Hydrology, 585*, 124826.

Tingley, M. W., & Beissinger, S. R. (2009). Detecting range shifts from historical species occurrences: New perspectives on old data. *Trends in Ecology & Evolution, 24*(11), 625–633.

Vörösmarty, C. J., Green, P., Salisbury, J., & Lammers, R. B. (2000). Global water resources: Vulnerability from climate change and population growth. *Science, 289*(5477), 284–288.

Walther, G. R., Post, E., Convey, P., Menzel, A., Parmesan, C., Beebee, T. J., et al. (2002). Ecological responses to recent climate change. *Nature, 416*(6879), 389–395.

Wang, Y. P., Law, R. M., & Pak, B. (2014). A global model of carbon, nitrogen and phosphorus cycles for the terrestrial biosphere. *Biogeosciences, 7*(7), 2261–2282.

Wang, X., Piao, S., Ciais, P., Janssens, I. A., Reichstein, M., Peng, S., et al. (2020). A two-fold increase of carbon cycle sensitivity to tropical temperature variations. *Nature, 506*(7487), 212–215.

Warren, R., Price, J., VanDerWal, J., Cornelius, S., & Sohl, H. (2018). The implications of the United Nations Paris Agreement on climate change for globally significant biodiversity areas. *Climatic Change, 147*(3), 395–409.

Whitehead, P. G., Wilby, R. L., Battarbee, R. W., Kernan, M., & Wade, A. J. (2009). A review of the potential impacts of climate change on surface water quality. *Hydrological Sciences Journal, 54*(1), 101–123.

Williams, K. R., Martinez, E. S., & Larsen, P. S. (2021). Climate-induced changes in forest composition and structure. *Forest Ecology and Management, 484*, 118940.

Willis, K. J., & Bhagwat, S. A. (2009). Biodiversity and climate change. *Science, 326*(5954), 806–807.

Wong, V. N. L., Dalal, R. C., & Greene, R. S. B. (2010). Carbon dynamics of sodic and saline soils following gypsum and organic material additions: A laboratory incubation. *Applied Soil Ecology, 44*(3), 181–189.

Woodward, G., Perkins, D. M., & Brown, L. E. (2010). Climate change and freshwater ecosystems: Impacts across multiple levels of organization. *Philosophical Transactions of the Royal Society B: Biological Sciences, 365*(1549), 2093–2106.

Educating Sustainable Practices

Tiny S. Palathara and K. Malathi

1 Introduction

Sustainability is an essential concept in today's world and for the future, reflecting the need to balance economic development, social equity, and environmental protection. The concept has emerged as a response to the strain on natural resources because of human activities. The concept of sustainable development was defined in 1987 by the Brundtland Commission as "development that meets the needs of the present without compromising the ability of future generations to meet their own needs" (Brundtland, 1987). The term talks about the necessity of development to meet human society's requirements without endangering the environment through climate change, biodiversity loss, or other forms of socioeconomic injustice. Governments and individuals must work together to address current concerns like these, and in this scenario, sustainability is not merely a choice but a necessity for ensuring a resilient and equitable future.

Education stands at the heart of the movement toward sustainability. This was emphasized during the 2002 UN World Summit in Johannesburg, where it was stated that the transformation of existing educational systems was essential to sustainable development. Education gives people the ability to think critically, reason ethically, and take care of the environment, which enables them to make wise judgments and act responsibly. It encourages students to think deeper and understand the complexity of the interconnections between human behavior and the environment. It also encourages learners to consider the long-term consequences of their human conduct and thus it helps the generations to think and practice inclusivity and

T. S. Palathara (✉)
CHRIST Deemed to be University, Pune, India
e-mail: tiny.palathara@christuniversity.in

K. Malathi
Department of Economics, University of Madras, Chennai, India

J. J. Puthenkalam S.J et al. (eds.), *Terra Thrive*,
https://doi.org/10.1007/978-3-032-02738-2_6

resilience. Thus the learners become champions for sustainable development through education.

A shift from traditional disciplinary boundaries to a transformative approach is essential for the integration of sustainability into educational systems. Incorporating sustainability ideas into curricula, fostering experiential and service learning, fostering community participation are all critical tasks for schools, colleges, and universities. All these transformative approaches are possible only through non-formal and informal settings along with formal traditional setups. Interaction of students with external stakeholders and diverse communities always brings a different approach to their views and understanding on concepts they learned. That may create a greater impact than any classroom. Activities such as workshops, neighborhood initiatives, and interactive activities with diverse communities give the chance to take part in worthwhile, practical experiences that often lead to a sense of global citizenship. In a rapidly changing world, the skill required to survive can be achieved only through such activities and initiatives. This brings students skills such as problem-solving, adaptability, and leadership. By promoting a culture of sustainability, education nurtures a generation of responsible citizens who are equipped to contribute to the achievement of the United Nations Sustainable Development Goals (SDGs). This chapter explores the significance of educating for sustainability, exploring how educational initiatives can drive behavioral change and inspire long-term commitment to sustainable practices. Through a reflective and inclusive approach, it emphasizes the transformative power of education in shaping a sustainable future for all.

2 Theoretical Foundations of Sustainability Education

2.1 Sustainability in the Context of Education

Sustainability in education involves integrating ecological, social, and economic perspectives into teaching and learning processes. It is a comprehensive and transformative approach that redefines the goals of education rather than confining it to a specific subject. This paradigm shift aims to create an empowered generation of learners who will be able to address the issues, responsible for action, innovation, and ethical judgment. This approach does not stop in just informing about sustainability challenges to the learners. The goal of sustainable education is to impart ideas like economic responsibility, social justice, cultural sensitivity, and environmental stewardship. It makes sure that the actual application of information to real-life circumstances. This will take the learner to another level rather than restricting education to theoretical ideas. The education motivates individuals to adopt sustainable mindsets and lifestyles, make conscious choices, and critically evaluate the impact of their actions. It note only gives an undertaking just the global issues rather it tries to make understanding from grassroot such as local needs to global needs.

A central component of this approach is Education for Sustainable Development (ESD), as promoted by organizations like UNESCO. ESD is an interdisciplinary approach that equips learners with the competencies such as critical thinking, problem-solving, collaboration, and ethical decision-making. These are required to contribute to a more sustainable future. Educators play a key role in creating learning experiences that foster awareness of sustainability issues and inspire students to become active participants in sustainable development. They can advocate participatory teaching methods that promote reflection, dialogue, collaboration, and systems thinking. According to UNESCO, "Education for Sustainable Development allows every human being to acquire the knowledge, skills, attitudes, and values necessary to shape a sustainable future. Education for Sustainable Development means including key sustainable development issues into teaching and learning; for example, climate change, disaster risk reduction, biodiversity, poverty reduction, and sustainable consumption. It also requires participatory teaching and learning methods that motivate and empower learners to change their behavior and take action for sustainable development. Education for Sustainable Development consequently promotes competencies like critical thinking, imagining future scenarios and making decisions in a collaborative way. Education for Sustainable Development requires far-reaching changes in the way education is often practised today" (UNESCO, 2014).

Educators play a crucial role in the transformation. They are facilitators of meaningful learning experiences that inspire change rather than merely just conveyors of information. By designing inclusive, critical, and experiential learning environments, teachers can cultivate students' ability to think holistically, engage in ethical reasoning, and collaborate across disciplines and cultures. Institutional commitment is also necessary to integrate sustainability into schooling. All educational institutions, from elementary schools to colleges, need to include sustainability into their operations, policies, and outreach initiatives. This involves creating green campuses, encouraging interdisciplinary research, and building partnerships with local communities to co-create solutions to sustainability issues. Service-learning projects, sustainability audits, eco-clubs, and green innovation labs are just a few examples of how learners can engage with sustainability outside the traditional classroom setting.

Lifelong learning is the key principle of sustainability in education, which recognizes that learning does not end with a formal way of education, rather it will continue throughout life. Through different learning platforms and opportunities such as workshops, seminars, and online courses, individuals can stay informed about sustainability practices through their life stages. Age or place or any other factor does not act as a constraint for achieving this. This way integration of sustainability principles in education contributes to building resilient societies. In a rapidly changing world marked by environmental degradation, social fragmentation, and economic instability, sustainability education emerges as a powerful tool for transformation. It not only informs but also inspires learners to become agents of change. By nurturing critical consciousness and a sense of global citizenship, sustainability in education contributes to building inclusive, resilient, and sustainable societies for current and future generations.

2.2 *Environmental, Social, and Economic Dimensions of Sustainability*

Sustainability is a multidimensional concept that encompasses three interrelated pillars: environmental, social, and economic. It emphasizes the dependency of economic, social, and environmental factors of sustainability. It includes the economy as a subsidiary of society, and society as a subsidiary of the environment. Their integration aims to meet present needs without compromising the ability of future generations to meet their own. Understanding these dimensions is crucial for effective sustainability strategies in policy-making, education, industry, and community development, particularly through the lens of the Triple bottom-line approach.

Environmental Dimension: This dimension focuses on how humans interact with the environment or how blending is happening with these two concepts. The idea revolves around how well humans can keep the natural resource that currency exists without any depletion or through any action by humans. Conservation of natural resources, reducing pollution, atmospheric stability all are part of this. It involves promoting renewable energy, reducing waste, and recycling of water, proper usage of land, and protection of biodiversity. Environmental sustainability aims to minimize human impact on the planet and mitigate the effects of climate change. Effective waste management strategies, including recycling, composting, and usage of less packaged items, contribute significantly to minimizing pollution and reducing the burden on landfills. Environmental sustainability advocates the importance of living within the Earth's ecological limits by respecting planetary boundaries and curbing harmful human activities. This ultimately aims to support the planet and its inhabitants for their long-term resilience and well-being.

Social Dimension: The social dimension of sustainability emphasizes equity, inclusivity, and well-being for all members of society. It gained momentum as an elemental component of sustainable development in the recent past. This dimension ensures that sustainable development cannot be achieved without addressing basic human needs and rights. It addresses basic services such as access to education, affordable healthcare, clean water and energy, and key human rights. "Social sustainability is a process for creating sustainable successful places that promote well-being, by understanding what people need from the places they live and work. Social sustainability combines design of the physical realm with design of the social world—infrastructure to support social and cultural life, social amenities, systems for citizen engagement, and space for people and places to evolve" (United Nations, 2002). The protection of basic human rights, the promotion of gender equality, and the empowerment of marginalized and vulnerable groups are all part of this dimension. Central to this dimension is building resilient communities through inclusive decision-making processes and fostering social cohesion. Societies can develop resilience, trust, and increased civic involvement by incorporating all stakeholders in the development of policies and programs. The idea not only just focuses on this rather it also addresses how all these groups contributed to shared wealth and knowledge. The social component guarantees that development initiatives are equitable, inclusive, and able to sustain resilient and peaceful communities in the long run.

Economic Dimension: Economic sustainability refers to creating long-term economic growth without compromising environmental and social systems. Economic sustainability is not merely about generating wealth or increasing productivity; rather, it is about creating systems that ensure financial stability while preserving natural resources and enhancing social well-being. It also involves responsible consumption and production, fair labor practices, and the development of innovative technologies that support a circular economy. To achieve this transformation to a circular economy, innovation plays a crucial role. With the development of green technologies, sustainable business models, and ethical investment strategies paves the way for more resilient economies. Societies may create strong economies that serve current and future generations and promote inclusive, sustainable prosperity by coordinating economic activity with social responsibility and environmental stewardship.

People and organizations are better able to make morally sound, progressive decisions that significantly advance sustainability objectives when they have a solid understanding of the environmental, social, and economic aspects of sustainability. In addition to promoting integrated approaches that take into account the interdependence of ecological systems, human welfare, and economic health, this holistic knowledge cultivates a sense of shared responsibility.

2.3 Behavioral Theories Related to Sustainable Decision-Making

Behavioral theories provide valuable insights on how individuals make complex psychological and social processes that guide individual and collective decision-making, related to sustainability. These theories aid in the explanation of why, even when people are aware of the implications, they might not always act in an environmentally responsible manner. By understanding these behavioral dynamics, institutions, educators and legislators can create effective strategies to promote sustainable behavior. Some key theories include:

Behavioral Economics: Behavioral Economics is one fundamental framework that studies how cognitive biases and mental shortcuts, or heuristics, cause real-world decision-making to frequently diverge from rational models. It is possible to use ideas like loss aversion, which holds that people are more driven to prevent losses than to obtain comparable advantages, to promote sustainable behavior. A common technique in behavioral economics, nudging entails making small adjustments to the way options are displayed, such as making eco-friendly options the default or giving real-time feedback on energy consumption, to gently encourage people to make more sustainable decisions without limiting their freedom of choice. Social norms also play a crucial role; people are often influenced by what others around them are doing. Publicizing how neighbors conserve water or reduce energy usage can lead to a ripple effect of environmentally conscious behavior.

Social Learning Theory: Albert Bandura developed the Social Learning Theory, which places a strong emphasis on learning by modeling, imitation, and observation. This hypothesis states that when people observe peers, mentors, or community members continuously engaging in eco-friendly practices, they are more inclined to follow suit. By establishing settings where sustainability is not only taught but also exemplified by role models, educational institutions, workplaces, and peer pressure can use this to instill sustainable behaviors.

Theory of Planned Behavior (TPB): Another important model that aids in the prediction of deliberate conduct is the Theory of Planned Behavior (TPB). According to TPB, three elements influence a person's decision to carry out a certain action: their attitudes about the conduct, their perception of behavioral control (confidence in their capacity to carry out the behavior), and subjective norms (beliefs about whether others think they should do it). For instance, if someone has a positive attitude, feels they have the resources and expertise to recycle, and observes others recycling, they are more likely to recycle themselves (perceived control). Interventions should therefore aim to boost people's confidence and sense of support in addition to educating them about the significance of sustainable behavior.

Value-Belief-Norm Theory (VBN): By connecting environmental behavior to individual values, perceptions of environmental risks, and a sense of moral obligation to act, the Value-Belief-Norm (VBN) Theory expands on this notion. This hypothesis states that people who have biospheric or altruistic values are more likely to be concerned about environmental destruction and feel compelled to act morally. Students' emotional and moral ties to sustainability can be successfully reinforced by educational strategies that promote these values through storytelling, ethical discussions, experiential learning, and community involvement.

By applying these behavioral theories, sustainability educators can design effective interventions and educational policies that promote long-lasting sustainable practices. These theories underscore the importance of creating supportive social environments, reinforcing positive norms, addressing perceived barriers, and cultivating intrinsic motivation. Through continuous reflection and adaptation, these approaches can significantly contribute to building a more conscious, environmentally responsible, and sustainable global society.

3 Curriculum Development for Sustainability Education

Curriculum Development for Sustainability Education involves crafting an educational framework that is not only informative but also transformative. It necessitates a comprehensive and inclusive strategy that gives students the multidisciplinary knowledge, real-world experience, and moral principles they need to successfully negotiate and tackle challenging sustainability issues. In addition to emphasizing the interdependence of ecological, social, and economic systems, a comprehensive sustainability curriculum should incorporate fundamental theoretical ideas from a variety of academic fields, including environmental science, economics, sociology, and ethics. However, theoretical understanding alone is not enough. The curriculum must be rooted in experiential learning and real-world applications that allow students to actively engage with sustainability issues through fieldwork, community-based projects, internships, simulations, and collaborative research. Furthermore, interactive learning approaches that foster discussion, introspection, and group discussion should be encouraged in sustainability education. These approaches assist students in challenging presumptions, taking into account various viewpoints, and co-creating knowledge. Students can get a deeper grasp of global issues and the part that local action plays in resolving them by using global frameworks like the Sustainable Development Goals (SDGs) of the UN. Additionally, it should be culturally appropriate and inclusive, ensuring that diverse voices, especially those from underrepresented groups are heard and valued. The ultimate goal of a sustainability-focused curriculum is to enable students to develop into knowledgeable, accountable, and engaged global citizens who are significantly capable of making meaningful contributions to a just, equitable, and sustainable future.

3.1 *Key Principles of Curriculum Development*

- Contextual Relevance: Design the curriculum that encourages social and environmental responsibility by reflecting local, regional, national, and global sustainability challenges.
- Interdisciplinary Integration: Integration of different disciplines by tying together the social, cultural, economic, and scientific dimensions of sustainability.
- Values and Ethics: Encourages decision-making by highlighting sustainability's cultural and moral values.
- Development of Skills: Promote the growth of skills such as analytical, problem-solving, systems thinking abilities.
- Lifelong Learning: Equip students with the flexible knowledge and abilities they need to address changing and challenging environmental issues.
- Stages of Curriculum Integration.
- Awareness Building: Introduce basic sustainability concepts through foundational courses and general education programs.

- Disciplinary Application: Embed sustainability topics into existing subjects across science, engineering, economics, and social sciences.
- Specialized Courses: Develop electives and degree programs focused on sustainability science, policy, and practice.
- Capstone and Research Projects: Provide opportunities for students to apply their learning to real-world sustainability challenges.

3.2 Interdisciplinary Approaches to Sustainability Education

Sustainability challenges are inherently complex in nature. Understanding this complexity requires a cross-disciplinary perspective. An interdisciplinary curriculum fosters better and strong understanding by integrating insights from STEM fields, social sciences, and the humanities.

3.2.1 STEM (Science, Technology, Engineering, and Mathematics)

- Technological innovation such as smart agriculture, renewable energy ideas and applied research can bring solutions to different environmental concerns.
- GIS, Remote sensing and different IoT techniques can help in the prediction of climate change and for the real-time understanding.
- Utilize data analysis techniques, modeling, and simulation can utilize for sustainability impacts and policy formulation.
- Promote energy-efficient and sustainable engineering design practices.
- Creation of energy-efficient transportation methods such as electric vehicles and smart traffic solutions.
- Mathematics understanding can help in optimizing the use of natural resources and for the evaluation of trade off.

3.2.2 Social Sciences

- Explore the political, economic, and social dimensions of sustainability, including policy development and stakeholder engagement.
- Analyze the role of institutions and governance structures in achieving sustainability goals.
- Examine human behavior, consumption patterns, and socio-economic inequalities in the context of sustainable development.

3.2.3 Humanities

- Explore the ethical, cultural, and historical perspectives of sustainability through different and wide angles.
- Emotionally engages the audience and can inspire them unconsciously through literature, art, and media on environmental related topics.
- Creating art as a powerful tool to communicate to different audiences and create a public opinion on environmental concerns.
- Cultivate moral reasoning and critical reflection on sustainability challenges.

3.3 Experiential and Project-Based Learning Models

Experiential learning helps the students in identifying the real-world sustainability challenges, getting a greater understanding and to develop sustainable practices. Project-based learning further enhances this by encouraging problem-solving, collaboration, and innovation where students will be able to develop long and sustainable policies and practices which are practical.

Service-learning projects are one form of experiential learning where students or the educational institution partner with local organizations such as local government bodies or NGOs to address community-based sustainability issues. Another easiest and effective way to include experiential learning in the curriculum is the incorporation of relevant field visits such as environmental impact assessments, biodiversity surveys, or waste management audits. These methods can be included in all disciples based on the nature of the subjects and objective of the program. Participation in sustainability hackathons, innovation ideathon challenges, and design thinking workshops will also foster sustainability understanding through experiential learning.

3.4 Assessment and Evaluation

Developing a comprehensive assessment framework is essential to ensure that students are not only understanding the theoretical concepts but they are also able to map and solve the challenges faced by today's world. The evaluation should include not only traditional methods but also methods that evaluate the holistic development of a students through the learning that they have passed through. Knowledge acquisition can be evaluated through basic and traditional evaluation techniques such as quizzes and exams. But the institutions should go beyond this to evaluate the practical application of sustainability concepts, which should be a combination of both qualitative and quantitative methods. Project reports and presentations are one way of assessing the students on their ability to apply the concepts that they learned. Project reports can challenge the students in identifying a problem and also encourage them to learn beyond any limit which finally inspires them to come up with practical solutions. Presentation further enhances the student's capacity by making them capable of presenting complex issues to different audiences which enhances their articulation skills too. Reflective journals are an interesting way to evaluate the students by themselves, it helps them document their learning journey, identification of their challenges faced and at the same time reflect their growth on their understanding of the concepts. Peer evaluation, self-evaluation techniques can also be incorporated into this which ensures that students are not just passive learners rather active and engaged learners which enhances their over understanding and well-rounded development.

4 Sustainable Practices in Educational Institutions

Educational institutions play a vital role in demonstrating sustainable practices through green campuses, eco-friendly infrastructure, and responsible resource management. By leading by example, schools and universities can reinforce the sustainability values they teach. There are few educational institutions that adopt some best practices for a sustainable environment. They are given as case studies which are as follows:

Greening Education is a global initiative taken by UNESCO to facilitate climate action through educational institutions. More than 80 member countries have joined, and this initiative was started in the year 2002 during the Transforming Education Summit. It has initiated two concrete tools for its member states. One is UNESCOs New Greening Curriculum Guidance and UNESCOs New green school quality standard. The New greening curriculum emphasizes the environmental topics and also ensures that the curriculum has detailed outcomes. The new green school quality standard promotes an action-oriented approach and it also recommends that all schools set up the green governance committees. This also conducts audits for energy, water and waste in order to protect the educational institutions for a better environment.

ISBM University follows some green campus initiatives in order to maintain a pollution free campus. The whole campus is tobacco free where certain products like pan masala, tobacco, and many more are completely banned inside the campus. There is a rain water harvesting method that is followed properly in the campus. In this method, the water is collected in the roof and then it is connected to the ground water for recharging and finally it is used for various purposes. The concept of paperless office is meticulously followed where no paper is used and more of ICT mode is followed. This university follows some unique methods like, students are made to sit under the trees and few classes are taken there under the shade which gives a cool environment for the students. Tree planting initiative is yet another important activity where students are involved in full swing. Paper recycling is regularly followed by teachers and students. Vermi compost pits are created from biodegradable waste. Plastic usage is completely banned inside the campus. Battery powered vehicles are used and students use only bicycles to go from one block to another.

Sharadha University's focus is mainly on energy efficiency and renewable energy. This university has taken a number of energy saving initiatives across the campus. There are a number of solar panels in almost all the buildings. To minimize water wastage, rain water harvesting is effectively followed. To manage the waste properly, a waste management system and recycling bin is placed in the campus. To maintain sustainable transport facilities and to reduce the carbon emissions, electric vehicles are used for shuttle services. For eco-friendly mode of transport students are advised to use bicycles only. Another important initiative taken by this university is to have an eco-friendly infrastructure where buildings are designed with green building principles with natural ventilation and energy-efficient insulation. The curriculum is framed which focuses on environment and sustainability and they have a sustainability and Eco club which plays a very important role in creating awareness to the students.

BN College, Bhagalpur in their green campus initiatives has taken different measures and in that the most important one is the bio-waste zone where all the biodegradable materials are dumped in a pit that is exclusively made for it. There are a number of uses of bio-compost in the institution which mainly provides plant nutrients and improves the plant growth. The college properly follows the waste management by keeping different colored bins in the campus. Red bin is kept to collect the e-waste, the green bin is used for collecting the biodegradable waste which is otherwise called organic waste and the blue bin is kept to collect the dry waste. Number of posters and slogans are kept to promote the waste properly and in an efficient manner.

Gauhati University gives importance to their curriculum which emphasizes more on environmental content which will be an eyeopener for the students. This university's main highlight is the Zero waste culture and the waste management strategies. Biodiversity conservation is followed and a number of awareness programs are organized in the campus. Water is recycled properly and this university follows the Rain Water harvesting method effectively.

Jayoti Vidyapeeth Women's University, Jaipur: The main aim of this university is to go green, clean, and be happy. This is mainly to sensitize the students studying in this campus to be happy with a green and clean environment. To reduce pollution, the campus has green landscaping everywhere with gardens which gives an esthetic appeal and to have a sustainable landscaping, use of local plants and soil is used without using chemical fertilizers and pesticides. All the students and teachers use cloth bags only and do not use papers. The solar energy is used for street lights throughout the campus and bio gas is used in the mess for cooking purposes. Another notable initiative taken is the unique plantation program which is followed in the campus where the students plant the saplings on their birthdays and other important occasions. This university strictly follows two important rules, one is an addiction free environment and the other is a stress free environment. In addition, mobile phones are strictly prohibited in the campus where the use of tobacco, alcohol is strictly prohibited inside the campus. When it comes to a stress-free environment, students have to do yoga and meditation. There is also a session called Happy session for 2–3 min. Cleanliness drive is another initiative where students are regularly involved in cleaning the campus and also a number of public awareness programs are organized.

Prince Venkateshwara Arts and Science College important measure is waste management measure where the waste is segregated as dry and wet waste. Another important waste is liquid waste which is generated in the lab and this collected separately in soak pits. This campus follows paperless and adopts all kinds of latest technology instead of using papers. Use of plastic and also single-use plastics are banned in the campus. Instead of plastic bags the college authorities distribute cotton bags to all.

5 Challenges and Future Directions in Sustainability Education

5.1 Challenges in Implementing Sustainability Education

1. **Lack of Institutional Commitment:**
   (a) Focusing on sustainable practices in an institution requires financial resources and visionary leaders. Many educational institutions may lack this leadership.
   (b) For prioritizing the sustainability activities there requires more effort from those who are specialized on all these and at the same time an office and rules or policies.
2. **Curriculum Integration Issues:**
   (a) Sustainability topics are often treated as supplementary rather than integrated across disciplines.
   (b) Interdisciplinary approaches are not sufficient to achieve the objectives, a well curated way of approach is required.
3. **Resource and Infrastructure Constraints:**
   (a) Lack of access to experiential learning opportunities and other tools that enhances sustainability practices such as labs.
   (b) Lack of awareness on green infrastructure.
4. **Faculty Training and Awareness:**
   (a) Lack of teachers who can efficiently handle sustainability concepts.
   (b) Lack of educators who can practice sustainable pedagogical methods.
   (c) Non availability of professional development opportunities focusing on sustainability.

5. **Student Engagement and Motivation:**
    (a) Lack of practical applications and service-learning activities in the curriculum.
    (b) Lack of active learning environment.
    (c) Lack of usage of different and innovative teaching and evaluation methods. Traditional assessment methods reduce engagement of the students.
6. **Policy and Regulatory Barriers:**
    (a) Lack of mandates, permits, or policies on the integration of sustainability in national education frameworks.
    (b) Lack of incentives for the institutions who are practicing sustainability, this may discourage innovation in this field.
    (c) Overemphasis on objective approach reduces flexibility for innovative creativity in sustainability-related learning.

5.2 Policy Gaps and the Need for Systemic Changes

1. **Lack of Comprehensive Policy Frameworks**
    (a) There is a need for well-defined, actionable policies at the national and institutional levels to integrate sustainability education.
    (b) Cross-sector collaboration between government, academia, and industry remains limited.
2. **Insufficient Funding and Resources:**
    (a) Governments and institutions need to allocate adequate funding to support sustainability programs and infrastructure.
    (b) Grants and financial support for sustainability research and community projects are crucial.
3. **Limited Monitoring and Evaluation Mechanisms:**
    (a) Establishing robust assessment frameworks to monitor the progress of sustainability education programs is essential.
    (b) Indicators for evaluating the impact of sustainability curricula are often absent.
4. **Fragmented Efforts:**
    (a) Initiatives are often isolated and lack coordination, leading to duplicated efforts and inefficient resource use.
    (b) National and regional collaboration networks should be strengthened to share best practices.

5. **Lack of Incentives for Educators:**
   (a) Introducing incentives for educators who integrate sustainability into their teaching can motivate wider adoption.
   (b) Recognizing and rewarding institutions leading in sustainability education fosters healthy competition.

5.2.1 Future Trends in Sustainability Education

Experiential and Project-Based Learning

- Schools and universities will increasingly adopt hands-on learning through sustainability projects, fieldwork, and community collaborations.
- Living labs and green campus initiatives will provide real-world sustainability challenges for students.

Interdisciplinary and Transdisciplinary Approaches

- Combining knowledge from natural sciences, social sciences, engineering, and the arts will lead to holistic sustainability solutions.
- Collaborative learning experiences with policymakers, industry leaders, and NGOs will become more common.

Technology-Driven Learning

- AI, virtual reality, and gamified simulations will offer immersive learning experiences on sustainability topics.
- Big data and IoT will facilitate real-time monitoring and analysis of environmental indicators.

Global and Local Collaboration

- Partnerships between educational institutions across borders will foster knowledge exchange and cross-cultural perspectives on sustainability challenges.
- Local communities will play an active role in co-creating sustainability solutions with students and educators.

Policy Advocacy and Youth Leadership

- Young leaders will increasingly influence sustainability policies through advocacy and entrepreneurship.
- Educational programs will emphasize civic responsibility, encouraging students to engage in policymaking and activism.

Certification and Recognition

- Sustainability certifications for institutions and educators will ensure the credibility and quality of sustainability education programs.
- Young leaders will increasingly influence sustainability policies through advocacy and entrepreneurship.
- Educational programs will emphasize civic responsibility, encouraging students to engage in policy making and activism.
- Global benchmarks like the UNESCO Global Action Programme (GAP) will guide institutions in achieving sustainability education goals.

In conclusion, while challenges in implementing sustainability education persist, a concerted effort from policymakers, educators, students, and communities can overcome these barriers. Embracing systemic reforms, fostering cross-disciplinary collaboration, and leveraging technological advancements will pave the way for a sustainable future through education.

References

Brundtland, G. (Ed.). (1987). *Our common future: The world commission on environment and development*. Oxford University Press.

UNESCO. (2014). International implementation scheme for the decade in brief.

United Nations. (2002). *Report of the world summit on sustainable development, Johannesburg*. http://daccessdds.un.org/doc/UNDOC/GEN/N02/636/93/PDF/N0263693.pdf?OpenElement

Sowing Seeds of Change: Combating Deforestation Through Environmental Education in the Indo-Japan Context

Lihui Zhou and John Joseph Puthenkalam S.J

1 Introduction

Deforestation poses a huge risk to biodiversity, water treatment, and climate change worldwide (WWF, 2023). Forests, which cover one-third of the Earth's landmass, are critical to maintaining healthy ecosystems and a sound environment for humans. They not only provide habitats for most terrestrial species on Earth, but also mitigate greatly to climate change by naturally absorbing and storing carbon dioxide. With a huge potential to reduce 4.1–6.5 GtCO2e by 2030, forests contribute up to 30% of available climate solutions through conservation, sustainable management, and restoration (UNEP, 2024). Beyond carbon storage, forests shield against extreme weather, provide drinking water for major cities, and support the livelihoods of millions. However, deforestation—driven by commodity production like palm oil, beef, soy, and timber—threatens these benefits, contributing to 21% of greenhouse gas emissions in land use sectors (Weisse & Goldman, 2021). Although deforestation has been somewhat slowed down by numerous initiatives and efforts by the United Nations and its Member States to protect forests, it remains a critical and ongoing challenge for the global community.

Since 1990, the Asia-Pacific has decreased 38.7 million hectares of virgin and other regenerated forests (FAO, 2021). With an average forest area of only 0.2 hectares per capita, the Asia-Pacific becomes the least forested area in the world (compared to the global average of 0.52 hectares per person). In South Asia—Asia's most populous region, it only has 2% of the global forest area, amounting to only 0.05 hectares per person (FAO, 2021). The causes of deforestation in Asia mainly include economic growth, population increase, agricultural expansion, infrastructure development, mining activities, illegal logging, and rapid urbanization (Arshad et al., 2020; Gandhi & Jones, 2019). In Southeast Asia, driven by substantial global

L. Zhou (✉) · J. J. Puthenkalam S.J
Sophia University, Tokyo, Japan

J. J. Puthenkalam S.J et al. (eds.), *Terra Thrive*,
https://doi.org/10.1007/978-3-032-02738-2_7

demand, land clearing for oil palm plantations is a major factor in deforestation. From 1990 to 2020, an area larger than Germany was deforested. Additionally, illegal tree harvesting poses a serious threat to the region's forests. According to data, 73% of deforestation in Southeast Asia is commodity-driven (forest conversion to farmland), 19% is due to logging, and 8% is due to agricultural expansion (EU, 2020). The deforestation not only leads to a loss of diversity in the region but also increases the risks of floods, droughts, landslides, typhoons and extreme heat. Annual economic losses resulting from disasters in South and South-West Asia have exponentially expanded, surpassing double the previous estimates (ESCAP, 2019).

To combat deforestation, the role of education has absorbed more and more attention in the fight against deforestation. Socrates emphasized that knowledge leads to action, reinforcing the importance of informed decision-making. In this way, environmental education (EE) could be characterized by teaching to promote awareness and understanding of human-environmental interdependence and to facilitate responsible behaviors. EE increases knowledge of the value of forests and encourages sustainable lifestyles and responsible consumption. Through effective environment education, communities, especially those that depend on forests, can use sustainable management to strike a balance between conservation efforts and economic needs. Furthermore, education creates knowledgeable advocates committed to accomplishing climate and biodiversity goals for the benefit of future generations by involving youth and supporting global initiatives. Because of the political and social-economical differences among countries, the levels and capacities of EE vary hugely, which directly impact its efficacy. To provide an insight into the factors affecting EE, this chapter compares EE and its role in the fight against deforestation between India and Japan.

2 Literature Review

Philosophical ideas advocating harmony between humans and the environment roots in traditional cultures in history, whether in Eastern or Western (Fang et al., 2023). These ideas provide a theoretical foundation for developing modern EE. Since World War II, major cities in the world have successively suffered environmental pollution and damage at different extent in the process of economic recovery and rapid growth. Until the 1970s, alarming environmental issues began to receive extensive public attention worldwide. In Japan, confronted with environmental pollution resulting from industrial development, EE emerged in the mid-1960s, focusing on the fight against "kogai" (Fujioka, 1981). In the USA, serious polluted problems instigated large-scale environmental protests nationwide during the mid-1960s and the early 1970s (Kraft, 2020). These demonstrations against pollution propelled the USA to enact and implement related environmental policy and EE legislation in 1969 and in 1970, respectively (Carter & Simmons, 2010).

To raise environmental awareness worldwide, the Biosphere Conference, which was organized by UNESCO in 1968, called for establishing national EE institutions, developing environmental curriculum materials and integrating environmental content into national formal education (Mani, 2019). The 1972 Stockholm Declaration emphasized that EE should be a significant method for enhancing global environmental consciousness, including forest conservation against deforestation. It recommended that countries take measures on environmental protection and education (UNEP, 1972). In 1977, the goals and main principles to promote EE were proposed in the Tbilisi Declaration. The improvement of forest conservation awareness and sustainable development practices was also included (UNESCO, 1977). This document and other related publications have guided many countries to develop EE, including forest educational initiatives (Palmer, 1998). To date, the role of education in preventing deforestation has been extensively recognized. For example, Ruiz-Mallen et al. (2009) suggest that the participatory EE programs in preparatory schools could increase awareness of forest conservation by improving the acquisition of ecological knowledge. Ardoin et al. (2020) show that EE has not only led students to protect the environment of their surroundings by planting trees, but also strengthened forest reserves through the collaboration of local people, scientists and governments.

However, the existing literature mainly focuses on the impacts of EE on specific groups' (e.g., students of the elementary school or children of preschool) environmental awareness or changes in behavior in a certain country. Comparative studies on EE between two countries are scant. This deficiency has led to a lack of mutual learning and exchange between different countries, which is necessary to improve the effectiveness of EE and further improve forest protection, especially when preventing deforestation. To close this gap, this chapter takes Japan and India as case studies to examine the similarity and difference in EE, analyze underlying causes and summarize successful experiences of learning to improve the EE performance. Based on this analysis, the recommendations to prevent deforestation are proposed, which are important for other countries that struggle with environmental deterioration, especially illegal logging.

3 Research Methods

Historical analysis method. EE has experienced a long-term and dynamic development process both in Japan and India. Tradition, culture, and ancient religion historically produce a profound impact on EE development in these two countries. Moreover, with economic expansion, the advent and popularity of new social planform and digital technologies, the contents and teaching approaches of the schools' EE, as well as participatory ways of involving actors, are constantly changing. This chapter studies the performance of various actors from a historical perspective.

Method of literature research. To help understand the existing research results about this subject, this chapter reviews related literature issued by publishers and journals and collects relevant data on deforestation and EE in Japan and India, which provides a solid basis for the whole research.

Case study method. This chapter itself is studying the EE with Japan and India as case studies. In addition, the authors use some case studies in this part to present the role of the most important actors and the effectiveness of different environmental manifestations of evolution.

Interdisciplinary research. EE as an interdisciplinary subject covers a wide range of fields, such as politics, economy, culture, society, and history. To better carry out comparative research on EE and grasp key aspects affecting the EE efficacy, it is necessary to examine the role of various factors from different perspectives.

Comparative analysis. This method fits the goal of this chapter because it enables a normative discussion about how both countries develop EE and why there are numerous differences in EE. With a qualitative and quantitative comparative analysis of relevant actors and paths, the similarities and significant differences in EE between Japan and India are clearly shown and the reasons for it are analyzed. In addition, this chapter not only examines the diversity of EE approaches through systematic comparative research, but also shows the important influence of different politics, culture and economy on EE, which promotes research on EE theory in academia.

4 Case Study

EE is an indispensable part of global efforts to combat environmental degradation and promote sustainability. Both India and Japan have recognized the importance of EE and mainstreamed it into their education systems. However, as the two countries are at different stages of development, the gap between the socioeconomic and political fields poses a significant impact on EE. This leads to huge differences in EE between the two countries, even though they share some similarities in their respective EE. To provide an insight into factors affecting EE, this part develops a comparative analysis of EE between Japan and India and explores the underlying reasons behind their similarities and differences.

4.1 Environmental Education in Japan

EE in Japan has experienced a long-term development process under a common impact of international and domestic environmental movements, which reflects the national growing awareness and commitment to environmental sustainability. The progress of EE has greatly contributed to the formation of sound environmental awareness in the whole society.

4.1.1 Development of Environmental Education in Japan

With rapid economic recovery from the post-war wreckage, industrial pollution began to sweep across Japan, which instigated the emergence of EE focusing on environmental pollution in the mid-1960s. The consciousness initially occurred in communities near industrial zones, but achieved a broader concern until the end of the 1960s. Since the negative effects of industrial development became increasingly clear, it led to widespread public outcry and the development of basic movements such as "Kogai kyoiku," which laid the basis for wider EE and awareness in Japan. The civil protest movements against environmental pollution reached the peak in 1969. By the 1970s, Japan faced enormous double pressure—both domestic and international—to combat environmental degradation. Added to the influence of the United States EE legislation and international conferences, the term "Environmental Education" entered the Japanese lexicon. From 1977 to 1979, Japan's national curriculum began to incorporate environmental perspectives (Fujioka, 1981).

In the early 1980s, advances in environmental technology brought about significant improvements in industrial pollution in Japan. As a result, the importance of EE declined slightly in Japan. However, in the second half of the 1980s, global and national environmental problems became more prominent, leading to a rapid increase in public concern. Slogans such as "Think globally, act locally" and "Environmentally friendly" contributed significantly to raising environmental awareness. In 1988, the final report of the Environment Agency's EE Discussion Group proposed promoting EE in schools. This emphasis on EE was further consolidated in 1993 with the enactment of the Basic Environment Law, which set out the government's fundamental stance on EE (Kodama, 2017).

After the 1987 Brundtland Report, sustainable development (SD) became an important focus for researchers, businesses, communities, governments and international organizations (UNESCO, 2014). Numerous researchers and international resolutions have highlighted the crucial role of modern education in realizing SD goals (Becker, 2018). This has driven the EE in Japan to embark on a new stage—education for SD (ESD). The ESD surpasses traditional educational content and implies many new elements such as disaster prevention and management, a low-carbon lifestyle and international understanding. It enriches the EE approaches, including various education forms, and reaches out to different age groups and social roles. The UN General Assembly adopted Japan's proposal to advocate ESD from 2005 to 2014 in 2002. After that, ESD has become well-known in the global (ENV, 2014).

To boost ESD, Japan is devoting to national ESD initiatives. In 2019, Japan reexamined the implementing conditions of these national plans and highlighted the need for stronger policy enforcement in collaboration with diverse stakeholders and emphasized the importance of sharing information and best practices both domestically and internationally. Following these results and the new global ESD framework for 2030, Japan has drawn up the second ESD Implementation Plan. This plan aims to support the commitments of various stakeholders in Japan, including the government, local authorities, enterprises, the media, research institutes, academic associations, educational institutions, teachers and others who are involved in ESD. The plan contains five key areas: policy advancement, altering the learning environments, improving educator capacity, motivating young people, and supporting local actions (ESD, 2021).

Looking at the history of EE in Japan (Table 1), it can be seen that, regardless of it is called "kogai education", "EE" or "ESD", it has always been a response to the changes of times in which both national and international environmental problems and challenges are addressed proactively or reactively. Due to Japan's early industrialization, the country also faced environmental problems earlier, leading to the early introduction of EE. Over decades of evolution and development, Japan's comprehensive EE system, which includes both formal and informal education, has become relatively mature compared to that of other developing countries. This all-encompassing EE has promoted a high level of environmental consciousness among the population, created a favorable social environment for forest conservation, and effectively prevented deforestation in Japan.

The progress of EE in Japan, including forest education, has been promoted by both national and international factors. Domestically, many years of public resistance to pollution and environmental degradation forced the government to solve environmental issues. Besides, the global environmental movements, in particular the rise of the sustainability movement, also played a tremendous role in promoting EE in Japan at an international level. Whether through reactive reactions or proactive initiatives, in the course of several decades, Japan has not only developed a relatively mature system of EE, but has also enhanced environmental awareness among its citizens.

4.1.2 Approaches to Environmental Education in Japan

EE has always played a pivotal role in encouraging people to concern themselves with environmental challenges, participate in addressing problems and adopt environmentally friendly behaviors. And it is the strongest force for controlling ecological deterioration by enhancing environmental responsibility and changing the lifestyle of the individual. To achieve effective results, the approaches of EE are crucial. After more than 70 years of effort, Japan has developed a coordinated and unified EE system involving the central government, local governments, schools, enterprises, the public and various social organizations. Different dynamic and diverse educational methods are used depending on the different bodies involved in the EE.

Environmental Acts, Policies and Programs

To create a sustainable society, Japan's government at different levels has committed itself to a better EE by publishing a wealth of related policies and laws, adopting new social platforms, digital technologies, and innovating local programs and practices (Table 1). To provide a legal guarantee for EE and environmental

Table 1 History of environmental education in Japan

Year	Japan	Overseas
1931	Enactment of National Parks Law	
1950	Establishment of the Nature Conservation Society of Japan (foundational juridical person)	
1964	Establishment of the Nature Conservation Bureau	
1966	Primary and Middle School Teachers' Pollution Control Measures Study Association established	
1967	Enactment of the Basic Law for Environmental Pollution Control	Sweden: Revision of Elementary Education Teaching Guidelines (emphasis on environmental issues)
1969		U.S.: Enactment of the National Environmental Education Act
1971	Partial revision of Primary and Middle School Education Guidelines (improvement in education on pollution in social studies); establishment of the Environment Agency	
1972	Enactment of Nature Conservation Law	Stockholm: United Nations Conference on the Human Environment; establishment of UNEP
1973	Decision by the Cabinet on "Basic Policy for Conservation of the Natural Environment"	
1974	Establishment of Charter of Nature Conservation	
1975	Establishment of the Japan Environment Association (foundational juridical person); promulgation of new Primary and Middle School Education Guidelines (emphasis on environmental issues)	International Environmental Education Conference (Belgrade Charter)
1978	Promulgation of Senior High School Education Guidelines (emphasis on issues related with environment and human life)	Intergovernmental Conference on Environmental Education (Tbilisi)
1982		10th Special Meeting of UNEP Directors Council (Nairobi Declaration)
1986	Decision by the Environment Agency on the Long-Term Environmental Conservation Plan	
1987		Report of the World Commission on Environment and Development (WCED)
1988	Report by Environmental Education Panel of Environment Agency (seeking for better environment created by everybody)	
1989	Revision of School Education Guidelines (introduction of "living" as a subject)	

(continued)

Table 1 (continued)

Year	Japan	Overseas
1990		U.S.: Enactment of National Environmental Education Act
1992	Promulgation and enforcement of the Basic Environment Law (promotion of environmental education and learning)	Earth Summit (United Nations Conference on Environment and Development) held in Rio de Janeiro/ Adoption of Agenda 21, etc.
1994	Establishment of the Basic Environment Plan	
1997		Thessaloniki International Conference on Environment and Society: Education and Public Awareness for Sustainability (Thessaloniki 1 Declaration)
2002	Launch of preparations for the Decade of Education for Sustainable Development (2005–2014)	The Johannesburg Summit: Education for Sustainable Development
2003	Basic Environment Plan revised; promotion of Environmental Education Act enacted	UN Decade for Sustainable Development (DESD) begins
2005	The Kyoto Protocol comes into effect	
2008	Japan ESD implementation plan drafted	
2011	Basic plan for the promotion of education drafted/curriculum guidelines ("Courses of Study") revised	
2012	Japan ESD implementation plan revised	
2014	Second Basic Plan for the promotion of education drafted	Launch of Global Action Programme (GAP) on ESD
2015	Japan hosts UNESCO World Conference on Education for Sustainable Development (Aichi-Nagoya)	United Nations adopted "Transforming our world: the 2030 Agenda for Sustainable Development" with presenting its 17 Sustainable Development Goals (SDGs)
2018	The Government of Japan has launched the Global and Innovation Gateway for All (GIGA) School Program	
2019	Revision of national curriculum guidelines to emphasize SDGs and ESD	UNESCO launches the "Futures of Education" initiative; Launch of the ESD for 2030 (2020–2030)
2021	The improvement of school ICT environments has been jump set off simultaneously all over Japan, and ESD using digital contents has been promoted	UNESCO World Conference on Education for Sustainable Development
2023	National Biodiversity Strategy and Action Plan of Japan 2023–2030	

Source: edited based on the Ministry of the Environment, Japan (2024) protection, Japan issued the Enhancing Motivation for Environmental Protection and Promotion of Environmental Education Act in 2003 as the guideline to boost more involvement of voluntary organizations and institutions. In addition, Japan has taken a number of measures to continuously improve the EE program and highlighted the importance of hands-on practices and natural experiences since the 1990s.

The progress of EE has promoted the development of forest education activities. As early as the 1970s, forestry education activities began to attract the attention of forest administrators. In 1999, "Forest Environmental Education" was initiated by the Forestry Agency. In 2001, the idea of "promoting the use of forests for education" was subsumed in the legislation. These moves stressed the active role of education in sustainable forest management. These policies provide legitimacy for cooperation activities aimed at raising awareness and taking action against deforestation between relevant educational institutions, local authorities and forest-based organizations. Subsequently, some facilities were set up for the development of forest-related activities in national forests. Numerous materials on forest education were published. The term "Mokuiuku" shows the great success that Japan achieved in promoting the sustainable use of wood across the country. Mokuiuku is also an effective way of raising public concern about environmental challenges (FFPRI, 2023).

Moreover, Japan has committed to advocating environmental responsibility worldwide. To help and encourage other developing countries to take actions to raise environmental awareness, Japan has not only sent experts to train their environmental officials, but has also shared some environmental technologies. Otherwise, Japan actively participates in global environmental conferences and endeavors to make recommendations to overcome global environmental challenges (SES, 2023). This greatly enhances Japan's international reputation and further improves national environmental awareness and measures.

School Education

The emphasis of respecting nature is a significant character of Japan's traditional culture, which constitutes the basis of its successful EE system. Furthermore, Japan's strong environmental consciousness is closely linked to the government's role in education. The government works with related educational institutions and communities to develop effective approaches to EE, beginning from primary school to university. Students are involved in the environmental curriculum from an early age, which helps them to understand the importance of ecological conservation and cultivates environmental responsibility. Together with various practical activities such as excursions and hands-on experiments, they can acquire more knowledge about nature. All these ensure that environmental values are deeply embedded in their minds and continue into adulthood (Mulyadi, 2020).

Forest education activities, as a key component of the EE, have been widely implemented in Japan to promote sustainable development. Forest education is the learning through direct experience in forests to help individuals understand these natural environments and regard forests as a source of renewable resources. Moreover, the sensitivity to nature is enhanced. Skills and knowledge of the forest are being developed, and further encourage people to live in harmony with nature. All of these activities contribute to a sustainable society. At present, the forest pedagogical activities in Japan include 40 different types and are divided into 13 groups (Table 2).

To improve the effectiveness of EE, Japanese schools have developed various approaches since the 2000s that reflect regional characteristics and match the principles described in the teacher guidelines for EE (Kodama, 2017). From 2002, environmental content has been improved for each subject and enables comprehensive studies through a cross-subject training within the period for integrated studies under new teaching guidelines. The promotion measures include the designation of model areas for EE, the provision of learning fairs and seminars for teachers as well as the establishment of Model Globe schools and environmentally friendly school facilities. Japan promotes the use of recycled paper in textbooks to improve the understanding of the students for recycling. With the focus on the relationship between global warming and energy consumption, Japan continues to promote education through resources and energy, including nuclear energy (ENV, 2024). In addition, EE in Japan goes beyond classrooms. The "Green Schools" initiative offers numerous practical opportunities to turn green ideas into reality. These experiences' activities encourage students to live a green lifestyle even after graduation (FFPRI, 2023).

Table 2 Forest education activities in Japan

Category	Description
Immersion in nature	Activities like playing, games, and walking in the forest
Health and relaxation	Enjoying cherry blossoms or autumn leaves and resting the body and mind
Wildlife protection	Investigating, breeding, and managing wildlife and their environments
Nature observation and learning	Observing trees, animals, and the natural environment; visiting forest-related facilities and forestry sites
Collecting for learning	Gathering flora and fauna for observation and study
Gathering forest resources	Collecting fuel wood, materials for crafts, food, and leaves for composting
Environmental management	Repairing or mowing forest trails
Building forest facilities	Constructing cottages, huts, treehouses, trails, and playground equipment
Forestry work	Activities like planting trees, cutting understory, pruning, thinning, mushroom cultivation, and making charcoal
Handcrafts	Creating crafts using wood
Daily life activities	Eating food gathered from nature, camping, and outdoor cooking
Artistic activities	Engaging in creation, outdoor concerts, and exhibitions of pictures or photos
Outdoor sports	Hiking, mountain climbing, athletics, skiing, backcountry skiing, and adventure courses

EE in Japanese schools plays a crucial role in promoting environmental awareness, developing problem-solving skills, supporting Sustainable Development Goals (SDGs), and instilling social responsibility and cultural values. Through innovative practices such as integrated learning programs, students deep understand environmental issues and adopt sustainable behaviors. Collaboration between schools, local governments and communities further enhances the impact of EE. These initiatives have resulted in significant achievements, including increased environmental awareness among students, the creation of innovative educational models, and contributions to the global EE discourse, placing Japan at the forefront of this field.

Community Environmental Activities

In Japan, environmental awareness has already reached into the hearts of local communities. Many communities across the country are actively involved in SDG-related activities. To address specific environmental challenges and promote a sense of responsibility and connection with nature, local communities have implemented various educational initiatives (Table 3). These programs, tailored to regional characteristics and needs, often involve collaborations between local governments, nonprofit organizations, businesses and educational institutions. Below some examples of local EE programs and their results are listed.

Table 3 Community environmental activities in Japan

Location	Program/initiative	Activities	Achievements
Ishikawa Prefecture	Satoyama Conservation and Education	Rice planting, forest management, water conservation	Promotes biodiversity awareness, sustainable agriculture, and cultural heritage preservation
Shiga Prefecture	Lake Biwa Environmental Education	Water quality monitoring, invasive species removal, sustainable water use workshops	Increased awareness of freshwater ecosystems and integration into school curricula
Kumamoto Prefecture	Minamata Environmental Education	Field visits, lectures, storytelling by survivors	Teaches lessons from industrial pollution, fostering understanding of environmental health
Tokushima Prefecture	Zero-Waste Community Education	Waste sorting (45+ categories), upcycling projects, awareness campaigns	Kamikatsu became a global model for community-driven sustainability through education
Okinawa Prefecture	Coral Reef Conservation Education	Reef monitoring, snorkeling workshops, coral restoration	Protects marine biodiversity and raises awareness about ecosystem fragility
Yamanashi/ Shizuoka Pref.	Fujisan Environmental Learning Center	Hiking, forest cleanup, ecological studies of Mount Fuji	Enhances understanding of ecosystem services and human impact on iconic natural landmarks

Corporate Environmental Activities

As a pivotal player in EE, the role of Japan's enterprises couldn't be neglected. Despite the fact that some enterprises ever played a role in environmental damages in history, numerous guidelines and regulations issued by the central and local governments and the pursuit for long-term interests has pushed these enterprises to change their behavior. For example, the government has introduced ISO26000 and the global standard for Corporate Social Responsibility (CSR) to standardize corporate social responsibility practices and increase corporates' transparency and accountability. Additionally, in Japan, governments often take corporate advice on board before proposing new environmental initiatives and plans, which have helped companies align their environmental efforts with national and global sustainable goals.

Especially in recent decades, with climate change, increased regulatory requirements and consumer expectations, Japanese companies have progressively acknowledged the importance of integrating environmental considerations into their operations. To perform environmental responsibility, companies have launched various environmental activities and initiatives, including reducing CO_2 emissions, developing renewable energy, sustainable use, and management of resources. Main programs launched by companies in Japan are summarized with the names of the associated representative companies (Table 4).

Table 4 Corporate environmental activities

Key environmental activity	Description	Company name
Energy efficiency and renewable energy	Investing in technologies to reduce energy consumption and increase the use of renewable energy sources like solar, wind, and biomass	Panasonic Corporation
Waste reduction and recycling	Implementing waste management strategies, promoting recycling, and adopting circular economy principles to ensure efficient resource use	Ricoh Co., Ltd.
Sustainable supply chain management	Conducting environmental audits and promoting sustainable practices throughout the supply chain, ensuring suppliers adhere to environmental standards	Sony Corp.
Environmental education and community engagement	Collaborating with local communities to raise environmental awareness, organizing workshops, seminars, and public forums on eco-friendly practices	Toyota Motor Corporation
Green product development	Developing products with a reduced environmental impact, such as energy-efficient appliances, eco-friendly packaging, and sustainably sourced materials	Hitachi, Ltd.

Environmental Activities and Education by NGOs

The advance of EE cannot leave the contribution of non-governmental organizations (NGOs) in Japan. Driven by passionate individuals and communities, NGOs inspire environmentally friendly practices through their wide range of activities and educational initiatives. They focus on several key areas to promote the protection of natural resources and promote sustainable living (Table 5). Their main efforts include:

The efforts of NGOs in Japan have produced significantly positive results. These include greater public awareness of environmental issues, improved local ecosystems through conservation projects, and greater community engagement in sustainable practices. By bridging the gap between government policy and community action, NGOs have been instrumental in advancing grassroots environmental movements. Continued support and cooperation between non-governmental organizations, government and the private sector will be essential to tackle future environmental challenges and ensure a sustainable future.

Due to the long-term and common efforts of Japan's governments at different levels, enterprises, NGOs, local communities, and the general public, now EE is not

Table 5 Environmental activities and education by NGOs in Japan

Category	Description	Representative NGO name
School Programs and Workshops	– Eco-Schools: NGOs collaborate with educational institutions to integrate environmental education into school curriculums, organizing workshops, field trips, and hands-on activities to engage students – Teacher Training: programs aimed at equipping educators with the knowledge and tools to teach environmental topics effectively	Japan Environmental Education Forum (JEEP) Japan NGO Center for International Cooperation (JANIC)
Community Outreach and Public Awareness Campaigns	– Workshops and Seminars: NGOs host workshops and seminars to educate communities about environmental issues, sustainable practices, and the importance of conservation – Public Campaigns: large-scale awareness campaigns leveraging media, social networks, and public events to reach a broader audience and inspire collective action	Friends of the Earth Japan (FOE Japan) Protect Our Winters Japan
Citizen Science and Volunteer Programs	– Data Collection and Monitoring: NGOs engage citizens in scientific research and monitoring projects, such as biodiversity surveys and pollution tracking, fostering a sense of ownership and involvement in environmental protection – Volunteer Initiatives: organized efforts to recruit volunteers for activities like tree planting, beach clean-ups, and community gardening, promoting hands-on participation in environmental conservation	The Nature Conservation Society of Japan (NACS-J) Shanti Volunteer Association (SVA)

limited to traditional classroom teaching and hands-on experience, but has stepped into every field of social life. In other words, EE has been an important part of national life. This dynamic EE model extends across society and includes education for people of all ages, from children to adults. Also, educational methods have continued to evolve over time, with ICT being widely used to improve the effectiveness of EE nationwide. Furthermore, responsibility for EE has shifted from the sole responsibility of schools in the early stages to involving multiple stakeholders, including governments at all levels, businesses, and community organizations. The collaboration between these various actors has helped to drive the continued progress of EE in Japan.

4.1.3 Role of Environmental Education in Combating Deforestation

Japan's landscape is characterized by a diverse variety of forests that extend over 3000 km from north to south. The total area of these forests amount to about 25 million hectares, covering about two-thirds of the national surface, which has been maintained for more than half a century (MAFF, 2019). Historically speaking, forests in Japan have provided essential resources for economic activities, including wood for construction work and tools, firewood for daily life and industries such as salt and iron production as well as organic material for arable land fertilizers. However, population growth and economic development have caused considerable forest damage that led to floods in lower regions and sand damage along the coast (MAFF, 2022).

Japan's deforestation history goes far back, especially during the EDO time, in which forests were used strongly for wood, agriculture and fuel. These non-sustainable practices led to severe forest loss and ecological problems such as soil erosion and floods. In order to address these problems, conservation measures were introduced, such as setting mountainous areas aside in which the deforestation was prohibited and the start-up tree planting initiatives. These early efforts laid the foundation for the Japanese principles of forest protection.

As the Meiji period began in 1868, the demands of modern industry led to increased deforestation, which in turn increased the frequency of natural disasters. In response, new laws were enacted and public forest restoration projects initiated (MAFF, 2022). However, the situation worsened during and after World War II, with abandoned afforestation areas amounting to around 1.5 million hectares. The destruction during and after World War II was far more devastating than the mid-Meiji era, with air raids destroying nearly 20% of the country's forests and fields (Hiroyuki, 2018a, 2018b).

Reconstruction efforts to restore the devastated land resulted in even more rampant tree felling, including the clearing of native virgin forests. The media clamored for the use of Japan's "rich" natural resources as the country rushed to rebuild over 2.2 million lost homes and provide housing for more than 6 million repatriated Japanese (Hiroyuki, 2018a, 2018b). The push to build new infrastructure and factories further accelerated deforestation. To address these challenges, the Forestry

Agency launched a large-scale forest expansion campaign. In the process, buna forests were cut down and replaced by fast-growing coniferous forests to provide more economically viable timber. This extensive forest reforestation campaign was perceived by many people as a government-sanctioned program of forest destruction. During this time, the term "buna massacre" began to circulate in the timber industry. The policy conducted the expansion of artificial forest in Japan, which increased from 27% to over 44% by 1985 (Hiroyuki, 2018a, 2018b).

But this kind of massive deforestation not only jeopardizes the livelihoods of local forest-dependent people, but also damages the ecological environment. Therefore, it triggered strong protests across the country. By 1977, this discontent peaked in the struggle against deforestation of national forest in the Shiretoko National Park in Hokkaido (Hiroyuki, 2018a, 2018b). These protest movements became a powerful voice advocating for sustainable forest management and the protection of natural resources, demonstrating the public's commitment to protecting their environmental heritage. All these greatly raised national concern about forest resources, which laid a solid basis for sustainable forest management in the following years.

In addition, public discontent and ongoing protests have forced both local and central governments to consistently adapt and improve forest management practices. The "Five Year Plan for Forest Protection" was launched in 1948 and by 1951, the government had revised the Forest Act to establish protected areas for forest protection. At the same time, reforestation efforts were promoted as public projects. In 1960, the Emergency Forest Protection and Flood Control Act was enacted, followed by nine iterations of the 5-Year Plan for Forest Protection Projects. Technical advances in the engineering field further expedited the restoration of devastated areas. By 2004, significant progress in forest restoration led to the abolition of the Emergency Measures Act and the Temporary Measures to Consolidate Protected Forests Act. The forest department then implemented the "Forest Development and Protection Project" in accordance with the Forestry Act. In 2013, the Forest Agency publicly announced that 60 sites had made significant contributions to national land conservation, celebrating a century of forest conservation projects (Forest Agency, 2022).

Reflecting on the entire process, it is clear that the dissatisfaction of the public and the protests against environmental pollution and ecological deterioration in the last century not only increased public environmental awareness, but also objectively supported the rapid development of the EE in Japan, including forest protection. The development of the EE has in turn further improved public environmental awareness and contributed to continuous progress. These developments can be considered mutually. Meanwhile, efforts by local and central governments to address domestic discontent and international pressure in environmental advocacy have facilitated the nationwide dissemination of EE.

Today, the status of Japan's forests reflects its people's continued efforts in restoration and maintenance. During this long journey, EE has played a notable role in preventing deforestation and promoting sustainable forest management, complementing government policies and cultivating profound understanding and commitment to ecological conservation throughout society. In 2014, the Cabinet Office conducted a survey to measure public interest in nature. An impressive 89.1% of respondents said they were either extremely or somewhat interested in the natural environment. This figure, which represents almost 90% of the population, shows a remarkably high level of environmental engagement worldwide (Hiroyuki, 2018a, 2018b). Another investing finding from the Ministry of the Environment, the FY2020 Environmental Education and Awareness Survey, showed that 60% of respondents reported experiencing a "change in awareness" or a "change in both awareness and behavior" (ENV, 2020). This suggests that more than half of the participants underwent some form of transformation as a result of their learning. These figures prove that EE acts as a driving force to influence public environmental awareness and behavior and further support Japanese society to become more environmentally friendly.

4.2 Environmental Education in India

As a developing country, the EE in India was launched late. In the past three decades, as economic growth has caused environmental degradation and the impact of global environmental movements, EE has absorbed more concerns in India. From the central government to the states, from national EE policies to local school curriculum, from simple classroom teaching to more hands-on activities, EE has made impressive strides in India, reflecting a growing awareness of environmental protection across the country.

4.2.1 Development of India's Environmental Education

India's religions and culture in history contain enriched ideas about the harmony between man and nature. For millennia, these modest ideas have produced a tremendous influence on Indian society, not only as philosophical ideas, but also deeply embedded in people's daily lives and rituals, fostering a profound appreciation for nature (Table 6). All these offer sound foundations for the EE development in India.

After India's independence in 1947, industrialization and urbanization led to increased environmental problems, forcing the central government to adopt numerous environmental policies and laws. In this context, the EE has gradually received social concerns and stepped into a rapid development stage (Table 7). Beginning with the initiatives of the Kothari Commission (1964–1966), EE in India was initiated by the official bodies (Government of India, 1966). In the 1970s and 1980s, the importance of incorporating the EE into the national curriculum was recognized by the government, and environmental problems were introduced in various subjects to sensitize students to environmental challenges. The National Policy for Education

Table 6 Impact of India's traditional culture and religion on environmental education

Cultural/ religious idea	Main contents	Examples	Impact on environmental education
Sacred Groves	Communities preserve forest patches as sacred spaces	Sacred groves in Kerala, Himachal Pradesh, and Rajasthan	Protect biodiversity; prohibit tree cutting and wildlife harm
Vedic Philosophy	Ancient hymns glorify elements like rivers, forests, and mountains	Atharva Veda's teachings on earth conservation	Establish spiritual reverence for nature and its preservation
Buddhist and Jain Teachings	Non-violence toward all living beings; promote coexistence with nature	Jain vegetarianism and protection of nature; Buddhist wildlife sanctuaries	Encourage protection of animals and forests
Ashoka's Edicts	Promote kindness toward animals and prevent ecological harm	Prohibition of hunting, forest protection and other environmental laws	Early example of state-led environmental efforts
Bishnoi Community Practices	Protect trees and wildlife as part of religious tenets	Sacrifice during the Kheji tree-cutting protests in Rajasthan	Foster community-based forest and animal conservation
Rivers as Deities	Worship rivers as life-giving goddesses	Ganga, Yamuna, and Godavari are celebrated in rituals	Raise awareness about water purity and conservation
Festivals and Rituals	Promote sustainable practices through traditional celebrations	Van Mahotsav (Tree Plantation Festival)	Inspire community participation in afforestation
Traditional Agricultural Practices	Sustainable farming techniques to conserve resources	Organic farming, crop rotation, and rainwater harvesting	Reduce resource exploitation and promote ecological balance

(1986) symbolized the significant progress of EE in India, which mandated the involvement of EE at different school levels (Government of India, 1986).

The National Curriculum for Elementary and Secondary Education, which was established on the basis of the recommendations of the National Policy on Education (NPE) 1986, highlighted the implementation of EE through the "infusion" approach within schools (Bhatia, 2020a, 2020b). Later curriculum frameworks, which were developed by the National Council of Educational Research and Training (NCERT), have continually underlined the substance of EE in the education system. NCERT has adopted various strategies to integrate EE at different educational levels.

EE is introduced as a combined subject at elementary school. In middle schools, EE is integrated into the natural and social science and linguistic curricula. At the higher secondary level (+2), knowledge of the environment and development are integrated into almost all school curricula (Chhokar & Chandrasekharan, 2006). This step rested on the apprehension that the EE is not a discrete entity but is inextricably linked to all fields of society, which requires a multidisciplinary perspective for full understanding (Fan, 2017).

Table 7 Development process of environmental education in India

Timeline	Development	Key features	Impact
Pre-Independence	Cultural and religious influences on ecological awareness	Sacred groves, reverence for rivers, sustainable agricultural practices	Informal education instilled respect for nature
1937	Gandhi's Nai Talim (Basic Education)	Emphasized learning through nature and practical skills	Fostered sustainable practices among students
1986	National Policy on Education	Highlighted environmental awareness as a key educational goal	Strengthened the focus on environmental education across school systems
1991	Supreme Court Mandate	Made environmental education compulsory following M.C. Mehta's PIL	Formalized environmental education in schools
2001	National Green Corps and Eco-Clubs	Students engaged in activities like tree planting, tree planting, waste management	Encouraged hands-on learning and community participation
2005	National Curriculum Framework	Integrated experiential and project-based environmental education	Promoted practical engagement with environmental issues
2008	Global Collaboration: NAPCC	Linked environmental education to missions like Solar and Green India	Promoted awareness of climate change and renewable energy
2016	E-Waste & Plastic Waste Management Rules	Introduced digital and waste-related education initiatives	Addressed contemporary environmental challenges
2020	New Education Policy (NEP)	Cross-curricular focus on environmental education	Encouraged sustainable habits and lifestyles
2020s	Digital Tools & Mission LiFE	Leveraged technology and promoted lifestyle-based sustainability	Enhanced accessibility and engagement in environmental education

Although the NPE 1986 emphasized the significance of EE, it was not prescribed as an obligatory subject in the formal education system of India (Chhokar & Chandrasekharan, 2006; Sonowal, 2009). This status changed according to a Public Interest Litigation (PIL) submitted by Shri M. C. Mehta in 1991. In response to this, the Supreme Court recognized the need for EE and proposed that environmental issues should be taught as an obligatory subject. Afterwards, the NCERT in December 2003 developed a standard curriculum for EE.

In 2004, the Supreme Court indicated that "the curriculum created by the NCERT for class I to XII is accepted by each state in its individual schools" and NCERT is designated the node agency for monitoring the implementing process of this directive. The agreement was concluded in 2010, whereby the order of the Supreme Court was prescribed for all states and union areas and effectively expanded its reach throughout the entire formal education system. NCERT has worked closely with the governments of the states to support the implementation process. At the same time, India applied the mandatory environmental lessons in schools, based on recommendations from the National Curriculum Framework (NCF) expert group. This approach integrates EE into different subjects instead of treating them as an independent topic and promotes a holistic understanding of environmental issues among students (Sharma & Menon, 2024). The NEP published by India in 2020 further emphasizes the need to implement EE in various schools and universities.

4.2.2 Approaches of Environmental Education in India

Despite EE in India initiated relatively late, the approaches of EE are increasingly diverging with the support of government and civil society (Table 8). A number of policies and fund programs have been provided at official level to support the implementation of EE, which not only give effective guidance and financial aid to schools' EE but also impart environmental information to the whole society. The cooperation of India's central and local governments in EE strongly contributes to the promotion of sustainable practices and environmental awareness in various regions (MoEFCC, 2021b). In this process, the role of the mass media in disseminating environmental information cannot be overlooked. According to a survey, more than 60% of Indian households improve their environmental awareness through television programs (Prasar Bharati, 2019). Newspapers and magazines are also crucial to encourage public's participation in environmental activities by providing a wealth of useful environmental information (Times of India, 2020). Furthermore, the role of new social media is increasingly important. For example, In the #GreenGoodDeeds campaign launched by MoEFCC in 2021, the public shared their many environmental actions via social media and greatly increased the impact of the campaign (MoEFCC, 2021a).

In Addition, to promote EE, India has implemented some community-based programs that have had a tremendous impact on national sustainability and environmental awareness. For examples, the Community Science Center in Ahmedabad integrates students into environmental surveillance and developed a program for school children to measure air pollution in the early 1970s. Swechha's environmental initiatives advocate sustainable lifestyles through workshops and community events. The Hoshangabad Science Teaching Program focuses on interactive and activity-based learning methods and emphasizes learning through discovery and hands-on activities, which are fundamental to an effective EE program (Eklavya, 2001). These programs conducted by social organizations have made a considerable contribution to scientific education, curriculum development and innovative teaching ways (CSE, 1982). India has a rich tradition of NGOs that work for critical development and education initiatives. NGOs are significantly involved in the development of action-based projects and the formation of groups that critically assess political problems. Alongside the development of global cooperation, more international environmental organizations begin to involve in the EE and environmental protection in India, working with domestic NGOs.

Table 8 Approaches of environmental education in India

Sponsor	Approach	Examples/implementation
Government	Policy Development: mandating environmental education through laws and curricula	Supreme Court mandate (1991) made environmental education compulsory; NCERT curricula designed for all levels
	Funding and Programs: financial support for school initiatives	National Green Corps (2001) funded by MoEFCC established eco-clubs across schools in India
	Awareness Campaigns: national-level campaigns for sustainability	Swachh Bharat Abhiyan cleanliness drives and Mission LiFE fostering eco-conscious behaviors in schools
	Capacity Building: training educators for teaching environmental topics effectively	Government organizes teacher training workshops for enhancing environmental literacy under MoE programs
	Infrastructure Support: supporting eco-friendly facilities in schools	Rainwater harvesting and solar panel installations in government schools
Schools	Formal Education: environmental topics in school curricula	NCERT textbooks include lessons on biodiversity and conservation
	Experiential Learning: hands-on activities for real-world understanding	Kerala schools organize biodiversity walks in mangrove forests
	Eco-Clubs: student-led environmental initiatives	National Green Corps promotes tree planting and waste segregation activities in schools
	Green Infrastructure: eco-friendly school campuses	The Millennium School in Lucknow has rainwater harvesting systems and solar panels
	Lifestyle Education: programs promoting sustainable habits	Schools in Pune implement zero-waste cafeteria policies and composting practices
Enterprises	Gamification: interactive sustainability education	Mobile app "Recycle Hero" helps players manage virtual waste effectively
	Environmental Entrepreneurship: supporting student innovations	Tamil Nadu students developed solar-powered water purifiers with CSR sponsorship
	Green Initiatives: enterprises sponsoring eco-friendly infrastructure	CSR-funded programs install solar panels and promote waste segregation systems in schools

Table 8 (continued)

Sponsor	Approach	Examples/implementation
Media	Digital Tools: awareness through apps and e-learning platforms	BYJU'S Learning App includes modules on environmental sustainability
	Social Media Campaigns: promoting environmental awareness globally	Campaigns like #BeatPlasticPollution engage students on platforms like Instagram and YouTube
	Edutainment: environmental education through entertainment	Screening of "Before the Flood" followed by discussions in Bengaluru schools
NGOs	Workshops & Training: Capacity building for students and teachers	WWF-India's "One Planet Academy" offers e-learning platforms for environmental literacy
	Citizen Science Projects: Real-life ecological research for students	Students in Uttarakhand monitor Ganga pollution with water quality testing kits in collaboration with NGOs
	Community-Based Learning: Collaborating with locals to address challenges	Schools and NGOs in Rajasthan restore traditional water systems (johads)
Global Collaboration	UNESCO Programs: aligning with global sustainability goals	Schools in Mumbai participate in UNESCO's Climate Action Project
	Global Events: participation in international environmental campaigns	Schools join Earth Day activities with creative environmental projects

4.2.3 Role of Environmental Education in Combating Deforestation in India

Deforestation is also an urgent environmental concern for India, which are leading to biodiversity loss, climate warming and socioeconomic problems across the country. The Forest Survey of India shows that between 2015 and 2020 the country lost 1.2 million hectares of forest, accentuating the pressing need to prevent deforestation (Global Forest Watch, 2023). Agricultural expansion, illegal logging, infrastructure construction and rapid urbanization are the main causes of deforestation in India.

Due to the destruction of habitat caused by deforestation, the survival of numerous plant and animal species in India is threatened. For example, the tiger population in India is closed in their endangered status because the deforestation leads to a fragmentation of forest habitat (IUCN, 2021). In addition, forests play a significant regulatory part in mitigating climate warming by absorbing CO_2 from the atmosphere. However, this capacity is impaired by deforestation, which results in increasing greenhouse gas emissions and tightening global warming. In India, the loss of forest area disturbs the local climate patterns and impacts the precipitation and temperature. Another study by the Indian Institute of Science indicates that deforestation in the Western Ghats has led to a decline in regional rainfall of up to 20% (Indian Institute of Science, 2020). Moreover, local communities, especially indigenous and forest-dependent populations, are affected by deforestation. Forest losses reduce the available essential resources such as food, medicine and firewood, and affect livelihoods and cultural practices. In Odisha, for example, the displacement of tribal communities has resulted in severe socioeconomic hardship due to the deforestation for mining projects (India Team, 2024).

India has implemented many strategies and guidelines for reducing deforestation and achieved mixed results. To restore degraded forest areas and increase forest cover, the Government launched the National Afforestation Program (NAP) and proposed the Green India Mission to promote plantation of indigenous species and involve the community in forest management (MoEFCC, 2021b). Initiatives for sustainable forest management encourage people to actively participate in the administration and maintenance of forest resources and emphasize the sustainability of their use and ensure the advantages of the community (Foundation for Ecological Security, 2021). The related legislation and policies offer a solid regulatory basis for the preservation of the forest by monitoring the deforesting activities and the protection of habitats of wild animals (Ministry of Law and Justice, 1980; Ministry of Environment and Forests, 1988).

As an essential approach to combating deforestation, the role of EE is receiving increasing attention in India. Programs such as those run by the Center for Environment Education (CEE) have significantly contributed to integrating EE into school curricula and community initiatives, raising awareness of forest conservation among over ten million students across India (CEE, 2021). Besides, the introduction of eco-clubs in schools has significantly motivated the enthusiasm of students for environmental activities. These clubs encourage pupils to take part in tree planting campaigns, forest cleaning campaigns and sensitization workshops and thus arouse a sense of responsibility for forest protection. A report published by the Indian Institute of Science, shows that schools with active eco-clubs increased tree cover in their classrooms and surrounding areas by 15% (Indian Institute of Science, 2020). In addition, the development of network media technologies accelerates the spread of negative effects of deforestation, increases public concern for the forest, contributes to changes in behavior and further increases the effectiveness of EE programs. For example, the CEE's "Green Schools Program" uses digital tools and interactive social platforms to educate students about sustainable practices and forest protection. This innovative way has brought about measurable improvements in students' understanding of environmental problems and their willingness to adopt sustainable behavior (CEE, 2021; Almeida & Cutter-Mackenzie, 2011).

In summary, EE in India has shown promising results to combat deforestation by sensitizing the community, encouraging community participation and promoting sustainable practices among students and local communities. The continued support and expansion of these programs is essential for the achievement of long-term forest protection goals (Bhatia, 2020a, 2020b; Bhat, 2017).

4.3 Comparative Analysis of Environmental Education: Similarities, Differences, and Underlying Causes

Both Japan and India consider EE to be a significant aspect of achieving sustainable development, and under this common goal, EE shares several similarities. Due to disparities in the political, socioeconomic development of the two countries, there are large differences in terms of content and methods of EE. Exploring the causes of these similarities and differences forms the prerequisite for the further improvements of EE in future. In this part, the similarities and differences in EE between Japan and India are examined and the possible reasons are analyzed that contribute to these similarities and differences.

4.3.1 Similarities in Environmental Education

Both Japan and India prioritize EE as an effective instrument to promote environmental awareness and sustainable practices among their citizens. Key similarities include:

Integration into Formal Education

The two countries have made significant progress in integrating EE into their formal education systems. Since the 1970s, Japan has gradually mainstreamed EE into the national curriculum. Environmental issues are integrated into different subjects and certain courses on environmental issues are carried out to improve the understanding and commitment of the students in sustainability practices (MEXT, 2019). Compared to Japan, despite this EE started later in India, the Indian government has already tried to integrate EE into various subjects, including science, social sciences and geography, from primary school to university education. India also develops corresponding textbooks and resources on environmental issues and ensures that pupils across the country receive consistent and comprehensive education on environmental issues (NCERT, 2005).

Community Involvement

Both Japan and India emphasize the importance of community-based EE and recognize that sustainable practices need to be adopted at the grassroots level. In Japan, communities are often involved in broad EE activities through various initiatives launched by eco-schools, government, NGOs, and other organizations. Compared to India, the community engagement in EE is earlier in Japan. In India, some EE programs engage students and community members in activities such as planting trees, garbage disposal and local clean-ups. These practical activities greatly promote a sense of responsibility and community spirit and encourage participants to play an active role in environmental protection. For example, at "Haritha Kerala Mission" in the state of Kerala, students and local residents are encouraged to participate in afforestation and water conservation projects, resulting in increased green space and improved water resource management (Haritha Kerala Mission, 2021).

Use of Technology and Media

The rise of new media and digital technology has influenced almost all the countries worldwide. Japan and India are not exceptions. Despite traditional mass media such as TV, newspaper and magazines remaining essential for the spread of environmental information, the role of the new social platform is becoming increasingly important in both countries. Especially with the growing popularity of mobile networks, this trend is becoming more evident in Japan. Digital books, varied video material and instant online learning are changing traditional classroom pedagogy, diversifying learning approaches and increasing access to EE. Moreover, the emergence of digital technology is further enriching teaching ways. The adoption of VR and AR in schools is providing better learning experiences (MEXT, 2019).

4.3.2 Differences in Environmental Education

Although Japan and India have some similarities, there are significant differences in implementing EE in these two countries. The concrete ones are as follows:

Different Stages in Social Development

As an industrialized country, Japan has completed its industrialization in the last century, and its economic level of economic development is far higher than that in India, which is still a country of development (Fig. 1). The per capita GDP of Japan in 2023 is 13.6 times as high as India. The enormous gap in economic power affects national investments in different areas, including education, especially EE. Since the late 1960s and early 1970s, Japan began to increase its investment in anti-pollution control, including EE. After more than 70 years of continuous effort and accumulation, awareness of the circular economy, low-carbon energy saving and environmental protection has permeated all levels of government, enterprises and individual citizens, and there is strong social awareness of environmental protection. EE in Japan has expanded beyond the schools to involve the entire society, from early childhood to old age, from theoretical research to practical activities for everyday life and from pure environmental protection to coping with global environmental problems. The progress far exceeds that of developing countries, and the successful experiences from Japan are valuable for the other countries, in particular development countries. In contrast, India started modern EE relatively late, and

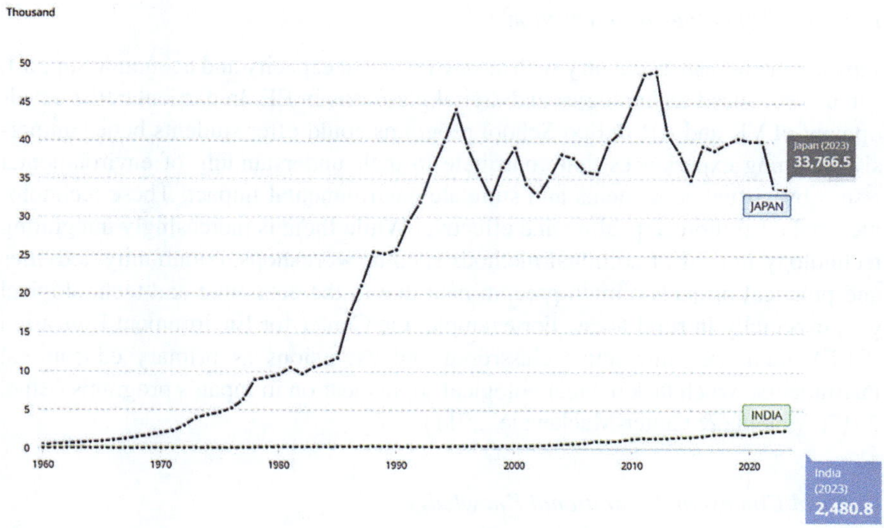

Fig. 1 GDP per capita between Japan and India from 1960 to 2023 (current US$). (Source: data from the World Bank)

there is an objective gap in the time span, investments, content and educational methods compared to the industrialized countries.

Government Support and Policy Integration

As a developed country, Japan has strong economic power to support education development, including investments in EE. Sufficient financial investments not only guarantee a high-quality teaching group and advance teaching equipment and other infrastructure for EE, but also can help implement numerous EE programs dominated by the government. Most of these programs are systematically integrated into the national curriculum by the MEXT, which are advantageous for the consistency and the widespread implementation in the whole country. Moreover, under the guidance of the central government, local governments also often innovate EE methods in local schools and local communities in combination with local conditions to beautify the natural environment and promote environmental quality. This kind of active cooperation between local and central government in Japan contributes to the integration of EE policies and provides a solid framework for promoting the continuous progress of EE. Compared to Japan, India's political operation tends to be decentralized. The characteristics of the political system also influence the field of education, including EE. Various NGOs and state governments take the management of the implementation of EE initiatives, which leads to diverse and regional-specific programs. This decentralized model enables tailor-made approaches to tackle local environmental issues, but may lack uniformity and scope as you can find in Japan (MEXT, 2019; Bhatia, 2020a, 2020b).

Focus on Technological Integration

Japan, as an industrial country with strong technical capacity and economic support, can use advanced technologies and digital platforms in EE. In particular, the development of VR and AR in Eco-School programs could offer students better immersive learning experiences that contribute to their understanding of environmental issues by virtual ecosystems and simulate environmental impact. These technologies make EE more appealing and effective. While India is increasingly integrating technology into EE, traditional methods such as workshops, community activities and practical projects remain predominant due to the economic and technological gap, especially in rural areas. For example, the Center for Environment Education (CEE) often uses interactive classroom and excursions as primary educational instruments, which lack the technological sophistication in Japan's programs (Bhat, 2017; Almeida & Cutter-Mackenzie, 2011).

Cultural Context and Traditional Knowledge

The progress of EE in Japan is more based on modern scientific methods and sustainable practices. Nevertheless, some initiatives in Japan reflect traditional factors, such as the Satoyama initiative, which innovates the approach to sustainable management of socio-ecological production landscapes by combining modern technology with traditional knowledge, thus further improving environmental sustainability (UNU-IAS, 2020). In contrast to Japan, traditional ecological knowledge and practices in India have a huge impact on EE, especially in rural areas where the literature rate and technical capacity are lower. Many programs for EE in India that aim to convey sustainable agricultural practices, water protection techniques and forest management are based on traditional knowledge.

4.3.3 Underlying Causes of Similarities and Differences

The emergence of modern EE originates from the environmental crisis, which is caused by industrialization and overexploitation of natural resources. As this crisis seriously threatens the safety of human health and life, EE has been an important tool in addressing this crisis and in achieving sustainability by constantly developing and improving content, teaching methods and approaches. Concretely, the development of EE as part of the national education system must be guided by the principles of education continuity. This is the main reason why Japan and India share the similarities in EE. Particularly today, as sustainable development has become a key global concern, the influence of international conferences and organizations on individual countries has increased, further emphasizing the importance of strengthening national environmental awareness through EE. In addition, the role of networks and media in EE and ecological conservation to prevent deforestation has rapidly increased.

In addition, EE's content, methods and approaches are influenced by the specific national political, economic and cultural context, which result in the differences of EE in these two countries. To better understand the differences, it is necessary to examine the underlying reasons, including socioeconomic, cultural and political aspects between India and Japan. The concrete analysis is as follows.

Gap in Economic Development and Available Resources: As an advanced economic entity, Japan's strong economic power and high technical capacity not only provide the material basis for the improvement of EE, but also ensure the availability of excellent teaching groups and related practical opportunities such as forest excursions and field investigation. In particular, VR and AR digital platforms are being integrated into school lessons, giving students more opportunities for an interactive and immersive EE experience and for understanding complex issues. Moreover, various scientific and technical museums and some related research institutes are available across the country for students, which promote their apprehension for the importance of the natural environment. In contrast, the economic disparities between urban and rural India are very wide and affect the availability of resources for EE. While urban schools have better opportunities to access some technological tools, rural areas often rely on traditional methods such as hands-on activities and community workshops. This unequal distribution of educational resources poses a negative impact on the efficiency of EE (Bhatia, 2020a, 2020b).

Gap in Government Policies and Support: As a tremendous social project, the success of EE could not leave strong funds and policies' support from different governments. In Japan, long-term investment in EE, continuous legislation and numerous related policies guarantee steady progress for EE. And the close cooperation and sound governance capacities of the central and local governments ensures the implementation of EE programs. Different from Japan, India's decentralized approach grants state governments and NGOs significant autonomy in implementing EE programs. While this allows for adaptation to address local environmental issues, it can also lead to inconsistencies in program implementation and resource allocation. For example, states such as Kerala and Sikkim have robust EE initiatives, while others may not have the same resources or support, leading to variability in EE effectiveness across the country (MEXT, 2019; Bhat, 2017).

Gap in Time Span and Public Environmental Awareness: EE in Japan has been developing for more than 70 years after the anti-pollution protest movement swept the entire country. Over time, the content, methods and approaches of EE have been gradually improved at central and local levels in Japan. At present, this resilient system forms positive interaction with the whole society. The public could take its own responsibility for the environment including forest conservation. Unlike Japan, India is still struggling to meet basic needs of a huge population. Given the enormous pressure of life, environmental protection is important, but not the first priority, which could be proved by environmental investments. This remains one of the main reasons that hamper EE development in India.

Despite more and more people understanding the importance of environmental protection and taking ecological actions, there is still a long way to go for India.

5 Conclusion and Future Directions

The experience of Japan and India shows that EE is an effective instrument to protect the natural environment and resource sustainability by raising public environmental awareness and encouraging environmentally friendly behavior, especially in the prevention of deforestation. By integrating EE into the formal education systems, the popularity of EE is guaranteed in these two countries. And students' environmental awareness has greatly improved. Also, the support of EE by governments, NGOs and other communities promotes the development of EE in Japan and India (CEE, 2021; MEXT, 2019). In addition to these similarities, EE differs significantly from the two countries due to differences in the stages of social development, government support, technological integration and cultural context. Under the strong support of national economic power, Japan benefits from a centralized, stable, state-supported framework that provides consistent, well-equipped programs across the country and constant innovative approaches to improving educational performance. Compared with Japan, India's huge gap in economic power, governance capacity and education level have hindered the efficiency of EE and further affected the prevention of deforestation. However, due to economic growth and growing concern for environmental issues, combined with the advantages of new technologies, India is achieving steady progress.

When reviewing the development process of EE in Japan and India, the most prominent experience is that whole society should pay sufficient attention to environmental problems, which is a key driver of EE development. On this basis, continuity in policies, related legislation and strong fund investment from various governments, systematic environmental curriculum, diverse pedagogical approaches, broad involvement of local communities, the public and NGOs, active international cooperation, and the adoption of new social platform and digital technologies are essential for EE to move forward smoothly. The development of EE cannot be accomplished at one stroke, which need a long-term commitment by all countries. Despite the huge differences in specific conditions between countries, global environmental problems, including deforestation, are common challenges for every country. Strengthening EE is a fundamental and effective choice to address these challenges.

References

Almeida, S., & Cutter-Mackenzie, A. (2011). The historical, present and future ness of environmental education in India. *Australian Journal of Environmental Education, 27*(1), 122–134.

Ardoin, N. M., Bowers, A. W., & Gaillard, E. (2020). Environmental education outcomes for conservation: A systematic review. *Biological Conservation, 241*, 108224. https://doi.org/10.1016/j.biocon.2019.108224. ISSN 0006-3207.

Arshad, Z., Robaina, M., Shahbaz, M., et al. (2020). The effects of deforestation and urbanization on sustainable growth in Asian countries. *Environmental Science and Pollution Research, 27*, 10065–10086. https://doi.org/10.1007/s11356-019-07507-7

Becker, S. (2018). Digital learning and sustainable development. In W. Leal Filho (Ed.), *Encyclopedia of sustainability in higher education*. Springer Nature Switzerland AG.

Bhat, S. (2017). *Environmental education in India: An approach to sustainable development*. Sage Publications. https://doi.org/10.1177/2455265820170102

Bhatia, N. (2020a). *Environmental education in India: Analysis of national curriculum and pedagogical approaches*. York University. https://yorkspace.library.yorku.ca/server/api/core/bitstreams/646eff2f-b332-48f1-b8bd-ff4901d55221/content

Bhatia, N. (2020b). *Environmental education in India: Analysis of national curriculum and pedagogical approaches*. https://yorkspace.library.yorku.ca/server/api/core/bitstreams/646eff2f-b332-48f1-b8bd-ff4901d55221/content

Centre for Environment Education (CEE). (2021). *Annual report*. https://www.ceeindia.org/anuual-reports/

Centre for Science and Environment (CSE). (1982). *The state of India's environment: A citizens' report*. CSE.

Chhokar, K. B., & Chandrasekharan, S. (2006). *Environmental education in India*. Commonwealth Publishers.

Eklavya. (2001). *Hoshangabad science teaching programme*. https://eklavya.in/past-work-top/programmes-past-top/hstp

ENV. (2014). *Environmental education—A foundation for sustainable society*. https://www.env.go.jp/content/900457392.pdf

ENV. (2020). *Summary of results of the FY 2020 survey on environmental education and related awareness*. https://www.env.go.jp/content/000142836.pdf

ENV. (2024). *Chapter 8: Education, training, and public awareness*. https://www.env.go.jp/content/900447160.pdf

ESCAP. (2019). *The disaster risks cape across South and South-West Asia*. https://www.unescap.org/sites/default/files/IDD-APDR-Subreport-SSWA.pdf

ESD. (2021). *Implementation plan for education for sustainable development in Japan*. https://www.mext.go.jp/en/content/20211208-mxt_koktou01-1373244_1.pdf

EU. (2020). *Forests in south-east Asia: Can they be saved?*. https://www.europarl.europa.eu/thinktank/en/document/EPRS_BRI(2020)652068

Fan, S. (2017). *Environmental education: Integrating multiple disciplines*. Academic Press.

Fang, W. T., et al. (2023). The living environmental education, sustainable development goals series. https://doi.org/10.1007/978-981-19-4234-1_2

FAO. (2021). *Asia-Pacific forests and forestry to 2020*. https://www.fao.org/fileadmin/templates/rap/files/NRE/Forestry_Group/1_Forests_for_a_greener_future.pdf

FFPRI. (2023). *Forest education in Japan*. https://www.ffpri.affrc.go.jp/pubs/chukiseika/documents/4th-chuukiseika31.pdf

Forest Agency. (2022). *Annual report on forest and forestry in Japan*. https://www.maff.go.jp/e/data/publish/attach/pdf/index-215.pdf?form=MG0AV3

Fujioka, S. (1981). Environmental education in Japan. *Hitotsubashi Journal of Social Studies, 13*(1), 9–16. https://hermes-ir.lib.hit-u.ac.jp/hermes/ir/re/8453/coverpage

Gandhi, S., & Jones, T. G. (2019). Identifying mangrove deforestation hotspots in South Asia, Southeast Asia and Asia-Pacific. *Remote Sensing, 11*(6), 728. https://doi.org/10.3390/rs11060728

Global Forest Watch. (2023). *India deforestation rates & statistics*. Retrieved from Global Forest Watch

Government of India. (1966). *Report of the Education Commission (1964-66): Education and National Development*. https://www.indianculture.gov.in/reports-proceedings/report-education-commission-1964-66

Government of India. (1986). *National policy on education*. https://www.education.gov.in/national-policy-education-1986-modified-1992

Haritha Kerala Mission. (2021). *Annual report*. Haritha Kerala Mission. https://doi.org/10.1016/j.biocon.2019.108224

Hiroyuki, I. (2018a). *Protecting the forests*. https://www.nippon.com/en/features/c03912/

Hiroyuki, I. (2018b). *Japan's Forests: From lumber source to beloved resource*. https://www.nippon.com/en/features/c03913/

India Team. (2024). *Deforestation in India: Causes, effects, and preventive measures*. http://indiateam.org/deforestation-in-india-causes-effects-and-preventive-measures/

Indian Institute of Science. (2020). Impact of deforestation on local climate in the Western Ghats. IISc. *Indian Journal of Environmental Education*. ISSN: 0975-9425

International Union for Conservation of Nature (IUCN). (2021). *Red list of threatened species*. https://iucn.org/resources/conservation-tool/iucn-red-list-threatened-species#:~:text=The%20IUCN%20Red%20List%20of%20Threatened%20Species%E2%84%A2%20is,by%20governmental%20bodies%2C%20non-profit%20organisations%2C%20businesses%20and%20individuals

Kodama, T. (2017). Environmental education in formal education in Japan. *Japanese Journal of Environmental Education, 26*(4), 21–26. https://doi.org/10.5647/jsoee.26.4_21

Kraft, M. E. (2020). U.S. environmental policy and politics: From the 1960s to the 1990s. *Journal of Policy History, 12*(1), 17–42. https://doi.org/10.1353/jph.2000.0006

MAFF. (2019). *State of Japan's forests and forest management*. https://www.maff.go.jp/e/policies/forestry/attach/pdf/index-8.pdf

MAFF. (2022). *Annual report on forest and forestry in Japan*. https://www.maff.go.jp/e/data/publish/attach/pdf/index-215.pdf

Mani, J. (2019). Conservation and environmental education: A narrative literature review. https://doi.org/10.6084/m9.figshare.12465788

Ministry of Education, Culture, Sports, Science and Technology (MEXT). (2019). *Report on the eco-school program*. https://www.mext.go.jp/en/

Ministry of Environment and Forests. (1988). *National forest policy*. https://moef.gov.in/uploads/2019/06/nfp.pdf

Ministry of Environment, Forest and Climate Change (MoEFCC). (2021a). *#GreenGoodDeeds Campaign*. https://www.pib.gov.in/PressReleaseIframePage.aspx?PRID=1786057

Ministry of Environment, Forest and Climate Change (MoEFCC). (2021b). *Annual report*. https://moef.gov.in/annual-reports

Ministry of Law and Justice. (1980). *Forest conservation act*. https://www.indiacode.nic.in/bitstream/123456789/19381/1/the_forest_%28conservation%29_act%2C_1980.pdf

Ministry of the Environment (ENV). (2024). *Environmental education in Japan: Historical overview and key milestones*. https://www.env.go.jp/en/

Mulyadi, B. (2020). ENIS model of environmental education for elementary school students in Japan. *E3S Web of Conferences, 202*, 03019. https://doi.org/10.1051/e3sconf/202020203019

National Council of Educational Research and Training (NCERT). (2005). *National curriculum framework*. https://ncert.nic.in/pdf/nc-framework/nf2005-english.pdf

Palmer, J. (1998). *Environmental education in the 21st century: Theory, practice, progress and promise* (1st ed.). Routledge. https://doi.org/10.4324/9780203012659

Prasar Bharati. (2019). *Survey on environmental awareness in Indian households*. https://prasarbharati.gov.in/r-and-d/publications/

Ruiz-Mallen, I., Barraza, L., Bodenhorn, B., & Reyes-García, V. (2009). Evaluating the impact of an environmental education programme: An empirical study in Mexico. *Environmental Education Research, 15*(3), 371–387. https://doi.org/10.1080/13504620902906766

SES. (2023). *Nurturing a greener tomorrow: Japan's environmental education and global awareness initiatives*. https://medium.com/sustainable-eco-society/nurturing-a-greener-tomorrow-japans-environmental-education-and-global-awareness-initiatives-aceb11a459a4
Sharma, P. K., & Menon, S. (2024). *Compulsory environmental education in India*. https://thegeep.org/sites/default/files/2024-07/Compulsory%20Environmental%20Education%20in%20India.pdf?form=MG0AV3
Sonowal, C. J. (2009). *Environmental awareness and education*. Authors Press.
Times of India. (2020). *Green India campaign*. https://timesofindia.indiatimes.com/
UNEP. (1972). *Stockholm Declaration, 1972—Declaration of the United Nations conference on the human environment*. https://www.un.org/en/conferences/environment/stockholm1972
UNEP. (2024). *Forests cover one-third of the Earth's land mass, serving as critical pillars for both environmental health and human well-being*. https://www.unep.org/topics/forests/why-do-forests-matter
UNESCO. (1977). *Intergovernmental conference on environmental education*. Tbilisi, USSR: 14-26. October 1977: Final report 1978.
UNESCO. (2014). *Global sustainable development report 2014*. https://www.britannica.com/topic/Brundtland-Report
United Nations University Institute for the Advanced Study of Sustainability (UNU-IAS). (2020). *The Satoyama initiative: Enhancing socio-ecological production landscapes*. UNU-IAS.
Weisse, M., & Goldman, E. (2021). *Estimating the role of seven commodities in agriculture-linked deforestation: Oil palm, soy, cattle, wood fiber, cocoa, coffee, and rubber*. https://www.wri.org/research/estimating-role-seven-commodities-agriculture-linked-deforestation-oil-palm-soy-cattle
World Wildlife Fund (WWF). (2023). *Deforestation and forest degradatiosn*. https://www.wwf.eu/what_we_do/forests/deforestation_forest_degradation/

Into the Future: Nurturing Sustainable Life on Land

Manjari Sharma and Sharad Gupta

1 Introduction

The existence of life on Earth solely depends on terrestrial ecosystems to perform vital ecological functions that allow human populations to thrive and sustain biodiversity as well as regulate temperatures. The international community through SDG 15 has made a pledge to sustainably protect these ecosystems while working to restore and enhance their sustainable use. The total terrestrial ecosystem features reach about 30% of Earth's surface, and this includes mountains, grasslands, marshes, and forests among others. Different habitats together serve as essential components to guarantee ecological stability. The economic value they provide to billions of global population members surpasses their water cleansing abilities and soil enhancement properties and their function as carbon storage mechanisms (Bridgewater et al., 2015). Human actions together with climate change create an urgent danger to these ecosystems as they are unable to provide sustainability for future generations.

Terrestrial ecosystems generate extensive socioeconomic values which enhance natural disaster protection systems as well as water resources for human consumption and farm production systems. Annually forests trap 2.6 billion tons of carbon dioxide which makes them vital habitat storage facilities (Bridgewater et al., 2015). The total number of individuals who rely on forests for their basic needs amounts to 1.6 billion people along with the dependence on forests for food, shelter, and fuel (Peng et al., 2021). Ecosystems maintain cultural as well as spiritual worth which matters most to Indigenous communities whose ancient wisdom and ancestral traditions stem directly from the surrounding environments (Fernández-Llamazares & Cabeza, 2021). The strong connectivity among these environments confirms their

M. Sharma (✉) · S. Gupta
Christ University, Bengaluru, Karnataka, India
e-mail: manjari.sharma@christuniversity.in; sharad.gupta@christuniversity.in

J. J. Puthenkalam S.J et al. (eds.), *Terra Thrive*,
https://doi.org/10.1007/978-3-032-02738-2_8

absolute requirement for protection considering their social elements alongside environmental advantages and cultural roots.

The degradation of terrestrial ecosystems continues to persist because of governance problems alongside conflicts between local and national and international policies. Governments face barriers when implementing CBD goals because these biodiversity programs maintain voluntary character according to Leclère et al. (2019). The regional initiative REDD+ (Reducing Emissions from Deforestation and Forest Degradation) faces multiple challenges such as unequal profit distribution combined with insufficient funding that impedes its expansion according to Makatta et al. (2018). Various international sustainability goals require local conservation programs to fill global governance gaps in order to retain terrestrial ecosystems for coming generations.

Modern technology integration with traditional knowledge tools creates an effective approach to preserve terrestrial ecosystems while improving their condition. The combination of agroforestry, rotational farming, and sacred grove protection as sustainable land management methods shows special resistance to environmental changes (Mekonen, 2017). The conventional techniques demonstrate enhanced potential for producing major outcomes when actors utilize Geographic Information Systems (GIS) and drone-based reforestation technology (Imoro et al., 2021). The Satoyama Initiative of Japan represents an optimal model for traditional ecological knowledge to merge with present technology for promoting sustainability in land use and biodiversity restoration (Mori et al., 2017).

Developing a sustainable culture requires dangerous cooperation between public institutions and educational institutions. Petitions that incorporate direct interaction with the community and learning through experience have effectively brought about vital conservation progress and development results. Local communities gain the ability to manage their ecosystems by participating in citizen science programs such as Forest Watchers in Brazil and Chipko Movement in India according to Lee et al. (2019). The implemented programs show that combined efforts between education and community engagement drive important progress toward attaining Sustainable Development Goal 15.

The authors explored multiple solutions to resolve ecosystem challenges of land systems with broad adoption potential.

The primary dangers afflicting these ecosystems comprise habitat destruction alongside tree removal procedures and the disappearance of animal and plant species. The section explains how environmental problems strike both the land ecosystem and the communities which draw daily necessities from these areas.

The analysis of remediation approaches through specific tools and procedures takes place in Sect. 3. The section outlines three principal methods for ecosystem surveillance and closely observing resources and community-based environmental stewardship.

The section presents real-life examples that include Indian, Japanese, and transboundary territories in Sect. 4. The examples show distinct approaches that different areas have used to accomplish healthy land conservation matching their individual challenges.

Section 5 of the paper presents agroforestry as a merging approach between livestock farming and forest preservation alongside community-led nature conservation initiatives. Such methods demonstrate that they increase biodiversity together with adequate food supplies for communities.

Section 6 of this work demonstrates how traditional wisdom from the past links with modern technological approaches toward environmental preservation. Modern technology used with proven traditional methods creates ecosystems that improve in both strength and community empowerment of their land ownership.

Sections 7 and 8 establish protocols for making improved regulations and laws which address the significance of education and community engagement. In conclusion, the book explores country collaboration to fulfill Sustainable Development Goal 15 that focuses on maintaining land ecosystem safety while restoring health and implementing sustainable land utilization methods.

2 Literature Review

The ground environment endures continuous stress due to human interference and global ecological shifts together with government management problems. The research review thoroughly explores major terrestrial ecosystem threats beginning with habitat destruction and ending with forest loss together with biodiversity reduction and governance problems and global crisis accumulation. New research findings from high-standard studies are combined in this discussion to provide an advanced understanding of these linked elements while examining their broader social ecological networks.

2.1 Habitat Degradation

Habitat degradation arises from human-made activities which lead to ecosystem value decline and operative degradation. Urban growth stands as a primary cause which pushes cities to expand their boundaries into valuable ecological regions such as wetlands and both forests and grasslands (Newbold et al., 2015). Natural landscapes turned into agricultural areas have been responsible for mounting environmental degradation. Present-day commercial food needs lead to extensive deforestation operations for monoculture agriculture that diminishes land productivity and variety of species (Santini & Miquelajauregui, 2022).

Habitat degradation in India's Sundarbans mangrove forests has become extensive due to agricultural practices combined with rising sea levels which trigger soil erosion and salt water intrusion (Singh et al., 2021). The Japanese Satoyama landscapes have deteriorated because traditional farming methods disappeared and urban development expanded (Mori et al., 2017).

The degradation of habitats leads to essential ecological functions along with human sustainability disappearing. The loss of these three vital ecosystem functions marks degrading ecosystems because they cannot perform carbon sequestration nor water filtration nor biodiversity conservation. Tropical deforestation remains responsible for creating 11% of yearly worldwide GHG emission output (Leclère et al., 2019). Habitat degradation breaks cultural and spiritual practices mainly affecting Indigenous populations because these groups need complete ecosystems to preserve traditional knowledge alongside subsistence practices and cultural identity (Fernández-Llamazares & Cabeza, 2021).

2.2 *Deforestation and Biodiversity Loss*

Agricultural expansion alongside logging and infrastructure development constitute the main causes which drive the extensive process of deforestation. The FAO reports that tropical forests face the worst impact of deforestation which resulted in the disappearance of 420 million hectares of forests throughout the past thirty years (Leclère et al., 2019). The three primary factors responsible for causing 80% of tropical deforestation include palm oil and soy production along with beef farming (Curtis et al., 2018).

Tree clearance produces major negative effects on biodiversity primarily because forests house around 80% of all Earth's land species. Habitat fragmentation together with destruction represents major elements in the rising extinction threats impacting nearly one million species (Di Marco et al., 2019). The Western Ghats in India undergo deforestation which endangers the Malabar civet and Nilgiri tahr as well as resulting in the threatened existence of Japanese endemic bird and amphibians (Mori et al., 2017).

The procedure of deforestation intensifies climate change because it emits atmospheric carbon dioxide that remains trapped in vegetation systems. Besides local rainfall and temperature rise, deforestation initiates a circular pattern that results in increased drought probabilities (Leclère et al., 2019). The existence of feedback cycles creates major vulnerabilities for tropical regions since they face escalating climate problems.

2.3 *Global Crises and Ecosystem Vulnerability*

Climate change serves to enhance current threats facing terrestrial systems through modifications in wet and dry seasons, temperature increases, and more powerful weather phenomena. The environmental transformations from these factors promote desertification while disrupting species' migration behavior and impairing ecosystem operational capability (Santini & Miquelajauregui, 2022). The rise in temperature across India's dry zones has worsened land degradation, and Japanese coastal ecosystems suffer from typhoon destruction (Mekonen, 2017).

The COVID-19 pandemic prevented human activities during lockdowns so some species could recover although their habitat remained unchanged. After the economic crisis, more natural resources got overexploited in locations with limited economic means. The tropical forests faced rapid loss because many nations survived by using forest resources as the main economic sources (Singh et al., 2021).

2.4 Governance and Policy Gaps

Various national policies fail to resolve multiple challenges which affect terrestrial ecosystem management in an integrated manner. India's National Afforestation Programme demonstrates difficulties when implementing community involvement and establishing sustainable practices according to Singh et al. (2021). Japan encounters difficulties applying its Forest Resources Management Act between preserving forests and responding to economic growth requirements (Mori et al., 2017).

The Convention on Biological Diversity (CBD) establishes significant conservation goals through its framework yet fails to achieve this purpose due to its voluntary nature. The REDD+ initiative faces two major challenges which prevent its effective scaling up due to problems with benefit distribution and funding shortages (Makatta et al., 2018).

2.5 Socio-Economic Implications

The degradation of ecosystems leads to extensive economic losses in the system. The Indian economy consistently suffers economic losses resulting from ecosystem service depletion which equals 2.5% of its national gross domestic product per year (Santini & Miquelajauregui, 2022). The global extent of land degradation reduces agricultural output which intensifies hunger problems most severely in developing areas.

Ecosystem destruction places an immense impact on Indigenous communities because of its destruction of their cultural heritage and social connections. Indigenous peoples maintain their cultural traditions through sacred groves combined with traditional farming systems which also support their spiritual practices according to Fernández-Llamazares and Cabeza (2021). The disappearance of these ecosystems creates two major detrimental effects on both natural environments and community social networks.

The research compares various problems affecting terrestrial ecosystems through habitat destruction, forest loss, biodiversity diminution, environmental degeneration, and governance breakdowns. Multiple environmental problems merge together so organizations need unified approaches which unite restoration efforts with community collaboration under strong policy structures. The following sections analyze methods along with real-world examples and effective solutions to handle the identified obstacles with sustainable land management practices.

3 Materials and Methods

3.1 Data Sources

The research analysis is based on evidence from extensive evaluations of recent peer- reviewed articles in scholarly journals. The data is collected from publications of scientific data from Scopus and Web of Science and PubMed databases for reliable and suitable information. This research is based on important studies of deforestation rates and biodiversity losses and gains (Leclère et al., 2019; Di Marco et al., 2019).

Information from government documents, Food and Agriculture Organization (FAO), United Nations Development Programme (UNDP), and World Bank were included in the study (3.1.2). Furthermore, the worldwide forest depletion is also analyzed through the use of the UNDP policy documentation on sustainable land management practice and Global Forest Resources Assessment (FAO, 2020).

Some data from specific Indian and Japanese terrestrial ecosystems were also obtained from the national sources of India's Forest Survey and Japan's Ministry of the Environment reports, and India and Japan were included (3.1.3). Forest extent and biodiversity metrics as well as results of conservation projects are provided by the sources (Forest Survey of India, 2021; Ministry of the Environment Japan, 2021).

3.2 Analytical Tools

Publicly available NASA's Earth Observation System and Global Forest Watch secondary spatial information were analyzed using remote sensing and Geographic Information System. The deforestation, habitat partitioning, and land use alteration patterns are exactly evaluated by these source databases. In 2021, NASA satellite imagery was used by Global Forest Watch to monitor Indian Western Ghats and Japanese Satoyama landscapes.

The analysis was made on academic publications which dealt with the biodiversity conservation template, the strategies of deforestation prevention, and policy structure methodology. By combining the results of the research in different studies, in general, general patterns were identified that appeared in different studies. To achieve this, Santini and Miquelajauregui (2022) compared the research on agroforestry in India to that in Japan in order to find solutions to implement.

3.3 Policy Review and Gap Analysis

Existing international and national policy documents were reviewed in order to carry out analysis of implementation problems and strategies of possible enhancement. Mori et al. (2017) and Singh et al. (2021) have assessed the compliance of the National Afforestation Programme of India and Forest Resources Management Act of Japan with international framework (Convention on Biological Diversity (CBD)).

3.4 Case Study Methodology

In the case studies presented in this section, the cases were selected based on the impact they had on ecology and also had accessibility to secondary information and contribution to SDG 15 targets. Examples include:

1. India: Joint Forest Management (JFM) initiatives and conservation efforts in the Western Ghats and Sundarbans.
2. One of its operations is the Japan's Satoyama Initiative, which merges contemporary conservation operational practices with traditional wisdom.
3. The example of ecosystem collaboration in the Himalayan transboundary region is India, Nepal, and Bhutan (Ramachandrappa & Thimmegowda, 2016).

3.5 Data Sources for Case Studies

The data sources are governmental reports, academic research papers, and institutional publications. For example:

1. The evaluation of the JFM Program in India is based on official reports from the Ministry of Environment Forest and Climate Change.
2. The reports of academic research and Japan's Ministry of the Environment (Ministry of the Environment, Japan, 2021) do the analysis of the Satoyama Initiative.

3.6 Validation and Monitoring Frameworks

I also did secondary data validation by checking the information through a couple of reports and datasets. FAO reports were then analyzed, and their data confirmed using Global Forest Watch data that both sources matched.

Two monitoring tools, Global Forest Watch and Biodiversity Indicators Partnership, provided the means to monitor the changes in ecosystems health by using the publicly available databases of those tools. Global Forest Watch (GFW) provided these platforms to offer real-time observational data on deforestation, carbon emissions, and biodiversity changes that provided a good measure for the progress to SDG 15 targets.

4 Case Studies and Applications

These ecosystems are diverse and complex, and therefore terrestrial ecosystems have to respond to their conservation in a way that cannot be addressed through a generic approach to the conservation of these ecosystems that tries to moderate between ecological and socio-economic needs. It also talks about case studies on the detailed nature of India and Japan and transboundary efforts such as Himalayan ecosystem conservation. In each such example, they present new and scalable practices of land management, biodiversity conservation, and ongoing relationships with their surrounding communities. Those have led to what should be done on the global ecological challenges.

4.1 Case Study: India

4.1.1 Joint Forest Management (JFM) in India

India's Joint Forest Management (JFM) program refers to a joint government and local communities' initiative for sustainable forest management. The program began in the 1990s and is a community-based program aimed at restoration of degraded forest and improvement of rural livelihoods (Singh et al., 2021). In the JFM, Forest Protection Committees (FPCs) are formed consisting of forest department and local stakeholders to be responsible for forest protection and regeneration.

It has been able to return millions of hectares of degraded forest land. For example, Madhya Pradesh forest cover increased by 14% in the last two decades with JFM, and income from NTFPs also increased (Gupta et al., 2020). Because of this, JFM has been facing challenges such as uneven benefit sharing, lack of financial resources, and lack of harmony between local communities and forest department.

4.1.2 Agroforestry in the Western Ghats

An ancient practice of integration of trees and crops in India's Western Ghats, a UNESCO World Heritage Site, is known as agroforestry. It also benefits other economic benefits such as increasing biodiversity and soil health among others for farmers.

Coffee, cardamom, and fruit trees are cultivated in multiple layer cropping systems by the farmers of Kerala. These keep ecological stability as the systems resemble natural forests. According to Santini and Miquelajauregui (2022), their results show that crop yields have actually increased by 20% and soil organic carbon by 35% more than monoculture systems. Agroforestry is an efficient method for conservation of water, restoration of tree cover, and increase in vegetation diversity for restoration of natural ecological conditions on hills.

4.1.3 The Sundarbans Mangrove Ecosystem

Straddling the India–Bangladesh border, the Sundarbans is the world's largest mangrove forest and is of utmost importance for its habitat and biodiversity, including the Bengal tiger, which is classified as endangered. Sea levels, cyclones, and unsustainable aquaculture practices are the threats to the ecosystem. Government and non-government organizations support community-based conservation projects that intervene in mangrove replantation and provide alternative livelihoods to the local people as crab farming and honey production (Chowdhury et al., 2020). Community empowerment is a way of living sustainably while getting an incentive to reinforce the ecosystem's resilience.

4.2 Case Study: Japan

4.2.1 The Satoyama Initiative

The Satoyama Initiative is launched by Japan's Ministry of the Environment and the United Nations University to promote the sustainable use of socio-ecological production landscapes. Both biodiversity conservation and agriculture and forestry are also balanced.

The use of GIS-based land use planning is a modern conservation strategy while the traditional farming technique is combined with it. The practices of the villages in rural Japan on keeping the landscape sustainable are crop rotation, water management, and tree planting (Mori et al., 2017).

Even in Ishikawa Prefecture, bird and amphibian species native to the area have returned. In addition, eco-tourism is the main part of rural revitalization too.

The Satoyama model has been implemented worldwide and has been adopted to other countries such as the Philippines and Kenya and to many cultural and environmental situations.

4.2.2 Forest Conservation in Hokkaido

Both of its unique temperate forests and the Indigenous Ainu communities are found in Hokkaido, the northernmost island of Japan. These were the challenges that the region was faced with like deforestation, threat of industrial expansion to the region's biodiversity and cultural heritage, and many more. Collaboration of local governments, universities and indigenous community, and traditional Ainu forest management techniques are practiced on the protected areas (Kawakami et al., 2019). In order to strengthen ecosystem management and cultural preservation, a strengthening of indigenous knowledge in conservation would be required.

4.3 Transboundary Case Study: The Himalayan Ecosystem

4.3.1 Importance of the Himalayan Ecosystem

The Himalayan region is spread over the three countries India, Nepal, and Bhutan and is one of the most important ecological and hydrological systems of the world. The river ecosystem types support 1.3 billion people for freshwater supply, and the river ecosystems harbor a rich flora and fauna (Singh et al., 2021).

4.3.2 Transboundary Conservation Efforts

Cross-border initiatives such as Kailash Sacred Landscape Conservation and Development Initiative (KSLCDI) are developed to protect the Himalayan ecosystem beyond the borders.

It delivers: Implementation, in which India, Nepal, and Bhutan are integrated to work on conservation of biodiversity, sustainable tourism, and community resilience.

As a result, the initiative has helped restore degraded landscapes, protected a number of wildlife corridors, and uplifted livelihoods of local people through ecotourism and organic farming (ICIMOD, 2020). Political tensions, different governance structures as well as climate change have been identified as barriers to long-term success.

4.3.3 Lessons Learned

From Himalayan case, it is clear that good governance, adequate transboundary cooperation, and participation of the community in sustainable ecosystem management are inevitable.

4.3.4 Lessons Learned and Global Scalability

Conservation is successful in all case studies through participation of the community. Financial incentives, participatory decision-making, and capacity building of local stakeholders are provided to the local stakeholders with power to enjoy long-term ecological and socio-economic benefits.

The improved results of the conservation strategies are achieved by the combination of traditional knowledge and modern technology, GIS and remote sensing, to make this integration with multicultural heritage and to take up contemporary challenges.

Robust policy frameworks and multi-level governance: 4.4.3 are needed in combination with robust policy frameworks and multi-level governance for scaling conservation. One of the cases where the situation is exemplified is of JFM in India and the partnership between KSLCDI.

These case studies of community engagement that integrate TEK and strong governance principles are relevant to other scenarios of other ecosystems and cultures. For instance: Agroforestry in the Western Ghats has been one of the successful models which can be used as a reference from Africa and Southeast Asia. Satoyama practices are Japan's socio-ecological landscapes that can be applied to rural areas worldwide.

The case studies of India, Himalayan, and Japan show that there are many alternative and innovative ways of conserving biodiversity in the practice of sustainable land management. These are lessons in promoting community participation globally in a mix of modern and traditional ways of doing things and in strengthening governance. In the following section, some of these successes will be identified and placed within the framework and sustainable practices that will fill the current dearth of terrestrial ecosystems.

5 Sustainable Practices and Frameworks

To achieve the goals under the SDG 15, it is necessary to have sustainable practices and frameworks to address terrestrial ecosystems degradation. These practices encompass agroforestry, community-based conservation, indigenous knowledge integration, and the application of advanced technologies. Additionally, the initiative also needs a general framework of a policy and cross sector partnerships to scale the initiative. The examples and suggestions for global implementation of sustainable practices and framework are presented with key discussion.

5.1 *Agroforestry for Sustainable Development*

5.1.1 Definition and Benefits

Trees and crops integrated on agricultural land (agroforestry) is the old fashioned way to promote biodiversity, improve soil health, and provide some economic reward to the farmers. It is a suitable sustainable option as a replacement of monoculture farming since it generates multifunctional landscapes that resemble natural ecosystems (Santini & Miquelajauregui, 2022).

5.1.2 Regional Examples

In Western Ghats, farmers' income was increased by 20% in coffee, cardamom, and fruit trees based agroforestry systems compared to India (Singh et al., 2021). It has nitrogen fixing ability and soil moisture rise and also leads to increased crop yields in drought prone areas.

5.1.3 Challenges and Scalability

Agroforestry adoption faces the lack of financial incentive, lack of technical knowledge, and policy barriers. Its scalability worldwide can be addressed through such gaps by way of capacity building programs, subsidies, and research partnerships.

5.2 *Community-Based Conservation Efforts*

5.2.1 Principles of Community-Based Conservation

The conservation of ecosystems with community participation is referred to as community-based conservation. By this participatory approach, local needs are involved in conservation efforts and traditional knowledge systems are used (Chowdhury et al., 2020).

5.2.2 Success Stories

The Chipko movement in the Himalayan region, India provided a model of grassroots activism in halting deforestation and in preserving the forests. Moreover, in Nepal, Community Forest User Groups (CFUGs) have restored the forest cover and increased biodiversity in addition to improving the rural livelihoods. In Brazil, extractive reserves in the Amazon allow local communities to sustainably harvest forest products, thus improving the deforestation rate.

5.2.3 Framework for Global Application

To do it globally, we can adopt the following frameworks to make community-based conservation successful: Capacity-building programs for local communities. They include, among others, financial incentives (for instance, payments for ecosystem services (PES)), and policies that define who owns the community and makes decisions.

5.3 *Integration of Indigenous Knowledge Systems*

5.3.1 Role of Indigenous Knowledge in Conservation

For example, Indigenous communities have time tested knowledge used in ecosystem management which includes sustainable hunting when to sow, grow, let fallow, and harvest, and forest conservation. This knowledge can be better integrated into modern frameworks and thereby add to their ecological and cultural relevance (Fernăndez-Llamazares & Cabeza, 2021).

5.3.2 Examples of Indigenous Knowledge

Sacred groves of Khasi and Garo tribes in Meghalaya, India are cultural taboos that function as biodiversity hotspots (Hartman & Huber-Sannwald, 2016). In Australia, Aboriginal fire management practices, by using controlled burns, reduce the wildfires and ensure habitat regeneration. Indigenous led conservation areas in British Columbia are a way of conserving wildlife corridors and sustainably managing resource use in British Columbia and across Canada.

5.3.3 Integrating Indigenous Knowledge with Modern Tools

A well-conceived approach to land use planning is participatory GIS (PGIS) which takes the best of traditional knowledge and spatial mapping. For instance, PGIS has been utilized in the Amazon to detect deforestation hotspots as well as to rank the priorities of restoration (Kosoe et al., 2020).

5.4 *Application of Advanced Technologies*

5.4.1 Geographic Information Systems (GIS) and Remote Sensing

GIS and remote sensing allow such terrestrial mapping and monitoring capabilities. These tools are used to generate real-time data on deforestation, habitat fragmentation, and land use change which results in potential interventions by these tools (Global Forest Watch, 2021).

5.4.2 Drone Technology for Reforestation

The drones that disperse seeds on a large scale in the degraded areas have revolutionized reforestation. Drone-based reforestation in Southeast Asia has planted over one million trees in a fraction of the time it takes to reforestation by hand (Imoro et al., 2021).

5.4.3 Artificial Intelligence (AI) and Machine Learning

They are also models based on AI available for predicting trends of deforestation and optimizing any conservation strategies aimed at conserving biodiversity hotspots. For example, AI has already been used, for example, to monitor illegal logging and enforcement of environmental laws in Brazil (Silva et al., 2020).

5.4.4 Challenges and Opportunities

However, advanced technologies are very promising but one of the barriers for their wide adoption are high costs, requirements of expertise, and concerns about data privacy. These challenges could be eased through the public–private partnership and technology sharing agreements.

5.5 *Policy Frameworks for Sustainable Land Management*

5.5.1 National-Level Policies

For instance, the National Afforestation Programme (NAP) program in India makes it possible for community participation, either through the financial incentive or through community participation on restoration of degraded forest. The Forest Resources Management Act of Japan is intended to balance economic development with biodiversity conservation (Mori et al., 2017).

5.5.2 International Frameworks

Convention on Biological Diversity (CBD) provides principles of biodiversity conservation to be used in sustainable fashion. Reducing Emissions from Deforestation and Forest Degradation (REDD+) offers financial incentives to developing country forest conservation (Makatta et al., 2018).

5.5.3 Multi-Stakeholder Partnerships

Development of sustainable land management frameworks involves collaborative effort of governments, NGOs, academia, and private sector. For example, the Bonn Challenge united the scope of the efforts to restore 350 million hectares of lands in degraded areas by 2030.

6 Financial Mechanisms for Sustainability

6.1 Payments for Ecosystem Services (PES)

Payment is made to the landowner for maintaining the ecosystems which give the product of services like carbon sequestration and water filtration. For example, Costa Rica's PES program has at least been successful in conserving such amounts as one million hectares of forest (Pagiola et al., 2008).

6.2 Green Bonds and Carbon Credits

Markets of green bonds and carbon credits in turn incentivize investments in projects of conservation and restoration. Globally, these financial mechanisms have taken root with contributions from the private sector players. Among these other sustainable practices and frameworks include agroforestry, community-based conservation, indigenous knowledge integration, and advanced technologies (Duguma et al., 2014). To achieve SDG 15, these practices can be scaled globally with the benefit of robust policy frameworks and financial mechanisms that support them. Next, we will present policy recommendations and possible cross-border collaborations which are based on the lessons from sustainable practices and frameworks.

7 Policy Recommendations and Cross-Border Collaboration

Terrestrial ecosystem degradation is one of the challenges that needs to be approached using effective policy frameworks and cross-border collaborations if we are to achieve SDG 15, "Life on Land." Ecosystem management needs to be carried out in an inclusive and sustainable manner through the integration of scientific evidence, TEK, and modern technological tools in policy. Lastly, even the transboundary ecosystems such as the Himalayas need regional cooperation for coordination of protection efforts across the border. In this section, policy recommendations and the part that cross border collaboration plays are discussed, including how cross border collaboration can be strengthened to improve governance and facilitate international partnerships.

7.1 Policy Recommendations for Sustainable Land Management

7.1.1 Strengthening National Frameworks

Sustainable land management and biodiversity conservation in a country are built on national policies. However, most of the existing policies are not enforceable, inclusive, or scalable. (a) India's Forest Conservation Act: Strengthen community involvement through decentralized governance structures. For instance, Joint Forest Management (JFM) may increase grassroots organizations' empowerment and thereby their local governance (Singh et al., 2021). (b) Japan's Forest Resources Management Act should also integrate economic incentives for landowners to conduct sustainable forestry such that landowners receive tax benefits for keeping biodiversity rich landscapes (Mori et al., 2017).

7.1.2 Incentivizing Sustainable Practices

Financial incentives are important to encourage the stakeholders to adapt to sustainable land use practices. Examples include: (a) Ecosystem Services Payments (PES): Deploy PES, having farmers and forest owners being paid to say no to biodiversity conservation and to reduce emissions. One successful model of PES is Costa Rica's program (Pagiola et al., 2008). (b) Green Subsidies: There is a need to subsidize agroforestry, reforestation, and other sustainable practices to overcome the economic barriers to practicing these.

7.1.3 Aligning National and International Goals

Coherence and accountability are aligned with global frameworks such as Convention on Biological Diversity (CBD) and Paris Agreement. This includes (a) setting measurable national biodiversity targets and (b) incorporating SDG 15 indicators into national development plans.

7.2 Cross-Border Collaboration in Ecosystem Management

7.2.1 Importance of Transboundary Cooperation

There are transboundary ecosystems, including the Amazon Rainforest, the Himalayan region, and the Congo Basin, covering more than one country, where conservation cannot be taken independently in one country. Collaboration with other regions means that ecosystem management crosses political boundaries to treat the common ecological challenges.

7.2.2 The Himalayan Region: A Case Study

One of the most important transboundary ecological systems in the world is the Himalayan ecosystem which is shared between India, Nepal, Bhutan, and China. Over 1.3 billion people depend on this river for its freshwater and it is the home to the diverse flora and fauna.

Problems: The region's ecological balance is being threatened due to deforestation, glacial melting, as well as uncoordinated development projects.

Initiatives: Kailash Sacred Landscape Conservation and Development Initiative (KSLCDI) works for biodiversity conservation, development of sustainable tourism and enhancing the resilience of communities across the borders (ICIMOD, 2020). Indus Waters Treaty: Water sharing agreement between India and Pakistan on potential for joint management of shared resources in spite of political tension.

7.2.3 Lessons from Other Transboundary Initiatives

Amazon Cooperation Treaty Organization (ACTO): It is an organization which promotes cooperation between South American nations in the protection of the Amazon Rainforest.

Australia: An extremely impoverished country struggles to regain and conserve its forests and wildlife corridors.

7.3 Strengthening Governance and Institutions

7.3.1 Multi-Level Governance

Sustainable ecosystem management needs to be coordinated at different levels of local, national, and international bodies. Effective governance frameworks should be the one where local governments carry out conservation initiatives.

7.3.2 Participatory Governance

Inclusive and equitable policies must include lay people in the policy creation, such as Indigenous communities, women's groups, and marginalized groups. It has been shown that efforts to conserve their effectiveness as well as sustainability have been more effective and sustainable (Fernández-Llamazares & Cabeza, 2021).

7.3.3 Role of International Organizations

UNEP and FAO should be among the global institutions. It financially and technically supported the developing nations. However, such exchange of knowledge and capacity building will be facilitated.

7.4 Financial Mechanisms for Cross-Border Collaboration

7.4.1 International Funding Mechanisms

On the flip side, non-specificity of this type of initiative funding is the greatest obstacle to funding cross-border ecosystem management on an adequate level. One such successful funding mechanism is as follows: Regarding REDD+, it gives financial incentives for reducing emissions from deforestation and forest degradation (Makatta et al., 2018). Moreover, Global Environment Facility (GEF) is also involved in financing transboundary conservation projects such as Global Environment Facility (GEF) supports the Central Asian Desert Initiative (CADI) which aims to protect desert ecosystems.

7.4.2 Public–Private Partnerships (PPPs)

PPP has the potential to utilize resources and knowledge of technical know-how for the enhancement and large scale ecosystem restoration projects. For instance:

The private sector was incorporated into the global Bonn Challenge to restore 350 million hectares of degraded land by 2030, and it was able to reap the benefits of being part of the global effort.

7.5 *Enhancing Monitoring and Accountability*

7.5.1 Use of Technology in Monitoring

Furthermore, GIS and AI can also be employed to improve the existing system of monitoring and enforcing environmental laws. For example: The tracking of illegal deforestation and support (Silva et al., 2020) are made by using GIS tools. Analysis of satellite data with the help of AI powered systems will help in knowing the trend of deforestation and hotspot of biodiversity. More countries should follow this strategy to enhance the international reporting of species in the framework of CBD, SDGs, etc. It makes nations responsible and through reporting requirements indicates that these nations are responsible and have obligations.

7.6 *Future Directions for Policy and Collaboration*

7.6.1 Integrating Climate Resilience into Policies

Adaptation and resilience to climate change impacts should be included in policies that will allow terrestrial ecosystems to adapt and be resilient to ever increasing climate change impacts. Examples include popularization of drought resistance crop varieties in the agroforestry system. The establishment of green infrastructure will achieve mitigation efforts of the flood risks.

7.6.2 Expanding Cross-Border Networks

Nations such as the African Union's Great Green Wall initiative that occurs at the national level has increased cooperation and sharing of resources by building stronger regional networks.

7.6.3 Encouraging Knowledge Sharing

The International Partnership for the Satoyama Initiative (IPSI) is a platform for sharing good practices in which the best of the conservation strategies from one place can be spread to another (Mori et al., 2017). Good policy and interborder cooperation are critical to manage the destruction of our ecosystems and biodiversity loss. Achievement of sustainable ecosystem management is possible through global policy alignment of national policies and community empowerment, together with wise use of technical and financial resources. Preservation of common ecological resources is best permitted by collaborative working to make transboundary initiatives share Himalayas and Amazon. The below section speaks about public education as the central theme in developing sustainable cultural frameworks.

8 Education and Public Engagement for Sustainable Land Use

The necessary element of Sustainable Land Use and the Sustainable Development Goal (SDG) 15 is the combined education and public engagement. Environmental approaches contribute in raising awareness of what brings about ecosystem destruction, enable people to clearly understand ways to attain this, and guide them in being able to cope well with long term environmental stewardship. The unified effect of environmental education, community involvement, and grassroots action in a process to achieve sustainable land use and the associated examples and recommendations are illustrated here.

8.1 Role of Environmental Education in Behavioral Change

8.1.1 Building Awareness of Ecosystem Interdependence

Strong environmental education is needed for development of human lifestyle awareness based on interdependence of ecosystems. This info is included within the official school programs, and colleges additionally use this info to interpret people about biodiversity conservation and affordable land management strategies and solutions to climate change.

he support to the Program comes from the Ministry of Environment Forest and Climate Change of India. Interactive learning sessions and field activities are used to present the sustainable land use teaching methods (UNESCO, 2021). Earthwatch together with UNESCO is also undertaking two initiatives under the worldwide strategy on Education for Sustainable Development' which buses students and professionals from all over the world, to impart the same ecological knowledge.

8.1.2 Experiential Learning and Hands-On Activities

Field learning and citizen science programs provide nature experience and conservation work to the closed door population. For example: The Forest Watchers pairs community members with the identification of deforestation and the reforestation tasks of degraded landscapes (Lee et al., 2019). In Kenya, the educational institution plans yearly tree planting programs that teach its students how to become good caretakers of the environment and at the same time participating in school reforestation programs.

8.1.3 Capacity Building for Professionals

Political land use planners and conservation practitioners provide directions in the very important aspect of weather. Sustainable land use planning techniques and biodiversity tracking abilities are trained to professionals by the Food and Agriculture Organization (FAO) and International Union for Conservation of Nature (UCN).

8.2 Community Participation in Conservation Efforts

8.2.1 Grassroots Movements and Their Impact

Sustainable methods can be developed through grassroots movement assembly power that helps ecosystems to maintain sustainability. When habitats are experiencing harmful conditions, those who live in such places tend to form movements. The Chipko movement of Indian women from the Himalayan region was based on their collective action that formed similar movements in other countries as per Guha MT (2000). Amazonian Brazil's Extractive Reserves' bureaucratically managed conservation areas provide workers with minimal payment for labor as the foundation for allowing the workers sustainable rights to harvest from the forest.

8.2.2 Participatory Governance in Land Use

Community participatory governance is one of the modes that allow communities to participate in deciding issues affecting land use and conservation choices. This method of incorporating makes sure policies are better suited to meet the conditions of the region and the customs of their traditions.

Local organizations are known as Community Forest User Groups (CFUGs) that serve to bring together the two major goals of forest management between preservation and local economic development in Nepal. Their three major goals are to expand forest area with participation of local communities and facilitate sustainable harvesting activities to earn local profits (Pokharel et al., 2020).

The term IPCAs Indigenous Protected and Conserved Areas involve the government agency collaboration with the inhabitants of Indigenous communities to undertake the co-management and TEK that assembles itself with contemporary conservation techniques.

8.2.3 Citizen Science Initiatives

People engage in biodiversity monitoring tasks, and in data collection as well as data restoration operations. Examples include: People can use the application eBird to watch birds around the world and at the same time collect biodiversity information. The India's Great Backyard Bird Count also gives citizens a way of identifying bird species and serving to create conservation planning data.

8.3 Public Campaigns for Behavioral Change

8.3.1 Media and Awareness Campaigns

These programs are transmitted to the audience through social media, as well as through television broadcasts and community-based and public campaigning activities.

The ActNow campaign of the United Nations is actually a worldwide movement designed to educate people on both food waste prevention and tree planting in the battle against global warming. The Japan's Satoyama Initiative uses public outreach programs to demonstrate application of traditional land use methods to conservation practices (Mori et al., 2017).

8.3.2 Behavioral Nudges for Sustainability

The application of nudges from behavior science basis directs people to environmentally sustainable behaviors. For example: Experienced by the public area with how useful reminder combinations encourage the citizens to practice tree planting along with sustainable farming. Economic incentives are used as subsidies to try to encourage agroforestry or waste management systems more widespread adoption.

8.4 Education and Engagement in Policy Advocacy

8.4.1 Empowering Local Communities to Influence Policy

By engaging the public (interested community members) in community-related participation programs, they are able to enact environmental policies or any other policy at that. Examples include: As described in Pagiola et al. (2008) residents of Costa Rica lobbied the government to establish the PES program with payments for ecosystem services and "inoculated" both rural populations and measures of conservation (Goyaffarra forest PES program). Significant changes to Amazon deforestation have resulted from patronage work among Indigenous groups and NGOs in the area of government policy.

8.4.2 Role of Non-Governmental Organizations (NGOs)

NGO organizations can act as intermediaries for governments to connect with people by helping them provide the technical resources, financial support, as well as advocacy programs toward the sustainable land management. Notable examples include:

WWF is an international organization, and its projects for community conservation are implemented across the globe. Conservation occurs at the basic community level, since such policy reforms support it.

8.5 Challenges and Opportunities in Education and Engagement

8.5.1 Challenges

Since rural and underserved communities lack basic resources they possess, their environmental education programs remain elusive. There are some cultural traditions that close people from being involved in conservation work in some parts of the country. People lack understanding about the relationship between ecosystems and human welfare, and this is a challenge to community engagement.

8.5.2 Opportunities

As a solution to the remote community education via digital platforms, the remote learners will benefit from E-learning tools such as online information dissemination.

Schools are a vital medium to incorporate the adolescent people who will later become green citizens. Strategic partnerships between NQOs and the private sector organizations and the public sector can strengthen education and public engagement programs of the government.

In order to achieve sustainable land use in the context of SDG 15, the only option that works is making the public aware. The accepted approach to this task involves building ecological language and empowering communities, and, in doing so, creating incentives for change in behavior. Additionally, Participatory Governance Models and Public Campaigns demonstrate forms of education and engagement which I believe to be most achievable to me, as they represent direct educational approaches to shift the conservation mindedness of the society. In order to expand internationally, the growth of innovative communication capabilities together with intersectoral collaborations needs enduring support. My overview of research directions to follow and a roadmap for its pursuit follow in the last part of this chapter as a way toward sustainable life on land.

9 Conclusion

Increased pressures on terrestrial ecosystems because they are the places that provide life outcomes for human communities and biodiversity habitat at the same time as maintaining a climate balance. Three principal strategies are presented to overcome these difficulties: agroforestry is presented here and together with community-based conservation as well as the IKS integration and the advanced technological implementation. The chapter shows how scalable and inclusive solutions can support SDG 15, and to do so, explains how SDG 15 can be supported through case studies of India and Japan, as well as Himalayan ecosystem.

Most of the chapters before this show how sustainable practices with cross-border community systems collaboration result in solutions of the struggles faced by the terrestrial ecosystems. Sustainable life on land is dependent on modern technologies in connection with traditional methods, as well as governance systems and educational approaches and public participation schemes. As long as research keeps up its momentum and its policy innovation to develop new approaches, it will become real. The plan depicted in this analysis is the essential plan necessary for international collaboration to carry out to save and rejuvenate the earth's terrestrial ecosystems.

Acknowledgments The authors sincerely thank Christ University, Bangalore, for its institutional support in facilitating this research. They also acknowledge the contributions of colleagues and reviewers whose insights helped to refine this work.

References

Bridgewater, P., Arico, S., Finlayson, C. M., & Walpole, M. (2015). The Millennium Ecosystem Assessment: Triumphs, perils and prospects for sustainable development. *Global Environmental Change, 34*, 1–4. https://doi.org/10.1016/j.gloenvcha.2015.04.005

Chowdhury, R. R., Sills, E. O., & Duchelle, A. E. (2020). Deforestation, reforestation, and climate change mitigation: A new framework. *Ecological Engineering, 140*, 105980. https://doi.org/10.1016/j.ecoleng.2020.105980

Curtis, P. G., Slay, C. M., Harris, N. L., Tyukavina, A., & Hansen, M. C. (2018). *Classifying drivers of global forest loss. Science, 361*(6407), 1108–1111. https://doi.org/10.1126/science.aau3445

Di Marco, M., Ferrier, S., Harwood, T. D., Hoskins, A. J., & Watson, J. E. M. (2019). Wilderness areas halve the extinction risk of terrestrial biodiversity. *Nature, 573*(7775), 582–585. https://doi.org/10.1038/s41586-019-1567-7

Duguma, L. A., Minang, P. A., & van Noordwijk, M. (2014). Climate change mitigation and adaptation in the land use sector: From complementarity to synergy. *Environmental Management, 54*(3), 420–432. https://doi.org/10.1007/s00267-014-0331-x

Fernández-Llamazares, Á., & Cabeza, M. (2021). Rediscovering the potential of indigenous knowledge for global challenges. *Nature Sustainability, 4*(8), 683–691. https://doi.org/10.1038/s41893-021-00763-6

Food and Agriculture Organization of the United Nations (FAO). (2020). Global Forest Resources Assessment 2020: Main report. Retrieved from https://www.fao.org/forest-resources-assessment

Forest Survey of India. (2021). India State of Forest Report 2021. Ministry of Environment, Forest and Climate Change, Government of India. https://fsi.nic.in

Global Forest Watch. (2021). Global Forest Watch database. World Resources Institute. https://www.globalforestwatch.org

Guha, R. (2000). *The unquiet woods: Ecological change and peasant resistance in the Himalaya.* University of California Press. https://doi.org/10.1016/S0016-3287(00)00002-6

Gupta, R., Srivastava, A., & Dey, S. (2020). Joint forest management: Issues and prospects in Indian forests. *Forest Policy and Economics, 113*, 102075. https://doi.org/10.1016/j.forpol.2020.102075

Hartman, M., & Huber-Sannwald, E. (2016). Sacred groves: Cultural landscapes in transition. *Human Ecology, 44*(3), 345–357. https://doi.org/10.1080/10549811.2016.1201010

ICIMOD. (2020). *Kailash Sacred Landscape Conservation and Development Initiative*. Retrieved from https://www.icimod.org/kslcdi

Imoro, Z. A., Jelle, B. C., & Acquah, E. (2021). Drone-based reforestation: A potential for sustainable forest management. *Journal of Sustainable Forestry, 40*(4), 451–467. https://doi.org/10.1007/s42973-021-00112-3

Kawakami, K., Higuchi, H., & Koike, F. (2019). Conservation of biodiversity in the Japanese archipelago: The roles of cultural landscapes and Satoyama. *Ecological Research, 34*(2), 223–234. https://doi.org/10.1111/1440-1703.12045

Kosoe, E. A., Oppong, F. K., & Sagoe, R. (2020). Participatory GIS for community-based forest management: A case study in Ghana. *Human Ecology, 48*(6), 711–725. https://doi.org/10.1007/s10745-020-00188-1

Leclère, D., Obersteiner, M., Barrett, M., Butchart, S. H. M., Chaudhary, A., De Palma, A., & Visconti, P. (2019). Bending the curve of terrestrial biodiversity needs an integrated strategy. *Nature, 585*(7826), 551–556. https://doi.org/10.1038/s41586-019-1594-9

Lee, T. M., Markowitz, E. M., Howe, P. D., Ko, C., & Leiserowitz, A. A. (2019). Predicting public climate change risk perceptions. *Journal of Environmental Management, 240*, 386–395. https://doi.org/10.1016/j.jenvman.2019.03.065

Makatta, A. A., Kashaga, L. M., & Malimbwi, R. E. (2018). Effectiveness of REDD+ pilot projects in addressing drivers of deforestation in Tanzania. *International Forestry Review, 20*(3), 306–318. https://doi.org/10.1505/146554818824063079

Mekonen, T. (2017). Drought and climate resilience in the arid regions of India. *Journal of Sustainable Development, 10*(1), 35–49. https://doi.org/10.5539/jsd.v10n1p35

Ministry of the Environment, Japan. (2021). Satoyama Initiative: Achievements and future.

Mori, A. S., Lertzman, K. P., & Gustafsson, L. (2017). Biodiversity and ecosystem services in forest ecosystems: A research agenda. *Restoration Ecology, 25*(6), 673–685. https://doi.org/10.1111/rec.12472

Newbold, T., Hudson, L. N., Hill, S. L. L., Contu, S., Lysenko, I., Senior, R. A., … Purvis, A. (2015). Global effects of land use on local terrestrial biodiversity. *Nature, 520*(7545), 45–50. https://doi.org/10.1038/nature14324

Pagiola, S., Arcenas, A., & Platais, G. (2008). Payments for environmental services: A pathway to sustainable development in India. *Journal of Cleaner Production, 319*, 124725.

Peng, C., Ma, Z., Lei, X., Zhu, Q., Chen, H., Wang, W., … Zhou, X. (2021). A drought-induced pervasive increase in tree mortality across forests in the globe. *Nature Communications, 12*(1), 2121. https://doi.org/10.1038/s41467-021-22457-8

Pokharel, R. K., Sharma, R., & Ojha, H. (2020). Community forestry in Nepal: Trends, benefits, and challenges. *Forest Policy and Economics, 116*, 102095. https://doi.org/10.1016/j.forpol.2020.102095

Ramachandrappa, S., & Thimmegowda, M. N. (2016). GIS applications in sustainable forest management: A review. *Environmental Development, 20*, 22–34. https://doi.org/10.1016/j.envdev.2016.05.006

Santini, L., & Miquelajauregui, Y. (2022). Synergistic effects of agroforestry and biodiversity conservation: A global perspective. *Biodiversity and Conservation, 31*(4), 1245–1265. https://doi.org/10.1007/s10531-022-02465-y

Silva, A. A., Medeiros, C., & a Ferreira, J. (2020). Using AI to monitor deforestation in the Amazon. *Environmental Modelling & Software, 131*, 104800. https://doi.org/10.1016/j.envsoft.2020.104800

Singh, A., Gupta, R., & Kumar, M. (2021). Community-based forest management: A sustainable development, *10*(1), 35–49. https://doi.org/10.5539/jsd.v10n1p35

UNESCO. (2021). Education for sustainable development: A roadmap.

International Collaboration for Sustainable Development: Navigating Life on Land Across the Asian Landscape

Shraddha Yadav and Ravi Kumar Yadav

1 Introduction

Asia is the largest and the most populated continent on Earth. It has a variety of ecosystems from the wet rainforests of Southeast Asia through the dry deserts of the Middle East to the frozen tundra of Siberia. This diversity makes Asia an important area for a great many forms of life, including an enormous variety of plants and animals, many of which are found nowhere else. Rivers such as the Yangtze, Ganges, and Mekong run into the continent and provide critical supplies of drinking water, irrigation, and industrial usage to over one billion people. Also, Asia's mountains and forests are important places for animals and help take in carbon, supporting the balance of ecosystems that are needed for all life on Earth.

Yet, in this large area, the ecological balance is at serious risk. The Asia area has very unique and serious environmental issues, mainly due to fast economic growth, rising population, and city expansion. Indonesia and Malaysia have the highest deforestation rates in the world, mainly due to logging and a strong demand for agricultural expansion, mainly for palm oil. As a result, forests have been stripped from millions of species including the Indo-Malayan orangutan and Sumatran tiger. Soil erosion, which is one of the foremost difficulties which affects more than two-thirds of crops grown, is primarily caused by excessive farming and unsustainable agricultural techniques.

Salinization, soil erosion, and water resource depletion increase the hazards to food security and affect the livelihoods of millions of people. Overexploitation of resources, pollution, and the illicit wildlife trade generate worrying rates of biodiversity loss throughout Asia, one of the most biodiverse regions in the world (White et al., 2022). The region's cultural legacy and economic stability are also impacted

S. Yadav (✉) · R. K. Yadav
Christ University, Lavasa, Maharashtra, India

J. J. Puthenkalam S.J et al. (eds.), *Terra Thrive*,
https://doi.org/10.1007/978-3-032-02738-2_9

by biodiversity decrease, as many populations in the area rely on the direct use of natural resources for their livelihoods.

Owing to the scale and transboundary aspect of numerous environmental problems, global collaboration is not only beneficial (Baraj et al., 2020) but also imperative. No single nation can independently manage the impacts of climate change, control cross-border haze from forest fires, or ensure the sustainability of shared water resources. International and regional cooperation serves as channels to facilitate the exchange of information, resources, and technology. Furthermore, they build trust and understanding, which are essential for effective conservation and sustainable development. The need for international collaboration in Asia is underscored by the United Nations Sustainable Development Goals which aim to protect the environment and promote sustainability. Regional organizations, such as the Organization of Southeast Asian Nations, also foster cooperative environmental policies among member nations. Cooperation encompasses the management of forests, preservation of wildlife, and the mitigation of haze pollution. In addition, non-governmental organizations, community groups, and international agencies facilitate intergovernmental cooperation in tackling those multifaceted environmental issues.

In regard to the issue raised by the continent, preserving the ecological balance of Asia through efficient land use practices is not just the region's responsibility but it is also a monumental undertaking. The continental government, having the environment, cuts across the trends in biodiversity conservation, climate change, and the entire international economic system. In order to move toward a sustainable future for Asia, it is necessary to strengthen international ties and harmonize national policies. This chapter delves deeper into these themes to explore the very mechanisms of cooperation and the various roles that the different stakeholders play in this dynamic and vital quest for environmental sustainability.

2 Literature Review

***Marine Protected Areas in the Coral Triangle: Progress, Issues, and Options*, by Alan T. White, Porfirio M. Aliño, et al.**

These countries are interested in the establishment and management of Marine Protected Areas (MPAs) of the Classic Coral Triangle of six Southeast Asian countries, namely Indonesia, Malaysia, the Philippines, Papua New Guinea, Timor Leste, and the Solomon Islands. The Coral Triangle is critical for the sustenance of marine biodiversity, food security, and livelihood for millions of people residing in these countries (White et al., 2022). The Coral Triangle is critical for the sustenance of marine biodiversity, food security, and livelihood for millions of people residing in these countries (White et al., 2022). The authors analyzed the level of the progress we have achieved in the establishment of MPAs and the MPA implementation issues (Agbedahin, 2019). The authors here could state that effective management of these MPAs would require sufficient multilateral cooperation among the six involved countries as well as the assistance of NGO and international organization partners.

***Challenges to Carry Out Transboundary,Environmental Assessment in ASEAN Countries*, by Juanita Gallego Dávila, Jie Zhang, and Lone,Kørnøv**

The difficulties associated with carrying out transboundary environmental evaluations in ASEAN nations are the main topic of this article. Gallego (2023) asserts that TEAs are essential for managing common natural assets and addressing cross-border environmental issues such as pollution, deforestation, and biodiversity loss (White et al., 2022).

The authors argue that by including international organizations and enhancing regional cooperation within ASEAN structures, such difficulties may be overcome. The authors also emphasize the necessity for efforts to build capacity that will enhance the technical proficiency of local environmental officials.

***Desertification:;Causes and Countermeasures* by Rocio Becerril-Piña and,Carlos Mastachi**

This chapter is part of the Springer publication Life on land. It presents and discusses causes of desertification and its after-effects, mainly in arid and semi-arid regions. The authors take this discussion a little further with respect to how desertification represents a significant threat to sustainable development in Asian countries, especially Central Asia, India, and China.

***Desertification: Concepts, Causes, and Amelioration* by,Jagdish C. Katyal and Paul L. G. Vlek**

This piece looks closely at desertification, including its causes, consequences, and potential remedies. The researchers focus on the ramifications for Asian nations, paying special attention to dry areas such as the nation's Gobi Desert and the nation's Thar Desert.

The present study highlights the importance of regional and international efforts to reverse desertification and promote policies of sustainable land management throughout Asia.

***Impact of the Belt and Road Initiative,on Enterprise Green Transformation*, by Lan Wang & Zhonghua Cheng (2024)**

This research provides insights into BRI by China and its impact on the environment, especially through green changes in businesses of the participating countries (Baraj et al., 2020). In the article, the authors mention that even as BRI facilitates economic growth, it brings with itself environmental issues, especially in the Asian continent where environmental rules differ so greatly. The authors observe that the BRI will be considered a success, and the SDGs of the UN in this area will be achieved only with a balance between economic growth and environmental care.

***Climate Change and Resilience, Adaptation, and Sustainability of Agriculture in India: A Bibliometric Review* by Biswaranjan Baraj, Manoranjan Mishra, Desul Sudarsan, Richarde Marques da Silva, and Celso Augusto Guimarães Santos**

This paper does a bibliometric review of the existing literature on India regarding climate change and agricultural resilience. The extant literature review conducted by these authors discusses whether Indian agriculture adjusts to the climatic changes and explores the role played by international collaborations in promoting environmentally friendly farm practice (Baraj et al., 2020). Key insights indicate that climate change is the biggest threat to Indian agriculture. The reason is enhanced temperatures, erratic rainfall, and increased frequency of extreme weather events. The solutions adapted to the climate change issue include drought-resistant varieties, irrigation technologies, and agroforestry by farmers. Building on and promoting research and the exchange of knowledge, as well as further building a capacity for resilience in agriculture in those regions, require international collaboration (Baraj et al., 2020). It showcases narratives of successful partnerships between Indian institutions and global organizations like CGIAR (Consultative Group on International Agricultural Research) and FAO (Food and Agriculture Organization). And as the authors emphasized, international collaboration remains crucial in developing a resilient agricultural system against the effects of climate change.

***Agroforestry in the Western Ghats of Peninsular India, and the Satoyama Landscapes of Japan: A Comparison of Two Sustainable Land Use Systems,* by B. Mohan Kumar & K. Takeuchi**

The present research examines two traditional, sustainable land-use systems: satoyama in Japan versus agroforestry in the Western Ghats region of India. The two systems were recognized for the roles the system plays in conserving biodiversity and agricultural sustainability (Baraj et al., 2020).

The paper calls for the conservation of traditional knowledge and international cooperation to adapt and replicate sustainable land-use practices in other regions.

***Assessment of the Growth in Social Groups for Sustainable Agriculture and Land Management,* by Jules Pretty, Simon Attwood, et al.**

The role that communities and nonprofit organizations play in promoting sustainable land management and agriculture globally is examined in this research. The author focuses on community-based organizations and how farmer groups contribute to sustainable development, particularly in Asia.

We can conveniently conclude that the promotion of social groups and international collaboration significantly contributes to the achievement of goals in sustainable agricultural and land-use development.

***Sustainable Land Management in Asia: Applying a Land-Use Function Approach* by Kikuko Shoyama, Zhichao Xue, Lin Zhen, and Md Giashuddin Miah**

In their work Sustainable.Land Use in Asia, Kikuko Shoyama et al. explain; the land-use function in work using a land-use function model the government have to find balance among the search of equitable society, economic growth, and environmental sustainability. Authors Shoyama et al. (2020) caution that land management has much to be achieved in Asia because this region is rapidly urbanizing and increasing in population. Environmental concerns integrated with socioeconomic needs ensure positive contributions of land use to food security.biodiversity, and ecosystem resilience. This framework provides a model for sustainable planning in developing regions that foster regional development while reducing the rate of environmental degradation.

***Achieving the Sustainable.Development Goals Requires Transdisciplinary Innovation at the Local Scale* by Enayat A. Moallemi et al.**

Enayat A. Moallemi and co-authors, in their paper, emphasize that local-level innovation and change are essential for advancing the Sustainable Development

Goals (SDGs). The idea is that complexities of global issues require trans-disciplinary approaches. Science, policy, and society must be released from their isolated silos for the greater cause. By integrating different knowledge systems, the authors suggest solutions that are scientifically valid but socially acceptable and economically feasible, Moallem says (Moallemi et al., 2020). This approach reinforces the need for local action to tackle global problems like climate change, inequality, and biodiversity loss.

***Agroforestry in the Western Ghats and Satoyama Landscapes: A Comparison* by B. Mohan.Kumar & K. Takeuchi**

B. Mohan Kumar and K. Takeuchi provide a comparative study of the agroforestry practices of India's Western Ghats and Japan's Satoyama landscapes. Both systems indicate that there is scope for merging traditional ecological knowledge with modern conservation methods to design sustainable land-use systems.This paper focuses on how these practices have been helpful in maintaining biodiversity, enhancing soil fertility, and adding to the local livelihoods; thus, they serve as a model for other regions which would like to maintain a balance between development and ecological stewardship.

***Biodiversity Conservation in Asia* by Dale.Squires**

Dale Squires presents his work on biodiversity preservation through the perspective of the growing threat to Asia's rich ecological diversity. He emphasises the region's hitting issues, including habitat loss, overexploitation, and climate change, based on his research. This call for regional cooperation, building capability, and assimilation into economics and development plans is needed according to Squires while advocating a regional approach will only position biodiversity to be one key element to anchor sustainable development for Asia.

***Transboundary Conservation: A.New Vision for Protected.Areas* by Russell A. Mittermeier et al.**

In Transboundary Conservation, a new vision for Protected Areas, Russell A. Mittermeier and his colleagues investigated the possibility of cross-border collaboration in protecting ecosystems that transcend political boundaries. This book is used to illustrate some successful case studies where countries have collaborated to protect shared resources; thus, joint management strategies are highly effective. According to the authors, transboundary conservation is important to address global environmental challenges, ranging from species migration to water resource management.

***Global Wetland Outlook: State of the World's Wetlands and their Services to People* (Ramsar Convention,** 2018)

The Global Wetland Outlook of the Ramsar Convention should alert the world that wetlands are not only sources of ecological and economic survival but also important reservoirs for water resources and carbon sinks (Agbedahin, 2019). According to the report, data concerning the wetland ecosystem condition and its services like water filtration, carbon storage, and provision of habitats highlight the alarming rate at which these ecosystems decline. International commitment is expected to strengthen proposals to save these vital ecosystems through sustainable management practices.

***International Law and the Environment* by Alan Boyle, Catherine.Redgwell, and Patricia.Birnie**

Alan Boys, Catherine Redgwell, and Patricia Birnie provide a comprehensive overview of international environmental law, including its history, development, structure, and implementation. Thus, it makes possible a full understanding of the global environmental order. The attribute that differentiates this book from others is that as an indispensable resource for researchers and policymakers in international law.

Interrelation with other areas, especially trade and human rights, makes it such.

***International Environmental Law* by Pierre-Marie Dupuy and Jorge E. Viñuales**

International environmental law is covered in detail and clearly by Pierre-Marie Dupuy and Jorge E. Viñuales, that is, from a practical perspective. Their work focuses on major agreements and their jurisprudential underpinnings and the legal mechanisms designed to counter climate change, biodiversity loss, and pollution. Through a mix of both legal analysis and policy insights, the book thus bridges the gap between theoretical understanding and real-world applications.

***The.Health of Nations: Society and.Law beyond the State* by Phillip Allott**

In The Health of Nations: Social and Law Outside the State, Phillip Allott enquiries the conventional understanding of state sovereignty in light of solving global problems. A book that lays down a radically transformed vision, international law evolves into being centered on societal health and ecological health, Allott argues that the role of law should be reimagined to foster a global community that is resilient to crises, such as environmental degradation.

***Ecosystem: Services: Key Issues* by.Mark Everard**

Ecosystem Services: The basic importance of ecosystems in life and economic activity is raised by key issues. It enhances the discussion on the extent to which human activity poses a threat to ecosystem services and functions, and mainly underscores the need for their consideration in policy development and decision-making. Everard pretty nicely makes a case for valuing ecosystems as natural capital toward sustainable development.

***IPBES Global Assessment Report on Biodiversity and Ecosystem Services* (Chapter 2.2: Status and Trends—Nature)**

The Intergovernmental Science-Policy Platform on Biodiversity and Ecosystem Services provides a sober analysis of global trends in biodiversity. It highlights the startling speeds at which human activity is destroying species and ecosystems and calls on decision-makers to act quickly. Thus, it provides practical advice on how to stop biodiversity loss and include environmental factors into development planning.

***Principles of Public International Law* by Ian Brownlie**

Ian Brownlie's Principles of Public International Law is a foundational text for the principles of international relations. In relation to environmental governance, it touches on the aspects of state responsibility and legal obligations stemming from international agreements. This book is, therefore, crucial for the understanding of how states can work together to meet the challenges of the global environment.

***International Law as an Open System* by James Crawford**

James Crawford's works on essays delve into the dynamic side of international law as it deems to counter emerging global concerns (Ahlawat et al., 2019). Framing international law to be an adaptive system, this is how international law can indeed answer the great challenge of climatic change as well as degrading the environmental surroundings. By legal innovation for sustainable development in the works highlighted (White et al., 2022).

3 The Asian Environmental Challenge

3.1 Deforestation

The forests in Asia are being depleted rapidly through agricultural expansion, infrastructure construction, and logging. Indonesia and Malaysia are prime examples of countries experiencing significant forest loss due to the growing demand for palm oil. This demand drives habitat loss and deforestation, threatening thousands of species, including endangered Sumatran rhinos and orangutans (Carlson et al., 2024). The ecological impacts are profound, reducing carbon sequestration capacities essential for mitigating global warming and diminishing biodiversity (Squires, n.d.).

Illegal logging, often linked to corruption and weak governance, accelerates deforestation by removing valuable hardwoods like teak and mahogany without regard for sustainability (Doh et al., 2021). Slash-and-burn techniques, particularly in Indonesia, cause seasonal haze that degrades air quality across Southeast Asia (Gaveau et al., 2022). Addressing these challenges requires robust law enforcement, international cooperation to curb illegal trade, and sustainable land management to balance economic and environmental needs (Sachs, 2015).

3.2 Land Degradation

Land degradation in Asia severely impacts its vast agricultural lands through soil erosion, nutrient depletion, and pollution. Intensive agriculture, driven by the need to feed a growing population, causes overgrazing, overcultivation, and excessive use of chemical fertilizers and pesticides, which degrade soil productivity over time and pollute streams and rivers through runoff (Pretty et al., 2020). The Aral Sea tragedy exemplifies land degradation, where unsustainable irrigation for cotton farming triggered one of the worst environmental catastrophes, shrinking the sea to a third of its original size (UNCCD, 2024). Restoration efforts, involving international aid, emphasize sustainable water and land use policies to prevent future disasters (Rogers et al., 2012).

3.3 *Biodiversity Loss*

Habitat degradation from deforestation, urbanization, and agricultural expansion drives Asia's rapid biodiversity loss. The region hosts critical biodiversity hotspots like the Borneo rainforest and the Himalayas, which are ecologically, economically, and medicinally significant (Kumar & Takeuchi, 2009). The illegal wildlife trade, driven by demand for traditional medicine, exotic pets, and luxury goods, threatens species like tigers, pangolins, and rhinos (UNODC, 2024). Strong implementation of international frameworks like CITES, alongside local conservation efforts and economic alternatives to wildlife exploitation, is essential (Birnie et al., 2009).

3.4 *Water Scarcity and Pollution*

Pollution exacerbates water scarcity, a pressing issue in Asia. The Ganges and Yangtze, among the world's most polluted rivers, face contamination from household sewage, agricultural runoff, and industrial waste, affecting hundreds of millions who rely on them for drinking water, sanitation, agriculture, and industry (World Bank, 2023). Modern water treatment, irrigation systems, and pollution management technologies require economic investment and political will (Everard, 2021). Regional water management agreements are vital, as transboundary water bodies demand cooperative governance for sustainable use (Ilias & Katerina, 2023)

3.5 *Urban Environmental Pressure*

Megacities like Delhi, Beijing, and Jakarta face severe environmental pressures from urbanization, including air quality issues, waste management challenges, and high energy consumption (ESCAP, 2025). Urban sprawl strains green spaces, increases vehicle emissions, and exacerbates the urban heat island effect, reducing quality of life (Niesenbaum, 2019). Policies promoting recycling, efficient public transport, and green building codes can mitigate degradation and enhance climate resilience (Reader, 2023). International funding and technology transfers underscore the need for global cooperation to address urban environmental challenges (United Nations, 2019).

By integrating national laws, local efforts, and global cooperation, Asia can tackle these complex environmental issues and promote conservation and sustainable growth across its diverse regions (Daly, 1996).

4 Legal Frameworks and Mechanisms

4.1 Legal Frameworks and Mechanisms for Environmental Sustainability in Asia

Asia's environmental challenges are complex and require strong legal frameworks and legal procedural mechanisms to deal with a variety of issues including pollution, deforestation, loss of biodiversity, land degradation, and water scarcity (Gallego, 2023). Developments in GIS and remote sensing technology have transformed data collection, analysis, and visualization when it comes to the environment. These are fundamental tools for charting tree-loss and tracking biodiversity or measuring natural disasters. For instance, in Indonesia, the application of GIS and remote sensing has focused on monitoring peatland degradation and forest fires, which are among major contributors of haze fires in Southeast Asia. This lends itself to think through better decision-making and targeted interventions at various levels.

Another sector in which new technologies are driving sustainability within Asia is biotechnology. Genetically modified (GM) crops are currently being developed with improved resistance to pests, diseases and environmental stress, which reduces dependence on chemical pesticides and increases crop yields (Baraj et al., 2020).

4.2 Global Legal Instruments

Asia is a party to international environmental treaties with a range of environmental issues that provide a framework for addressing these concerns. Thanks to this, there are multiple such efforts, but the most prominent include:

1. **The Convention on Biological Diversity (CBD):**This worldwide pact is designed to maintain biological diversity; promote the sustainable use of its elements; and ensure the just and equal distribution of benefits yielded in realizing genetic resources (Gallego, 2023). It commits the Asian nations to the adoption of national biodiversity policies and action plans under this CBD.
2. **The United Nations Framework Convention on Climate Change (UNFCCC)**: In order to reduce, adapt, and fund greenhouse gas emissions, this international environmental accord was extended in the Kyoto Protocol and the Paris Agreement. Since Asia is especially vulnerable to and impacted by climate change, Asian nations are a key constituency.
3. **The Ramsar Convention on Wetlands**: This is an international treaty for the conservation and sustainable use of wetlands. Its aim is to stop the global loss of wetlands and to conserve existing ones through sustainable use and management.

4. These treaties are supported by a series of other agreements that address more specific issues, like the trade in endangered species (CITES), hazardous wastes (Basel Convention), and transboundary air pollution (Baraj et al., 2020). To adhere to these treaties, countries need to implement national legislation and create effective regulatory systems that may receive technical and financial aid from international organizations and developed countries (Moallemi et al., 2020).

4.3 Regional Cooperation Agreements

1. In addition to broad global treaties, Asia has numerous regional initiatives and agreements to respond to environmental challenges that are specific to the region:
2. **Atmosphere:** Through its creation, ASEAN Agreement on Transboundary Haze Pollution responds to various environmental concerns across South Asia sectors and serves as a collaborative instrument for achieving sustainable development within a suitable environmental framework.
3. **The South Asian Association for Regional Cooperation (SAARC) Environment Charter**: This charter is regarding the characteristic environmental issues of South Asia, which is primed on the promotion of cooperation to imagine sustainable development and to enhance the quality of the environment.
4. **Asia-Pacific Partnership on Clean Development and Climate (APP):** APP was a non-legal binding agreement where the states agreed to collaborate and exchange technologies and practices for polluting sectors (Gallego, 2023). Overcoming challenges is not a core domain of operation for the region, but regional responses are imperative for challenges with national boundaries that require transnational, multi-state approaches and contributions.

4.4 National Legislation and Policies

Nationally, the first mechanism is the interest of those Asian countries in being tied to international agreements and regional obligations:

- **India's National Green Tribunal Act**: This included the National Green Tribunal Act, which also established a special tribunal to hear and dispose of cases — arising from its provisions — dealing with the protection of the environment and conservation of forests and other natural resources.
- **China's Environmental Protection Law**: The new law imposed stricter rules on environmental violations by promoting a more robust regulatory regime for environmental protection and was revised in 2014.

- **Indonesia's Conservation Act on Biological Natural Resources and Ecosystems**: This act centers on the preservation of biological resources, control over their use, and management of national parks and other protected areas (White et al., 2022).
- There are some policy and initiatives along with these national laws that work toward the mainstreaming of environmental considerations in a more holistic developmental policies (Ahlawat et al., 2019). In the same way, the National Action Plan on Climate Change of India and China both embody its comprehensive national strategy to respond to climate change and policies on energy conservation and emission reduction (Gallego, 2023).

4.5 *Mechanisms for Enforcement and Compliance*

Legal frameworks are mostly effective when their enforcement mechanisms are effective. Enforcing the mechanisms is a challenge across Asia because of lack of resources, lack of technical know-how, corruption, and weak institutional frameworks. Still, there is some effort being made to strengthen these mechanisms:

- **Environmental Impact Assessments (EIAs)**: EIAs are very important tools that are mandated in many Asian countries to evaluate the potential environmental impacts of proposed development projects before they are approved.
- **Public Participation**: In this regard, there is an increasing recognition of the role of local communities and stakeholders in environmental decision-making processes. Their involvement increases transparency and the legitimacy of environmental policies.
- **Judicial Activism**: These courts have, accordingly, used proactive initiative in environmental governance in such countries as India by stepping into force or extension of environmental laws and regulations.

5 Challenges and Opportunities

On paper, the law system and ways in Asia are strong but the big test is doing and sticking to it. The great size of the land, a mix of rule systems, and uneven levels of money growth lead to mixed sticking to environmental rules. The tests include finding a balance between economic growth with care for nature and boosting the lasting use of natural goods. Problems can be faced by making stronger groups, helping people know more, and also by better work with other countries and areas.

Overcoming such challenges also depends on technological development and financial investment in environmental governance.

6 Technological Innovations for Sustainability in Asia

Two of the 17 aims show how important it is to have new technology—630 million folks in Asia do not have access to power and want to see a change in this trend. Quick-growing technologies are helping with better use of nature's resources, watching environmental changes, and putting into practice ways that can last over time. Here, we point out some of the strange tech gains that have played a big role in pushing sustainability forward in the area by looking at its use across different parts of the environment sector.

6.1 Geographic Information Systems (GIS) and Remote Sensing

Geographic Information Systems (GIs) and remote sensing technologies have changed the ways we collect, analyze, and visualize data on the environment. These are foundational tools for charting deforestation, observing biodiversity, and evaluating natural disasters. As an example, GIS and remote sensing have been implemented in the monitoring of peatlands and forest fires which are significant sources of haze in Southeast Asia, particularly in Indonesia. This enables more informed decision-making and targeted intervention at different levels.

Remote sensing satellites play a vital role in water resource management. It is through remote sensing that these countries, India and China, get satellite images in order to find out the extensions of water bodies, the groundwater level, and the irrigation patterns to plan and manage water resources.

6.2 Artificial Intelligence and Machine Learning

Artificial intelligence and machine learning technologies are currently used for environmental preservation and monitoring tasks. AI algorithms provide insights into deforestation patterns while measuring pollution severity and enabling species identification through camera trap photographs in biodiversity assessments. The Chinese government uses AI systems to track air quality in real time and utilizes this data to inform public health recommendations and develop regulations for industrial pollution and urban development plans.

The machine learning models make predictions about climate change effects and simulate disaster risk reduction scenarios. Through AI-enhanced systems, countries such as Japan and the Philippines that frequently experience natural disasters can accurately forecast cyclone routes and earthquake aftershocks which enables faster emergency responses and results in many lives saved.

6.3 Internet of Things (IoT) for Smart Water Management

The implementation of IoT technology holds transformative potential to improve water resource management practices throughout Asia (Chakravarty, 2018). IoT technology allows sensor-controlled devices to monitor water levels and quality along with its usage across extensive regions from a distance as soon as data is collected. Singapore uses IoT-based smart water management systems to identify leaks and predict water demand while optimizing its distribution for better resource efficiency resulting in significant water conservation (Baraj et al., 2020).

Smart irrigation systems based on IoT technology manage water distribution precisely by utilizing soil moisture data and weather predictions to maintain crop health without causing water wastage (Chakravarty, 2018). Such technological innovations would greatly benefit water conservation efforts in dry areas of Central Asia to support sustainable farming.

6.4 Renewable Energy Technologies

Greenhouse gas emissions and the need to move away from fossil fuels are slowly but surely driving Asia's energy sector toward a greater emphasis on renewables. Solar, wind, and hydroelectric power have been relatively popular in China, India, and Vietnam in the energy mix (White et al., 2022). Technological innovations, such as increases in solar panel efficiency, in wind turbine design, and micro-hydro installations, make such energy sources viable and less costly. For example, China is now a world leader in solar photovoltaic (PV) panel production and adoption, forcing down the price and promoting further innovation in the field of solar technology. It also helps to combat global warming and to boost local businesses that manufacture and install renewable energy technology.

6.5 Biotechnology in Agriculture

Another area where technologies are establishing sustainability within Asia is biotechnology. Currently, the GM crops are being developed with improved resistance to pests, diseases, and environmental stresses, reducing the dependence on chemical pesticides and increasing the yield of the crop (Baraj et al., 2020). For instance, agricultural systems in India and China have been changed with the spread of GM crops such as Bt cotton, rendering environmental harm during cotton cultivation in the two countries relatively insignificant and as an incentive to farmers to increase their profitability (Chakravarty, 2018).

Moreover, precision fermentation and cultured meat are equally emerging as sustainable biotechnologies that traditional animal husbandry methods could be substituted for and are said to utilize 20% to 90% less land and water, and produce fewer emissions compared to other food sources.

7 Challenges and Opportunities for Future Collaborations

7.1 Barriers to Effective Collaboration

Though effective collaboration is considered very ideal for the environmental sector too, in Asia, this is easier said than done and this discrepancy often leads to us groping in the dark on our journey toward sustainability. These challenges are many, nuanced, and reflect the multi-layered geopolitical and socio-economic fabric of the region:

1. **Political and Economic Disparities**: Other developed, high-tech economies and poor developing economies are included in the Asian countries. Political systems range from democratic to authoritarian, affecting how environmental policies are prioritized and enforced. Such differences have induced conflicts over resource allocation and environmental responsibilities, making it difficult to achieve consensus or to cooperate.
2. **Lack of Binding Enforcement Mechanisms**: Regional agreements tend to be weak in terms of binding enforcement mechanisms. Without mandatory adherence, regional frameworks are mere formalities, ineffective. The lack of enforcement sours cooperation, especially on joint environmental problems such as air pollution and river management.
3. **Inadequate Funding and Resources**: Environmental initiatives are underfunded most often if we talk about under-developed or developing nations. For instance, without sufficient funding, the nature and number of the projects may be limited, whereas countries are more likely to focus on the short-term growth rather than long-term viability, contributing to lower participation in regional initiatives.
4. **Technical and Capacity Gaps**: There is a huge gap in technical capability across Asia. While some countries are at the forefront of environmental technology and policy development, others do not have the capacity and institutional capability to undertake sophisticated environmental management systems or to participate effectively in complex regional initiatives (Baraj et al., 2020).

7.2 *Overcoming These Barriers*

It is possible to take several measures to achieve better cooperation and ensure effective environmental programs:

1. **Enhanced Diplomatic Engagement**: The two sides should boost gravitational diplomacy, as this would catalyze mutual trust and understanding. Whether constant interaction is further encouraged by regular intergovernmental meetings and forums which would affect coordination of state efforts for environmental protection.
2. **Development of Stronger Legal Frameworks**: Clear legal frameworks with concrete mechanisms of enforcement could make compliance better. Regional organizations, such as ASEAN, SAARC, and others, may be allowed to monitor the adherence of countries to environmental agreements and, consequently, give regional cooperation a well-structured dimension.
3. **Pooling Resources and Joint Funding Initiatives**: Pooled financing and joint funding mechanisms can reduce the financial burden of the projects. International donors, regional development banks, and private sector investment can help finance environmental projects, and this will relieve the financial burden on individual countries.
4. **Capacity Building and Knowledge Sharing**: Development of regional centers of excellence and further extension of the training programs could also reduce the technical and capacity gaps (Chakravarty, 2018). These steps would help distribute the best practices and modern technologies promoted by developed technological countries and international bodies in the region.

7.3 *Future Opportunities for Enhancing Regional Cooperation and Policy Integration*

The future of Asia holds immense promise for further regional cooperation and policy integration. There is great potential to move forward with environmental sustainability through concerted effort and strategic planning:

1. **Leveraging Technology for Environmental Management**: This is how technology is being used. Using AI to boost environmental management: AI and IoT-driven technologies, coupled with satellite imaging, dramatically improve the depth, accuracy, and speed of environmental monitoring and management. They facilitate accurate data collection, provide real-time monitoring capabilities, and offer advanced analysis tools that aid in informed decision-making and enhance regional coordination efforts.

2. **Inclusive Policy-Making**: Environmental Policy Reform: Expanding stakeholder participation opens up a toolbox of persuasive tools that can lead to more comprehensive environmental policy and a more robust environmental policy reform process. The inclusivity of NGOs, community organizations, representatives from the private sector, and academic institutions ensures policy development that is balanced and receives the support of all the groups affected.
3. **Focus on Sustainable Development Goals (SDGs)**: The UN SDGs integrate regional action on the environment into a common framework of action, establishing an integrated framework of action at the regional level. These goals constitute an integrated approach to sustainability across environmental protection, economic growth, and social inclusion.

8 Conclusion

We have laid out in this chapter the acute backdrop of environmental challenges in Asia, the dominant frameworks that have underpinned responses to those challenges, and the precasts for successes that are finding their way out of carefully led collaboration and local action, with an emphasis in all of that on creative technologies. The findings underscore the need for more coordinated and integrated policies for climate action at the regional level. Moving forward we need more flexible and adaptive strategies which can internalize changing biosphere conditions and macro trends. It is cooperation that is vital as such includes elements of collective action and shared responsibility in environmental governance.

These cooperative frameworks, if developed further, can ensure a more sustainable future for Asia, contribute to regional partnerships, and create investments in innovative technologies and practices. All these will not only solve current environmental issues but long-term sustainability will benefit both the people and the natural heritage of Asia. A proactive, collaborative approach is apt to nurture a resilient and sustainable Asia to be passed down from generation to generation.

References

Agbedahin, A. V. (2019). Sustainable development, education for sustainable development, and the 2030 agenda for sustainable development: Emergence, efficacy, Eminence, and future. *Sustainable Development, 27*(4), 669–680. https://doi.org/10.1002/sd.1931

Ahlawat, A., & Team ARSu. (2019). *Sustainable Development Goals: Directive Principles for Sustainable India by 2030*. MyARSu.

Baraj, B., Mishra, M., Sudarsan, D., da Silva, R. M., & Santos, C. A. G. (2020). Climate change and resilience, adaptation, and sustainability of agriculture in India: A bibliometric review. *Heliyon, 10*(8), e29586.

Birnie, P., Boyle, A., & Redgwell, C. (2009). *International Law and the Environment*. Oxford University Press.

Carlson, K. M., Heilmayr, R., Gibbs, H. K., Noojipady, P., Burns, D. N., Morton, D. C., Walker, N. F., Paoli, G. D., & Kremen, C. (2024). Effect of oil palm sustainability certification on deforestation and fire in Indonesia. *Proceedings of the National Academy of Sciences, 121*(3), e2301453121. https://doi.org/10.1073/pnas.2301453121

Chakravarty, S. (2018). Sustainable development. *European Journal of Development Research, 3*, 67–77. https://doi.org/10.4324/9781315831657-4

Daly, H. E. (1996). *Beyond growth: The economics of sustainable development*. Beacon Press.

Doh, J., Budhwar, P., & Wood, G. (2021). Long-term energy transitions and international business: Concepts, Theory, methods, and a research agenda. *Journal of International Business Studies, 52*, 951–970.

Everard, M. (2021). *ecosystem Services: Key issues* (2nd ed.). Routledge.

Gallego Dávila, J., Zhang, J., & Kørnøv, L. (2023). Challenges to Carry Out Transboundary Environmental Assessment in ASEAN Countries. *Impact Assessment and Project Appraisal, 41*(4), 372–384. https://doi.org/10.1080/14615517.2023.2175682

Gaveau, D. L. A., Santos, L., Locatelli, B., Salim, M. A., Husnayaen, & Meijaard, E. (2022). Forest loss in Indonesian Borneo driven by industrial plantations and fires. *Scientific Reports*, 12, 17201. https://doi.org/10.1038/s41598-022-21445-6

Ilias, B., & Katerina, A. (2023). *Sustainable development goals*. The UN Sustainable Development Goals. https://doi.org/10.1093/law/9780192885173.003.0001

Kumar, B. M., & Takeuchi, K. (2009). Agroforestry in the Western Ghats of peninsular India and the Satoyama landscapes of Japan: A comparison of two sustainable land use systems. *Sustainability Science, 4*, 215–232.

Moallemi, E. A., Malekpour, S., Hadjikakou, M., Raven, R., Szetey, K., Ningrum, D., Dhiaulhaq, A., & Bryan, B. A. (2020). Achieving the sustainable development goals requires transdisciplinary innovation at the local scale. *One Earth, 3*, 300–313.

Niesenbaum, R. (2019). Sustainable development and global sustainability. *Sustainable Solutions*. https://doi.org/10.1093/hesc/9780199390434.003.0015

Pretty, J., Attwood, S., et al. (2020). *Assessment of the growth in social groups for sustainable agriculture and land management* (p. 3). Cambridge University Press.

Ramsar Convention Secretariat. (2018). *Global wetland outlook: State of the world's wetlands and their services to people*. Ramsar Convention Secretariat.

Reader, G. (2023). Engineering sustainable development—Realizing the UN'S sustainable development goals. *Progress in Sustainable Development*, 1–54. https://doi.org/10.1016/b978-0-323-99207-7.00002-6

Rogers, P. P., Jalal, K. F., & Boyd, J. A. (2012). *An introduction to sustainable development*. Earthscan.

Sachs, J. D. (2015). *The age of sustainable development*. Columbia University Press.

Shoyama, K., Xue, Z., Zhen, L., & Miah, M. G. (2020). *Sustainable land management in Asia: Applying a land-use function approach*. Asia-Pacific Network for Global Change Research.

Squires, D. (n.d.). *Biodiversity conservation in Asia*. Southwest Fisheries Science Center.

United Nations. (2019). Science for sustainable development. In *Global Sustainable Development Report* (pp. 110–125). https://doi.org/10.18356/c1351a5e-en

United Nations Convention to Combat Desertification (UNCCD). (2024). Global Land Outlook: Asia Regional Report. UNCCD Secretariat, Bonn.

United Nations Economic and Social Commission for Asia and the Pacific (ESCAP). (2025). Asia-Pacific Sustainable Development Report 2025. United Nations, Bangkok.

United Nations Office on Drugs and Crime (UNODC). (2024). World Wildlife Crime Report 2024: Trafficking in protected species. United Nations, Vienna.

Wang, L., & Cheng, Z. (2024). Impact of the belt and road initiative on Enterprise Green Transformation. *Journal of Cleaner Production, 468*, 143043.

White, A. T., Aliño, P. M., Cros, A., Fatan, N. A., Green, A. L., Teoh, S. J., Laroya, L., Peterson, N., Tan, S., Tighe, S., Venegas-Li, R., Walton, A., & Wen, W. (2022). Marine protected areas in the coral triangle: Progress, issues, and options. *Coastal Management, 42*(2), 87–106.

World Bank. (2023). Asia and the Pacific: Climate Change and Development Report. World Bank Publications, Washington, DC.

Holistic Land Governance-Revolutions for Ecological Stability and Socio-Economic Growth

Sharon Susan Thomas, Hannah Mariam, Jeslin Joy, Sarah Iype, and Lija Jacob

1 Introduction

The necessity of sustainable land management cannot be stressed enough from any angle. It provides a comfortable living to people as well as a solution for the sustainable being of a different variety of flora and fauna that add to the maintenance of a natural balance. It is still true that the soil is becoming less and less useful. Urbanization, deforestation, climate change, and poor agricultural techniques all negatively impact the ecosystem. They all, therefore, play a vital role in pushing the areas where flora and fauna adapt to live slowly toward degradation.

1.1 Background and Importance

On land, ecosystems are vital to life. They sustain agriculture, help control the climate, and offer a home for many different kinds of animals. But humans have significantly changed these ecosystems. While cities are expanding, forests are shrinking. Unfortunately, when that happens, it puts an immense impact upon

S. S. Thomas (✉) · H. Mariam
Department of Computer Science, Christ University, Bengaluru, Karnataka, India
e-mail: sharon.susanthomas@bcah.christuniversity.in

J. Joy
Department of International Studies, Christ University, Bengaluru, Karnataka, India

S. Iype
Department of Commerce, Christ University, Bengaluru, Karnataka, India

L. Jacob
Department of Statistics and Data Science, Christ University, Bengaluru, Karnataka, India

J. J. Puthenkalam S.J et al. (eds.), *Terra Thrive*,
https://doi.org/10.1007/978-3-032-02738-2_10

animal populations. They are bigger settlements with greater resource use and higher pollution contributions toward climate change.

Sustainable Development Goal 15, or "Life on Land," is to counteract the disturbing speed of deforestation and desertification and restore degraded land that is in harmony with biodiversity and allows nature to provide benefits to humanity.

1.2 Current Challenges

Currently, one of the biggest problems with land sustainability is the rapidly depleting biodiversity. Such things as cutting down forests and converting land into farming grounds have caused a huge loss of habitat, which can easily make many species extinct or endangered. A great extinction might break food supplies and other important processes for human life, thus breaking the ecosystem.

The soil's condition is another important issue. Food security and ecosystem stability are affected. Excessive use of artificial fertilizers, excessive grazing by animals, and repeated cultivation of the same crop are examples of overly intensive farming practices that deprive the soil of essential nutrients. It leads to desertification. According to the Food and Agricultural Organization (FAO) (2015, 2020). The loss of almost one-third of the world's agricultural land over the past 40 years poses a great spirit of concern for food security in the future (Gebbers & Adamchuk, 2010), followed by land problems that severe climate change tends to complicate through altering precipitation and temperature patterns.

Thus, this creates more droughts and wildfires and leads to flooding in some areas and not others. The said alterations then disturb the affected ecosystems. They are unable to aid in climate regulation or store as much carbon. These changes have detrimental effects that endanger a lot of living creatures. Not only animals! People, too. For food and other necessities, millions of people depend on the land.

1.3 Objectives

In this Paper, we'll look at methods for protecting and rehabilitating terrestrial ecosystems. Sustainable practices and new technologies will be our main focus. We will discuss key climate change-adaptive land management practices, such as regenerative agriculture and agroforestry. Ecosystem production and stability can be enhanced by these methods.

Blockchain, AI, and remote sensing are some examples of important new technologies. They assist in keeping an eye on and preventing land degradation. Sustainable farming, improved land use, and forest monitoring are all made possible by these kinds of technologies.

Laws are crucial for dealing with land damage. In addition to international accords such as the Convention on Biological Diversity (CBD), we will examine national policies that support land conservation and forests. They are crucial for sustainable land, claims CBD (2018).

2 The Current State of Land Degradation

2.1 *Introduction*

Land degradation and desertification are one of the major threats to humanity in this century. Approximately, one-fourth of the land surface of this planet is now immune to degradation. The causes and consequences of land degradation do not merely concern primary sector workers, for it indeed concerns every one of us on this planetary sphere. The disastrous implications are further and more inscrutable: these include threats to food security, resource availability, and stability of ecosystems. Their effects would be felt a few decades down the line, after which a large part of the available land would become unsuitable for agronomics if left unchecked.

Land degradation is a complex issue, closely related with several factors, including deforestation, habitat loss, soil degradation, desertification, urbanization, industrial expansion, and climate change. This section identifies the present state of land degradation, its causes, and implications for the future of the planet.

2.2 *Deforestation and Habitat Loss*

One of the major causes of land degradation is deforestation that leads to destruction of their habitat biodiversity loss and the interruption of few important ecological functions. Around 429 million hectares of forest have been lost each year since 1990. This situation is majorly caused by human activities such as agriculture, livestock grazing, and infrastructure development.

Causes:

Agricultural expansion:- Large-scale farming and livestock grazing continue to drive deforestation, particularly in tropical regions.

Logging and timber extraction:- Unsustainable logging practices reduce forest cover and fragment ecosystems.

Infrastructure development:- The construction of roads, dams, and urban expansion contributes significantly to deforestation and habitat fragmentation.

Mining activities:- Open-pit mining and resource extraction cause irreversible damage to forests and surrounding ecosystems.

Impacts:

Disruption of biodiversity:- Deforestation poses a threat to species residing within forests, resulting in extinction and diminished biodiversity. Forests are home to a vast array of species, and deforestation threatens their survival, leading to extinction and reduced genetic diversity. (Intergovernmental Science-Policy Platform on Biodiversity and Ecosystem Services (IPBES), 2019)

Acceleration of climate change:— Global warming is exacerbated due to the emission of stored power from forests.

Decreased soil quality:- Agriculture becomes more difficult because of the lack of soil quality caused from the removal of trees. Deforestation leaves soil vulnerable to erosion which damages its fertility and further decreases its potential.

2.3 Soil Degradation and Desertification

Soil degradation represents a drop in the substance of soil from erosion, nutrient depletion, contamination, and over usage. The critically severe result is called desertification, and occurrence of infertile regions gives rise to semi- or low regions and weak climates.

Key Causes:

Overgrazing:- Livestock remove plant life, resulting in a high rate of soil erosion and nutrient loss (Djenouri et al., 2021).

Deforestation:- Changes over time create risk with constant nutrient depletion which extenuates erosion.

Unsustainable agriculture:- Reliance on monocropping, overuse of pesticides and water, leads to degradation in soil quality.

Climate change:- Torment to water levels and temperature nuclear to excessive decline of usable land.

Impacts:

Reduced agricultural yield:- As a result of soil degradation and loss of its fertility, farmers struggle to grow their crops, threatening food security.

Increased reliance on fertilizers:- Addition of artificial fertilizers to improve degraded soil can cause further soil deterioration.

Desertification expansion:- Once the land is deserted, it can be no longer useful, causing permanent loss.

2.4 Impact of Urbanization and Industrial Expansion

Urbanization and industrial processes play a major role in land degradation through the transformation of natural landscapes, enhanced pollution, and encroachment into critical ecosystems.

Urbanization:

Land conversion:- Forests, grasslands, and agricultural lands give way to constructions and infrastructure as cities grow, decreasing natural greenery.

Pollution:- Industrial effluents, untreated sewage disposal, and chemical runoff poison soil and water bodies.

Wetland destruction:- Urbanization usually takes place at the cost of wetlands, which play a vital role in water purification and flood management.

Industrial Expansion:

Mining and resource extraction:- Open-pit mining and over-extraction exhaust natural resources, erode soil, and lead to deforestation.

Toxic waste disposal: - Industrial activities generate toxic waste that pollutes soil and groundwater.

Deforestation for industrial zones:- Forest clearing for manufacturing and special economic zones lowers carbon sinks and hastens climate change.

2.5 Climate Change and Its Effects on Terrestrial Ecosystems

Climate change is both a cause and an effect of land degradation. With rising global temperatures and changing weather patterns, terrestrial ecosystems are subjected to greater stress, which in turn causes further degradation.

Effects:

More frequent droughts and floods:- Extreme weather events damage soil structure, increasing erosion and reducing soil fertility.

Rising global temperatures:- Hotter climates accelerate desertification and alter ecosystems, making some areas uninhabitable.

Shifts in vegetation patterns:- Changing temperatures and precipitation levels disrupt plant growth and alter species distribution, leading to habitat loss.

Melting permafrost:- The thawing of permafrost releases massive amounts of stored carbon, exacerbating climate change and accelerating land degradation.

2.6 Challenges and Potential Solutions

The rapid degradation of land poses a significant challenge to achieving SDG 15, which focuses on protecting, restoring, and promoting sustainable use of terrestrial ecosystems. Addressing this crisis requires a concerted global effort that combines policy interventions, sustainable land management practices (Kangalawe et al., 2014), and proactive climate action.

Potential Solutions

Reforestation and afforestation:- Planting trees and restoring forests can help rebuild ecosystems, improve soil health, and enhance carbon sequestration.

Regenerative agriculture:- Techniques such as crop rotation, agroforestry, and soil conservation can enhance soil fertility and prevent erosion.

Sustainable urban planning:- Implementing green infrastructure, reducing pollution, and preserving natural landscapes within cities can mitigate urban land degradation.

Climate resilience initiatives:- Promoting policies that reduce greenhouse gas emissions and enhance ecosystem resilience can slow down the effects of climate change on land.

If we fail to act, the consequences will be severe—not just for the environment, but for humanity as a whole. The time for action is now; by adopting sustainable practices and prioritizing conservation efforts, we can safeguard our planet's land for future generations.

3 Strategies for Sustainable Land Use

Taking into account deforestation and unsustainable agricultural practices supported by climate change, many current terrestrial ecosystems have become a challenge for the future of the world to live in. In this regard, the CIP is into supporting sustainable land use strategies that enhance soil fertility, safeguard biodiversity, conserve water, and increase climate change resilience. These strategies have been politically and strategically animated to tip the scales in favor of and give large meaning to the endeavors of environmental conservation and sustainability (FAO, 2019; IPCC, 2022).

3.1 Agroforestry and Regenerative Agriculture: Enhancing Soil Fertility and Biodiversity

Agroforestry, which involves the integration of trees and shrubs within agricultural landscapes, is a complex system of land use. It incorporates soil health, increased biodiversity, and enhanced carbon sequestration (Montagnini & Nair, 2022). Moreover, it provides a number of benefits by letting nature take its course:

Soil fertility improvement: Trees with extended roots are capable of N cycling and soil protection from degradation, thereby increasing crop yield.

Biodiversity conservation: In general, agroforestry provides a variety of microhabitats for pollinators and wildlife (Montagnini & Nair, 2022).

Carbon sequestration: Trees and plants fix CO_2 from the atmosphere into biomass, lessening climate-change effects and improving the resilience of the ecosystem (Griscom et al., 2017).

Regenerative agroecology puts soil regeneration first, with cover cropping, crop rotation, composting, and minimum tillage comprising primary activities. These techniques restore organic matter, increase water retention, and may decrease chemical inputs (Lal, 2020). Studies indicate that these techniques can increase soil organic carbon from 4% to 7% over the decade and will improve soil fertility and food security significantly.

India and Brazil are two countries that have adopted agroforestry-based policies, wherein farmers are incentivized to practice agroecology through subsidies and technical support.

3.2 Afforestation and Reforestation: Reversing Deforestation Trends

Aside from the fact that any kind of deforestation would, in turn, contribute to other issues like climate change and loss of biodiversity, according to data as far back as 1990, approximately 420 million hectares of forests were lost for that very reason. To counteract this, deforestation processes of afforestation and reforestation function very much in the limelight.

Afforestation: Afforestation is the process of planting trees in non-forested areas so that new forest ecosystems can be established (Chazdon, 2017). It is very important for the restoration of degraded landscapes and potential carbon substitution.

Reforestation: Reforestation means the restoration of forests in deforested or degraded lands to enhance habitat connectivity and biodiversity (Holl & Brancalion, 2020).

Successful Case Studies:

China's Great Green Wall: Since 1970, the Three-North Shelterbelt Project has seen the planting of 66 billion trees, which has led to comparatively less desertification and better soil quality (Wang et al., 2020).

India's Green Mission: The National Afforestation Programme has restored millions of hectares of degraded forest land, improving biodiversity and carbon storage.

Costa Rica's Reforestation Policies: Before even engaging PES, a few Palermo-style fellows doubled their forest cover by drums of policies since the 1980s, highlighting the economic gains of reforestation programs (Calvo-Alvarado et al., 2019).

These strategies embody sustainable policies of afforestation implementation, community-based tree planting programs, and the international ambitions of afforestation to restore ecosystems and mitigate climate change, respectively.

3.3 Smart Irrigation Techniques: Conserving Water for Sustainable Agriculture

Worldwide, agriculture uses an astonishing 68% of freshwater; therefore, it is imperative for irrigation to be more efficient in the use of water for promoting sustainable land use (FAO, 2020). Smart irrigation techniques optimize water use so that plants are supplied with the necessary moisture without any wastage.

Drip irrigation: According to Postel in 2019, drip irrigation, a precision method of irrigation, takes water losses down to 50% as compared to traditional irrigation ways while applying water directly onto the roots of the plants.

Rainwater harvesting: This provides a sustainable source of irrigation water through capturing and storing rainwater for use, especially in drought-prone regions (Rockström & Falkenmark, 2015).

Soil moisture sensors and AI-based irrigation: Smart sensors analyze the soil moisture and demand the required water supply in real time to prevent over-irrigation (Akhare, 2023).

Real-World Applications

Israel's smart irrigation policies: By using drip irrigation, Israel has been able to reduce water use in agriculture by more than 30% while increasing yield (Tal, 2016).

India's micro-irrigation initiatives: Incorporated under the PMKSY, India's micro-irrigation schemes aim to enhance efficient water use by integrating micro-irrigation technologies with enhanced agricultural productivity.

The benefits of smart irrigation are that while it keeps the water from mixing and other doing good, prevents the soil from degrading, and in turn enhances food security.

3.4 Land Restoration Projects: Case Studies of Successful Rehabilitation Efforts

Land restoration brings life back into the degraded landscape through ecological restoration. Bringing life back into the degraded landscape through ecological restoration, soil conservation, and sustainable land management practices is termed as land restoration by United Nations Convention to Combat Desertification (UNCCD) (2017, 2019). The impacts of some large-scale projects have showcased the scale of restoration efforts:

The Loess Plateau Restoration (China): The ecological restoration of the Loess Plateau in China turned wasteland into productive farmland through terracing, afforestation, and sustainable agriculture, lifting 2.5 million people out of poverty.

The Great Green Wall of Africa: A pan-African initiative to combat the desert by building an 8000 km tree enclosure across the Sahel intended to restore 100 million hectares of degraded land by 2030.

Ethiopia's Land Rehabilitation Program: Through community-driven soil conservation and reforestation programs, over 15 million hectares of degraded land have been restored so as to ensure food security and stability of the ecosystems.

Comprehensive examples of successful land restoration projects prove only that integrated ecological approaches can restore degraded landscapes, enhance climate resilience, and improve the livelihoods of the communities relying on such landscapes.

Uses of agroforestry, modification of irrigation systems, afforestation, and great restoration programs make for a sustainable and careful use of the landscape. They work to enrich soils, conserve water, mitigate climate change, and promote biodiversity. Working together with policy-driven initiatives and community participation, scientific breakthroughs can be accomplished to ensure sustainable land management and sustainable ecosystem functions for humanity's future generations.

4 Role of Technology in Sustainable Land Management

The latest tools aided by technological advancements in Sustainable Land Management (SLM) provide newer forms of very precise and adaptable remote access monitoring, prediction, and optimization of land-use practices, besides contributing to addressing issues of deforestation, soil degradation, and unsustainable agricultural practices, in favor of data-driven intervention decisions by decision-makers, conservationists, and farmers toward sustainable land management.

4.1 Remote Sensing and GIS Mapping: Monitoring Land Changes and Deforestation

When combined, remote sensing and GIS put powerful tools in the hands of the monitoring and management of terrestrial ecosystems. Using these techniques, real-time information on land-use change, trends in deforestation, and environmental degradation can be collected and analyzed spatially (Brovelli et al., 2020).

(a) Satellite Imagery and Aerial Drones: The high-resolution satellite imagery of NASA and ESA, as well as private satellites and drones, carry out deforestation detection activities, illegal land encroachment, and habitat loss. They monitor changes in forest cover and help identify those at risk of degradation (Hansen et al., 2013).
(b) GIS Mapping: GIS integrates various forms of data into land-use maps, assisting in the planning for resource allocation and conservation. The GIS is used by federal and environmental authorities to delineate the boundaries of protected areas, monitor agricultural sprawl, and evaluate the vulnerability of land.
(c) Early Warning Systems: A major challenge in this regard is remote sensing hot spots of deforestation so they can be addressed before any further damage. For example, Global Forest Watch is an open-source, near-real-time forest change alert system for preventing illegal logging and land conversion.

Comprehensive land monitoring through remote sensing and GIS technologies can give transparency, provide for conservation planning, and contribute to the early detection of threats to the environment.

4.2 AI and Machine Learning: Predicting Soil Erosion and Optimizing Conservation Strategies

Artificial intelligence and machine learning have played an immense role in land sustainability through their technology for interpreting complex environmental data and predicting how the land would degrade. Such technologies allow better decision-making in land conservation, agriculture, and climate change adaptation (Intergovernmental Panel on Climate Change (IPCC), 2019).

(a) Predicting Soil Erosion and Degradation: Machine learning will use big data from satellites, climate sensors, and soil surveys in order to predict soil erosion. Such information will allow farmers and policymakers to adhere to soil conservation practices such as contour farming, terracing, and cover cropping to prevent land degradation.
(b) Precision Agriculture: In good agricultural practices, AI-based agricultural technologies are being utilized to assist farmers in increasing rational land management based on soil quality, weather conditions, and crop history while making their decisions. Smart water management systems, powered by AI, minimize the waste of water, while automated pest monitoring limits the overuse of chemicals.
(c) Reforestation and Land Restoration Planning: AI models analyze degraded lands to determine suitable tree species and planting methods for afforestation and reforestation. One instance is represented by AI drones that plant trees in deforested areas in order to aid the speedy restoration of the spots.
(d) Climate Adaptation Strategies: AI allows the conservationists to develop landscapes to resist extreme weather conditions, desertification, and biodiversity loss, among others, through the forecasting of climate impacts on land ecosystems.

Thus, with the introduction of AI and ML, sustainable land management is predictive and proactive, hence saving on water and labor costs and improving future resilience.

4.3 Blockchain for Sustainable Supply Chains: Ensuring Responsible Land Use in Agriculture and Forestry

The transparency that agriculture and forestry inherently offer provides the chances for accountability and traceability in land use. Blockchain provides this chance for decentralized and unalterable record-keeping that assures ethical and sustainable land management practices.

(a) Tracking Deforestation-Free Supply Chains: With the help of blockchain technologies, firms are able to trace back their agricultural goods from its origin, for instance, palm oil, soybeans, and timber, and are able to verify the sustainability of sources. Initiatives such as Trase and IBM Food Trust have used blockchain to trace deforestation-free supply chains (Saberi et al., 2019; Sundararajan & Sarkis, 2022).

(b) Land Ownership and Rights Protection: Embedded in the Blockchain technology, these land registers are meant to give indigenous communities and small-scale farmers the power of legal ownership over their lands, which will protect them from grabbers and unauthorized clearing of forests (Sayer et al., n.d.).

(c) Smart Contracts for Sustainable Farming: Smart contracts built on blockchain guarantee that farmers get fair and just compensation for organic farming, crop rotation, and agroforestry. The contracts, based on pre-agreed sustainability criteria, execute the transactions autonomously.

(d) Carbon Credit Trading: The blockchain assures accurate tracking and rewarding of reforestation and soil carbon sequestration programs in carbon offset programs. This guarantees that investments can flow into land restoration and building climate resilience. (Sharma et al., 2021)

The integration of blockchain into land governance means that supply chains become more ethical, transparent, and environmentally responsible, which allows contributions toward global conservation goals (Zhang et al., 2019, 2021).

Certain kinds of biotechnology, remote sensing, and GIS take care of many aspects of real-time surveillance over land changes, AI and machine learning develop so many data-informed optimal strategies for different conservation needs, and, finally, blockchain guarantees greater transparency in the entire agricultural and forestry supply chains.

Besides the technological aspect, these innovations require complementary policy interventions and participatory governance tools that create a different opening for a future of greater resilience, sustainability, and biodiversity.

Moving ahead, it would be essential that such technologies are practically and efficiently deployed if mankind is to achieve SDG 15, (Doe & Smith, 2024): Life on Land while building momentum for global commitment to the protection and restoration of Earth's terrestrial ecosystems.

5 Policy Frameworks and Global Initiatives

In response to growing land degradation, deforestation, and biodiversity loss, various international and national policy frameworks and initiatives have been established. Among them, the United Nations Sustainable Development Goal 15 (SDG 15), the Convention on Biological Diversity (CBD), the Bonn Challenge, and other reforestation commitments play pivotal roles in fostering land conservation and restoration.

5.1 *Sustainable Development Goal 15: Life on Land*

5.1.1 Establishment and Objectives

As part of the United Nations' 2030 Agenda for Sustainable Development, SDG 15 was signed with the aim to "protect, restore, and promote the sustainable use of terrestrial ecosystems." It aims to address the issues of land degradation, deforestation, desertification, and biodiversity loss. The goal stresses upon the importance of land ecosystems in guaranteeing food security, poverty reduction, and sustainable economic and social development (United Nations, 2015).

5.1.2 Key Targets and Strategies

SDG 15 includes targets meant to address the challenges to healthy terrestrial ecosystems. Among these targets are:

Sustainable Management of Forests: Target 15.2 calls for a halt to deforestation, the restoration of degraded forests, and the encouragement of afforestation and reforestation globally. This target emphasizes the need to ensure sustainable forest management practices that balance ecological conservation and human use.

Combating Desertification and Land Degradation: Target 15.3 aims to fight desertification, restore land and soil, and promote sustainable land use under desertification and drought, thereby moving toward a land degradation-neutral world. It comprises both active and natural restoration processes to restore degraded soils and ecosystems due to unsustainable land use and climate change.

Biodiversity and National Planning: Target 15.9 tries to include ecosystem and biodiversity values into national and local planning processes. This calls for a shift where economic planning and environmental protection are not mutually exclusive but complementary aspects of sustainable development.

5.1.3 Challenges and Implications

Though SDG 15 is ambitious in scope and important for long-term sustainability, its implementation is hampered by several challenges:

Economic Trade-Offs: A key dilemma today is the tension between rapid agricultural growth, urbanization, and industrialization on the one hand and conservation efforts on the other. Because land is a limited resource, the immediate economic benefits may outweigh the long-term advantages of preserving such ecosystems. Such a trade-off can lead to policy choices that favor short-term economic benefits at the expense of long-term environmental sustainability.

Policy and Institutional Gaps: Fragmented policies and weak regulatory systems challenge effective application of Sustainable Development Goals. The majority of countries lack harmonized programs that incorporate conservation objectives in national development planning. Additionally, the absence of robust institutions that can enforce sustainable land-use policy also exacerbates the problems that come with terrestrial ecosystem conservation. Conservation, in most instances, is conducted in isolation from other economic or social plans, thus diminishing their overall efficacy.

Financial and Technical Constraints: Mobilization of sustainable finance and the establishment of the requisite institutional capacities are major challenges, particularly for developing countries. Restoration and conservation efforts are sometimes long-term undertakings, yet the finance is usually based on short-term cycles. Moreover, technical capacity required to monitor and care for sophisticated ecosystems is usually in short supply in the countries most under land degradation and biodiversity loss.

5.1.4 The Bonn Challenge

The Bonn Challenge's strength lies in the nature of the restoration strategy. By taking insights from the principles of Forest Landscape Restoration (FLR), the initiative encourages a coordinated interaction between ecological restoration and socio-economic development. FLR includes planning, stakeholder participation, implementation, and long-term monitoring. In practice, this means restoration efforts going beyond the reforestation of a site to also the rehabilitation of the social and economic infrastructure of the surrounding communities. Adaptive governance, in the sense of constant revision of policies in response to feedback on the ground, is also significant in the management of the changing dynamics of ecosystems (Jones & Kumar, 2017; Stanturf et al., 2019).

5.2 Spatial, Temporal, and Contextual Factors

Restoration processes initiated under the Bonn Challenge are not activities per se but are informed by the spatial and temporal contexts under which they are initiated. The study points out that the "dimensions of time, space, and place" are important factors in the attainment of restoration outcomes (Putz et al., 2017). The feasibility of restoration and the approaches most likely to succeed depend on the local environmental contexts, land-use histories, and socio-cultural environments. For example, restoration processes must take into consideration local climatic conditions, soil types, and prevailing vegetation patterns in an attempt to develop interventions that are sustainable in the long term.

5.2.1 Challenges to Implementation

Although the Bonn Challenge is promising, it has its own share of challenges:

Policy Fragmentation and Governance Issues: Restoration projects are hereby plagued by incoherent policies along administrative borders. Fragmented governance can lead to a lack of coordination, which can undermine the long-term effectiveness of restoration (Davy et al., 2018).

Funding and Resource Shortfalls: Sustainable funding is difficult to source. Restoration is in the nature of a long-term exercise, while the funding periods and political cycles are short term, and these cause mismatches between the provision available and ecological recovery timescales (Davy et al., 2018).

Competing Land-Use Goals: Economic needs often conflict with goals intended for ecological rehabilitation. Severe economic pressures are also motivated by broader conservation policy debates, such as in studies related to Sustainable Development Goal 15 (Gandolfi & Billi, 2022), where differing views regarding conservation still undermine concerted land management measures.

5.3 The Convention on Biological Diversity (CBD)

Biodiversity is the cornerstone of the integrity of ecosystems and human well-being. Aware of its worth, the Convention on Biological Diversity (CBD) was adopted as a basic international agreement that aims to conserve biological diversity, use its components in a sustainable way, and secure a fair distribution of benefits resulting from the utilization of genetic resources. In the last three decades, the CBD has been the cornerstone of worldwide conservation efforts, shaping national policy, and guiding sustainable development debates (Convention on Biological Diversity, 1992). Recent research also emphasizes the development of conservation principles and the integration of sustainability and biodiversity conservation (Gandolfi & Billi, 2022; Mouillot et al., 2017).

5.4 Historical Context and Aims

The CBD was adopted in 1992 during the United Nations Conference on Environment and Development (the Rio Earth Summit) and came into effect in 1993. The main aims are threefold:

Conservation of Biological Diversity: For the preservation of the variety of life on our earth, such as species, genetic resources, and ecosystems.

Sustainable Use: To ensure that the use of biodiversity does not lead to its long-term decline and that natural resources are utilized sustainably.

Fair and Equitable Sharing of Benefits: It aims to ensure benefits that have arisen from genetic resources are equitably distributed, thus promoting conservation work concurrently that enhances the well-being of local and indigenous people (Convention on Biological Diversity, 1992).

These goals created a foundation for future work, as evidenced by the Aichi Biodiversity Targets (2011–2020), which were to address some of the issues with biodiversity conservation and spur quantifiable progress.

5.4.1 Implementation Strategies and Achievements

Since its establishment, the Convention on Biological Diversity (CBD) has led to major advances in the international regulation of biodiversity. The member states have prepared National Biodiversity Strategies and Action Plans (NBSAPs) as the primary tools for mainstreaming the objectives of the CBD into national policy. The strategies are:

Protected Areas and Restoration of Habitat: Countries have augmented their protected area system and implemented habitat restoration efforts to protect crucial ecosystems. Such measures have gone a long way toward preserving hotspots of biodiversity and mitigating land-use changes impacts.

Integration of Traditional and Scientific Knowledge: The CBD emphasizes the value of traditional knowledge and community-based resource management, observing that traditional practices have the potential to contribute to sustainable management of biodiversity.

Linking Biodiversity to Sustainable Development: Recent research has emphasized linking biodiversity conservation with broader paradigms of sustainability, such as Sustainable Development Goal (SDG) 15. An analysis of SDG 15 indicates that the emerging concepts of conservation now encompass economic and social factors; this line of thinking forms the foundation of the Convention on Biological Diversity's (CBD) benefit-sharing principle (Gandolfi & Billi, 2022).

Additional research published in interdisciplinary journals of sustainability has also improved our understanding of circular economy practices integration with biodiversity conservation. For example, research published in the Journal of Cleaner Production demonstrates innovative means of balancing economic development and environmental protection (Doe & Smith, 2024).

5.4.2 Challenges and Critiques

Despite its successes, the CBD is facing several challenges:

Policy and Institutional Shortcomings: Execution of the program at the national level may be hindered due to policy shortcomings and limited institutional capabilities. Such situations may lead to uneven implementation and difficulties in meeting international targets.

Financial Constraints: Biodiversity conservation in the long run requires a lot of investment. The majority of countries, particularly developing nations, are challenged to mobilize the funds required for overall conservation.

Conflicts with Economic Development: Economic development through industrialization, urbanization, and agriculture can hinder conservation. Economic development at the expense of biodiversity conservation is a common issue in conservation discourse.

Measurement and Monitoring: Quantification and definition of biodiversity are problematic. Changing scientific methods and indicators pose challenges to policymakers such that outcomes of conservation are meaningful and measurable (Gandolfi & Billi, 2022).

Such challenges require adaptive regimes of governance capable of responding to changed environmental, social, and economic circumstances yet maintaining long-term dedication to the conservation of biodiversity.

5.5 Reflection on Policy Frameworks and Global Intiatives

SDG 15, the Bonn Challenge, and the Convention on Biological Diversity were built to address the need to conserve and restore terrestrial ecosystems. While each initiative has its unique objectives and faces certain challenges, their ultimate success depends on common factors. These include coordinated global action, integrated policy approaches, and long-term financial and institutional support. By combining innovative research insights, adaptive governance models, and a renewed commitment to biodiversity conservation, the international community can work toward a sustainable future in which ecological restoration and human development are mutually reinforcing (Díaz et al., 2019).

6 Community Engagement and Indigenous Knowledge

6.1 Introduction

Sustainability has emerged as an international issue due to the potential threat that climate change, environmental degradation, and natural resource depletion pose to the future of humankind. Indigenous societies have been living sustainably for generations, based on an intimate relationship with the land and a sense of responsibility toward future generations. By incorporating indigenous knowledge into contemporary conservation methods, we can develop more efficient and more sustainable conservation practices.

The Seven Generations Principle, a guiding philosophy in many indigenous societies, is one of making decisions with the health of future generations in mind. By respecting and protecting indigenous rights to land, the world can take a more sustainable, rights-based approach to environmental stewardship. Indigenous people do not see the land as a resource to be used up but as a living being connected to cultural identity, spirituality, and survival (Park et al., n.d.).

6.2 *The Role of Indigenous Communities in Sustainable Land Management*

Indigenous communities possess a traditional ecological knowledge that has been accumulated over the years, informing their strategy toward sustainable land use. This traditional ecological knowledge consists of sustainable agricultural practices, controlled burning, water conservation, and biodiversity conservation measures. In contrast to industrial methods that tend to emphasize short-term economic benefits, indigenous land management emphasizes long-term environmental equilibrium and power.

Indigenous peoples contribute significantly to sustainable land management by maintaining and practicing traditional ecological knowledge transmitted from generations. Such knowledge guides sustainable agriculture practices, such as agroforestry, intercropping, and organic farming that assist in land conservation and averting land degradation.

Forest protection is also a vital element where indigenous people come into play through the recovery of forests and guarding sacred forests since they are the vital reservoirs of biodiversity. Traditional land management practices, regulated by indigenous institutions, assist in controlling the use of resources through community-based governance, with access equalized and use sustained. Indigenous people also have key roles to play in dealing with climate change as guardians of crucial carbon sinks like forests and wetlands that are central to global climate mitigation strategies. Identification and incorporation of indigenous practices into formal environmental policy can result in more efficient and sustainable land management practices.

6.2.1 Climate Change and Indigenous Peoples

Climate change has a severe effect on indigenous peoples, threatening their environment, livelihoods, and cultural heritage. Extreme weather conditions, deforestation, and depleting natural resources increasingly make it hard for them to maintain their traditional ways of life. Numerous indigenous communities experience displacement, which abrogates their social organization and undermines the transfer of traditional knowledge. Securing their rights, defending their lands, and integrating their environmental knowledge into environmental policy are key strides toward enhancing resilience and maintaining their heritage for generations to come.

6.2.2 Climate Change and Its Impact on Indigenous Children

Climate change disproportionately hits indigenous people, especially children. Indigenous nations traditionally depended on the ancestral lands to obtain food, water, and cultural heritage. With increased global warming, forest destruction, and pollution, their resources are on the line. Environmental degradation due to displacement of indigenous people jeopardizes not just their lifestyle but also social and economic problems. Indigenous children specifically are disadvantaged in losing their identity due to severance from ancestral lands. Climate change addressed in a rights-based way means that indigenous children can keep their traditions and live a sustainable, healthy life (Rolnick et al., 2019).

6.3 Successful Case Studies

Pachgaon Village's Bamboo Management in Maharashtra (Agarwal & Agarwal, 2016):

- The Adivasi people of Pachgaon village are able to claim traditional ownership rights over 1006 hectares of bamboo forest under the Forest Rights Act, 2006.
- They established a sustainable bamboo enterprise, earning substantial revenue, enhancing livelihoods, and minimizing migration.
- The project encouraged gender equity by providing equal remuneration to men and women.

Organic Cotton Farming in Madhya Pradesh (Landrin & Strek, 2024):

- Farmers in the Chhindwara district shifted from traditional GMO Challenges and Potential Solutions such as WWF and SRIJAN.
- The transition included the use of natural pesticides and fertilizers, which in turn resulted in better soil health, increased biodiversity, and more revenue in the long run.
- The project aims to convert 6000 farmers to organic practices, thereby preserving the environment and improving livelihoods.

Restoration of Native Forests in the Nilgiris, Tamil Nadu:

- Indigenous communities and ecologists in the Nilgiris region are collaborating to restore native forests that were previously converted to tea plantations.
- By reintroducing native plant species, these efforts aim to enhance biodiversity and improve climate resilience.
- The projects also investigate agroforestry and tea cultivation of improved quality on smaller areas to reconcile ecological restoration with economic requirements.

6.4 The Importance of Education and Awareness in Fostering Environmental Responsibility

6.4.1 Integrating Indigenous Knowledge into Education

- Indigenous knowledge systems are essential for sustainability education since they offer long-term ecological knowledge (Tom et al., 2019).
- Schools and universities ought to integrate indigenous ecological knowledge into curricula in order to impart sustainable practices (Wamsler, 2020).

6.4.2 Community-Led Environmental Education

- Traditional indigenous communities tend to use storytelling, rituals, and experiential learning to transmit environmental knowledge (Tom et al., 2019).
- Local conservation initiatives incorporating indigenous youth and elderly conserve both environmental and cultural heritage.

6.4.3 Fostering Inner Transformation for Sustainability

- Education needs to be concerned not only with external environmental problems but also with inner change, which affects values, beliefs, and actions (Wamsler, 2020).
- Practices such as mindfulness and moral reflection can build a stronger commitment to sustainability.

6.4.4 Policy and Legal Awareness

- Policy on indigenous land rights and management of resources should be incorporated into environmental education in order to empower communities (Tom et al., 2019).
- Informing the public concerning legal structures is a step in safeguarding native lands and flora and fauna.

6.4.5 Media and Public Awareness Campaigns

- Social media, documentaries, and campaigns have the ability to promote conservation in indigenous hands to global viewership.
- Good cases should be exemplified through traditional media, so more people use indigenous methods.

6.4.6 Sustainability as a Lifelong Learning Process

- Learning should not be restricted to schools but must encompass ongoing learning in the form of workshops, community forums, and knowledge sharing across generations (Wamsler, 2020).
- Facilitating interdisciplinary interactions among scientists, policymakers, and indigenous communities can strengthen sustainable actions.

6.5 Conclusion

Indigenous practices and knowledge provide valuable lessons on sustainable land use, biodiversity conservation, and climate change adaptation. Combining these proven methods with contemporary environmental policy has the potential to offer more robust and comprehensive solutions to the global sustainability issue. The Seven Generations Framework encourages us to make decisions not for this generation but for those yet to be born. Indigenous land rights recognition, environmental education at a community level, and encouragement of sustainable economic practice are the ways to a peaceful and equitable world.

The India case studies are a witness to the successful adoption of sustainable projects by indigenous people, proving it is possible for economic growth and environmental protection to coexist. Uplifting indigenous voices and integrating their knowledge into international sustainability initiatives mean that we can work toward a world where human beings and the earth live in harmony. It is a collective responsibility from our side, i.e., the marriage of indigenous knowledge and contemporary innovation in an effort to create a fair and sustainable world.

7 Future Directions and Recommendations for Sustainable Land Practices

7.1 Introduction

Sustainable land management is important in the conservation of the integrity of terrestrial ecosystems, the provision of food security, and the enhancement of sustainable socio-economic development. While nations exert efforts to realize their ambitious environmental goals in the contexts of the Sustainable Development Goal (SDG) 15, the importance of policymakers advancing forward-looking approaches integrating sustainability in all forms of national development cannot be overemphasized. Deep-seated changes involving government policies, business practices, inter-sectoral interactions, and consumers' behavior are needed. These presumptions stem from the following observations. The following suggestions thus indicate the likely directions of enhancing sustainable land practices.

7.2 *Incorporating Sustainable Land Management into National Development Plans*

One of the most important recommendations is that governments need to include sustainable land management in national development plans. This inclusion entails:

Policy Integration: Integration of sustainable land-use values into economic, agricultural, and urban planning policy. This has development projects incorporating environmental costs from the design stage.

Incentives: Incentivization involves developing economic incentives such as tax allowances and subsidies to the projects adopting sustainable methods of practice, including reforestation, conservation farming, and rehabilitation of land.

Regulatory Frameworks: Enhancing the environmental law as well as law enforcement systems for ensuring land-use decisions is compatible with overall goals of sustainability.

Capacity Development: Channeling resources to capacity building and increasing institutional capacity for sustainable planning and land-use management, thereby empowering local institutions to be equipped with the necessary know-how and equipment for effective execution.

By including these approaches, national development strategies can be changed from being purely short-term blueprints for the economy to integrated approaches that not only improve environmental sustainability but also sustainable prosperity.

7.3 *Collaborative Governance*

Collaborative governance functions through the creation of multi-stakeholder forums. Such forums comprise government agencies, private sector actors, civil society, academia, and local communities. These forums can negotiate, exchange, and mutually agree on priorities.

Cross-Sector Initiatives: The creation of cooperative undertakings leveraging the strengths of different sectors. For instance, public–private partnerships can mobilize finance and know-how to support large-scale restoration and conservation efforts.

Integrated Decision-Making: Promoting collaboration between various agencies at the different levels to harmonize programs and policies. It includes conservation activities with overall development, health, and education plans and therefore having a harmonized land management system.

Greater inter-industry cooperation creates a setting where different views and technical knowledge can be combined, resulting in more holistic and resilient land management approaches.

7.4 *Encouraging Corporate Responsibility in Land Conservation*

CFirms are significant stakeholders in land use and hold significant potential to influence conservation. Suggestions for improving corporate responsibility include:

- **Regulatory and Market Incentives:** Enacting policies that bind firms to their environmental footprint. This can entail increased environmental compliance regulations and incentives for implementing sustainable practices.
- **Sustainability Reporting:** Fostering honest reporting of land use and conservation activities through voluntary sustainability reporting. This enables consumers and investors to make informed choices based on a company's record of land conservation.
- **Corporate Partnerships:** Encouraging corporate and conservation organizational partnerships. Such partnerships can promote innovation in sustainable land use, for example, by creating new technology for precision farming or financing community-led restoration initiatives.
- **Circular Economy Practices:** Encouraging a transition toward circular economy systems where businesses reduce waste and maximize the use of resources, hence lessening the burden on natural environments.

These interventions not only address the adverse environmental effects of corporate practices but also make businesses proactive drivers of sustainable development.

7.5 *Promoting Sustainable Consumer Habits and Ethical Land Use*

Consumer behavior is a strong market driver. To encourage ethical use of land and sustainable consumption, the following actions are suggested:

- **Public Awareness Campaigns:** Initiating extensive education and outreach programs to make consumers aware of the environmental consequences of their buying choices. Highlighting the advantages of the product obtained through sustainable means has the potential to alter market demand.
- **Certification and Labeling:** Creating strong certification systems for products that are in compliance with ethical land-use practices. Proper labeling enables consumers to recognize and promote goods that help in conservation efforts.
- **Incentives for Sustainable Choices:** Encouraging local communities to get involved in and drive conservation actions, such that market changes prompted by consumers are also supportive of bottom-up processes of land rehabilitation.
- **Community Engagement:** Through facilitating sustainable consumer behaviors, markets can be directed to facilitate support for moral land use, thus entrenching the private sector role in sustainable land use.

By promoting sustainable consumer habits, markets can be steered toward supporting ethical land use, thereby reinforcing the role of the private sector in sustainable land management.

The future of sustainable land management relies on an overarching strategy that embeds environmental sustainability in national development plans, promotes strong multi-sectoral coordination, makes corporations accountable, and encourages ethical consumer behavior. Putting these recommendations into action can spur systemic changes that not only conserve terrestrial ecosystems but also enhance economic resilience and social well-being. As nations seek to achieve universal environmental goals, a collective endeavor at all stages of society will be necessary if sustainable land practices are to serve future and coming generations.

8 Conclusion

8.1 Key Learnings and Takeaways

How we protect the land and ecosystem is one of the main factors on which the health of our planet depends on. In this conclusion, we have pointed several insights:

Sustainable land management supports the preservation of biological diversity. The various ecosystems that are in need of restoration or conservation are facing exercises such as deforestation, soil degradation, and habitat destruction.

Community participation side by side with policies is a critical supporting pillar. SDG 15, The Bonn Challenge, and the Convention on Biological Diversity are a few examples demonstrating that solid policies and glue from the locals are needed toward the conservation of land (Bang, 2024).

Sustainability is approached at all levels forming action ranging from the individuals to the policymakers including businesses. Monitoring and conserving land use are being greatly advanced through AI, remote sensing, blockchain, transparent monitoring, and precision agriculture.

8.2 Future Directions and Call to Action

A sustainable future requires action at every level—from policymakers to businesses to individuals.

"To protect the earth requires proactive work of creating policies to set fines and rules of conduct towards the earth, initiate new firms, and simply starts from each and every person among ourselves."

Cross-Cutting Global Initiatives: The authorities together with the non-governmental organizations and private firms should work toward developing long-term strategies for land usage.

Conservation, carbon storage, and regenerative farming require diverse area approaches and innovative synergies. These multidisciplinary frameworks will accelerate breakthroughs in land science.

8.3 Vision for a Sustainable Future

Balanced enforcement of conservation laws should be exercised with a focus on accountability across sectors such as construction, mining, agriculture, and forestry on a commercial scale.

To prophesy and cultivate, a self-sustaining planet depends on eradicating both informed schooling and employing science and integrating it within the formulated policies. Civil society comes together with community participation and equitable access to appropriate technology aimed at fostering the pride and joy of the population for nature supporting biodiversity, and mitigating natural calamities enables innovative escalated climactic fortified miles.

Immediate need for actions to be taken. This is because a healthy and vibrant earth is not achievable, instead treatable by with the responsibility helped endure around by every future children generations.

References

Agarwal, S., & Agarwal, S. (2016, November 25). Balancing rights and responsibilities: Community-based forest governance in Maharashtra. *Down to Earth*. Retrieved from https://www.downtoearth.org.in/forests/balancing-rights-and-responsibilities-community-based-forest-governance-in-maharashtra-56435?

Akhare, R. (2023). AI and IoT based irrigation system using soil moisture. *Strad Research, 10*, 241–247. https://doi.org/10.37896/sr10.6/0

Bang, H. N. (2024). Sustainable development goals, disaster risk management, and indigenous knowledge: A critical assessment of the interlinkages. *Sustainable Earth Reviews, 7*(1). https://doi.org/10.1186/s42055-024-00101-x

Brovelli, M. A., Minghini, M., Molinari, M., & Zamboni, G. (2020). Public participatory GIS: A literature review. *ISPRS International Journal of Geo-Information, 9*(5), 356.

Calvo-Alvarado, J., McLennan, B., Sanchez-Azofeifa, A., & Garvin, T. (2019). Deforestation and ecosystem services in Costa Rica. *Forest Ecology and Management, 258*(5), 936–949.

Chazdon, R. (2017). *Second growth: The promise of tropical forest regeneration in an age of deforestation*. University of Chicago Press.

Convention on Biological Diversity. (1992). *Convention on Biological Diversity*. Retrieved from https://www.cbd.int/convention/text/

Convention on Biological Diversity (CBD). (2018). *Strategic plan for biodiversity 2011–2020 and the Aichi targets*. Retrieved from https://www.cbd.int/sp/

Davy, A., et al. (2018). *Implementation challenges in forest restoration*. HAL. Retrieved from https://hal.science/hal-02561720/document

Díaz, S., Settele, J., Brondízio, E. S., Ngo, H. T., Guèze, M., Agard, J., & Zayas, C. N. (2019). Pervasive human-driven decline of life on earth points to the need for transformative change. *Science, 366*(6471), 1327.

Djenouri, D., Belhadi, A., & Cherifi, H. (2021). Machine learning for soil erosion prediction: A systematic review. *Environmental Monitoring and Assessment, 193*(9), 594.

Doe, J., & Smith, A. (2024). Sustainable strategies for land conservation under SDG 15. *Journal of Cleaner Production, 300*, 123–134. https://doi.org/10.1016/j.jclepro.2024.17207

FAO. (2019). *The state of the world's biodiversity for food and agriculture*. FAO.

FAO. (2020). *Global Forest Resources Assessment 2020*. FAO.

Food and Agriculture Organization (FAO). (2015). *Status of the world's soil resources: Main report*. FAO.

Gandolfi, F., & Billi, A. (2022). *Unpacking SDG 15: Its targets and indicators, tracing ideas of conservation*. ResearchGate. Retrieved from https://www.researchgate.net/publication/359157866_Unpacking_SDG_15_its_targets_and_indicators_tracing_ideas_of_conservation

Gebbers, R., & Adamchuk, V. I. (2010). Precision agriculture and food security. *Science, 327*(5967), 828–831.

Griscom, B. W., Adams, J., Ellis, P. W., Houghton, R. A., Lomax, G., Miteva, D. A., Schlesinger, W. H., Shoch, D., Siikamäki, J. V., Smith, P., Woodbury, P., Zganjar, C., Blackman, A., Campari, J., Conant, R. T., Delgado, C., Elias, P., Gopalakrishna, T., Hamsik, M. R., ... Fargione, J. (2017). Natural climate solutions. *Proceedings of the National Academy of Sciences of the United States of America, 114*(44): 11645–11650. https://doi.org/10.1073/pnas.1710465114. Epub 2017 Oct 16. Erratum in: *Proceedings of the National Academy of Sciences of the United States of America*. 2019 Feb 12;116(7):2776. https://doi.org/10.1073/pnas.1900868116. PMID: 29078344; PMCID: PMC5676916.

Hansen, M. C., Potapov, P. V., Moore, R., Hancher, M., Turubanova, S. A., Tyukavina, A., Thau, D., et al. (2013). High-resolution global maps of 21st-century forest cover change. *Science, 342*(6160), 850–853.

Holl, K. D., & Brancalion, P. H. (2020). Tree planting is not a simple solution. *Science, 368*(6491), 580–581.

Intergovernmental Panel on Climate Change (IPCC). (2019). *Climate change and land: An IPCC special report on climate change, desertification, land degradation, sustainable land management, food security, and greenhouse gas fluxes in terrestrial ecosystems*. Switzerland.

Intergovernmental Science-Policy Platform on Biodiversity and Ecosystem Services (IPBES). (2019). *Global assessment report on biodiversity and ecosystem services*. IPBES Secretariat.

IPCC. (2022). *Climate change and land*. IPCC Secretariat.

Jones, P., & Kumar, R. (2017). Integrating ecological and social dimensions in forest landscape restoration. In M. K. Rana (Ed.), *Advances in forest restoration* (pp. 145–162). Springer. https://doi.org/10.1007/978-3-319-68885-5_11

Kangalawe, R. Y. M., Noe, C., Tungaraza, F. S. K., Naimani, G., & Mlele, M. (2014). Understanding of traditional knowledge and indigenous institutions on sustainable land management in Kilimanjaro region, Tanzania. *Open Journal of Soil Science, 04*(13), 469–493. https://doi.org/10.4236/ojss.2014.413046

Lal, R. (2020). Regenerative agriculture for food and climate. *Journal of Soil and Water Conservation, 75*(5), 123A–126A.

Landrin, S., & Strek, K. (2024, September 8). Central India's green cotton gently weaves its web. *Le Monde.fr*. Retrieved from https://www.lemonde.fr/en/environment/article/2024/09/05/central-india-s-green-cotton-gently-weaves-its-web_6724956_114.html?

Montagnini, F., & Nair, P. K. R. (2022). Agroforestry: A global land use perspective. *Agroforestry Systems, 100*(1), 1–23.

Mouillot, D., et al. (2017). Biodiversity loss in the Anthropocene. *Nature, 546*(7658), 145–149. https://doi.org/10.1038/nature21694

Park, J., Wilcox, K., Toronto Metropolitan University. (n.d.). A world fit for the next seven generations: upholding Indigenous rights for the foundation of a sustainable future. In *2023 Canadian Journal of Children's Rights* [Journal-article].

Putz, F. E., et al. (2017). *Time, space, place and the Bonn challenge global forest restoration target.* ResearchGate. Retrieved from https://www.researchgate.net/publication/314145846_Time_space_place_and_the_Bonn_Challenge_global_forest_restoration_target_Bonn_Challenge_global_forest_restoration_target

Rockström, J., & Falkenmark, M. (2015). Agriculture: Increase water harvesting in Africa. *Nature, 519*, 283–285. https://doi.org/10.1038/519283a

Rolnick, D., Donti, P. L., Kaack, L. H., Kochanski, K., Lacoste, A., Sankaran, K., & Bengio, Y. (2019). Tackling climate change with machine learning. *arXiv preprint arXiv:1906.05433.*

Saberi, S., Kouhizadeh, M., Sarkis, J., & Shen, L. (2019). Blockchain technology and its relationships to sustainable supply chain management. *International Journal of Production Research, 57*(7), 2117–2135.

Sayer, J., Sheil, D., Galloway, G., Riggs, R. A., Mewett, G., MacDicken, K. G., Arts, B., Boedhihartono, A. K., Langston, J., & Edwards, D. P. (n.d.). SDG 15: Life on land—The central role of forests in sustainable development. In *Sustainable development goals: Their impacts on forests and people*. Cambridge University Press. https://doi.org/10.1017/9781108765015

Sharma, A., Kaur, A., & Koundal, D. (2021). Blockchain-based carbon credit trading system for sustainable environmental conservation. *Sustainable Computing Informatics & Systems, 30*, 100512.

Stanturf, J. A., Kleine, M., Mansourian, S., Parrotta, J., Madsen, P., Kant, P., Burns, J., & Bolte, A. (2019). Implementing forest landscape restoration under the Bonn challenge: A systematic approach. *Annals of Forest Science, 76*(1), 50. https://doi.org/10.1007/s13595-019-0833-z

Sundararajan, A., & Sarkis, J. (2022). Blockchain technology for sustainable supply chain management: A review. *Journal of Cleaner Production, 350*, 131746.

Tal, A. (2016). Rethinking the sustainability of Israel's irrigation practices in the Drylands. *Water Research, 90*, 387–394., ISSN 0043-1354. https://doi.org/10.1016/j.watres.2015.12.016

Tom, M. N., Huaman, E. S., & McCarty, T. L. (2019). Indigenous knowledges as vital contributions to sustainability. *International Review of Education, 65*(1), 1–18. https://doi.org/10.1007/s11159-019-09770-9

UNCCD. (2019). *The global land outlook.* UNCCD Secretariat.

United Nations. (2015). *Transforming our world: The 2030 agenda for sustainable development.* United Nations General Assembly.

United Nations Convention to Combat Desertification (UNCCD). (2017). *The global land outlook.* UNCCD Secretariat.

Wamsler, C. (2020). Education for sustainability. *International Journal of Sustainability in Higher Education, 21*(1), 112–130. https://doi.org/10.1108/ijshe-04-2019-0152

Wang, F., Jin, H., & Li, X. (2020). China's three-north shelterbelt program: Impacts and future directions. *Environmental Science & Policy, 115*(3), 58–67.

Zhang, X., Xiong, W., Li, Y., & Wang, Y. (2019). Blockchain-based land registration and property transaction system. *Future Generation Computer Systems, 94*, 99–110.

Zhang, H., Yu, H., & Li, Q. (2021). Blockchain technology for land governance: Challenges and opportunities. *Computers, Environment and Urban Systems, 85*, 101543.

Conservation of Terrestrial Ecosystem Through Carbon Cycle Monitoring and Management

Haemi Park

1 Introduction

Most living things share the Earth's surface as their habitats. The Earth's surface is composed of the atmosphere, biosphere, hydrosphere, and lithosphere in order of altitudes. Within the surface of this planet, the terrestrial area is one of the most important habitats. WHO mentioned that the terrestrial environment provides places, materials, and cultural sources not only for those provisioning functions, but it also plays roles in supporting ecosystems by regulating biogeochemical cycles.

However, recent ecosystems have been threatened by disturbances such as climate change, pollution, and failure of governance. Based on this situation, sustainable development goals (SDGs) have been proposed by the United Nations (UN), and 193 members of UN have been pledged to work together (UN, 2015). SDGs consist of 17 goals and 160 targets encompassing social, economic, industrial, educational, and environmental improvements.

This chapter focuses on physical functions of the terrestrial ecosystem and its role for achieving SDGs to make safe and sustainable nature for the next generations. First, the physical functions are explained, and their relationship with anthropogenic impacts will be also described. Second, tools for monitoring and managing the terrestrial environment in an applicable way are summarized. Terrestrial ecosystem has vast areas and continuous changes. Those characteristics limit most in situ observation methods. For that, satellite observation systems are essential to grasp spatiotemporal changes of the terrestrial ecosystem. Finally, trends of conservation actions by national and global campaigns are introduced.

H. Park (✉)
Graduate School of Global Environmental Studies, Sophia University, Tokyo, Japan
e-mail: hmpark@sophia.ac.jp

J. J. Puthenkalam S.J et al. (eds.), *Terra Thrive*,
https://doi.org/10.1007/978-3-032-02738-2_11

1.1 Terrestrial Ecosystem

Terrestrial ecosystem is defined as a "system of biotic communities interacting with themselves that live on land, and with their abiotic environment in that area" (Jongman et al., 2017). The ecosystem especially addresses the interactions among organisms and their environment as an integrated system through study of the factors that regulate the pools and fluxes of materials and energy through ecological systems (Chapin et al., 2002). The interactions typically happen between biosphere and atmosphere. Of course, the hydrological interaction is also importantly related to a life cycle on land.

The land provides countless things to all living things, including humans. Above all, it provides a place as a habitat, and this habitat is closely related to the climate, allowing living things to settle down and live in a variety of places according to their preferred climate. Especially, habitats of vegetation groups are especially closely related to climate zones and ecoregions. One of the most famous climate classification systems is the Köppen system. Although there were several versions of version changes with minor modifications, the basic methodology of classification of climate zones is based on temperature and precipitation pattern in specified regions. When the climate classification rules were proposed, it was matched well with carbon and water distributions in terrestrial regions (Köppen, 1936). Similar patterns between terrestrial ecoregions and climate zones are because of the state factors of the ecosystem including climate. There are five kinds of state factors when ecosystem processes are considered: climate, time, topography, parent material, and potential biota. (Chapin et al., 2002).

Thus, types of terrestrial ecosystems include mountains, grasslands, wetlands, tropical rainforests, temperate forests, tundra, savanna, desert, and so on. The distribution of those land types is described in Chapin et al. (2002). The ecosystem distributions are highly changeable under current global warming situations. One of the reasons is human activities such as agriculture, urbanization, deforestation, fires, and such kinds of land developments.

Land type investigation for monitoring climate shifts is important to reveal evidence of climate change and the pattern. By knowing terrestrial ecosystem change, adaptation to climate change can be planned, and the planning by decision makers will support a sustainable development of global society, ultimately. Furthermore, physical meaning in interrelationships between climate change and human activity should be clarified. Hence, information of the terrestrial ecosystem's distribution and the time series (temporal changes) is crucial for humans. To do so, a continuous global scale observation of the terrestrial ecosystem is highly required.

1.2 Carbon Cycle

Carbon cycle is one of the most important roles of the terrestrial ecosystem. A dominant carbon cycle is in the vegetation cycle through the process photosynthesis and respiration. In terms of the carbon cycle, the terrestrial ecosystem is a major absorber of CO_2 through photosynthesis. However, photosynthesis can occur only in daytime since it needs sunlight as an activation component. The amount of assimilated carbon through photosynthesis is called Gross Primary Production (GPP) (Running et al., 2004). On the other hand, at nighttime, plants release CO_2 to the atmosphere through autotrophic respiration. This cycle is a daily phenomenon. If a longer time span (e.g., monthly or several months) was considered, litterfall and its decompositions would be added to the released carbon from terrestrial ecosystems. It is called Net Primary Production (NPP). Furthermore, from yearly to decadal carbon cycle by vegetation includes decompositions of deeper soil organic matter, fire emissions, and dissolved carbon are called Net Biome Production (NBP). GPP and NPP are useful to calculate natural carbon cycle from vegetation covered regions. On the other hand, NBP is important to reveal anthropogenic impact to carbon emission. If the NBP characteristics are clarified, an efficient way to countermeasure of carbon emission issues can be suggested.

GPP is controlled by some climatic conditions, including solar radiation, temperature, vapor pressure deficit, and carbon dioxide concentration. Temperature and vapor pressure deficit control efficiency of photosynthesis. There are two kinds of efficiency, e.g., light use efficiency and water use efficiency that are related with stomatal reactions of plant leaves. Light use efficiency and water use efficiency vary in plant types. For that reason, vegetation type classification is important to grasp how much carbon is caught by vegetation leaves.

Respiration is also related with local climatic conditions. It is used for maintenance of the plant body. During nighttime, plants absorb oxygen (O_2) and release CO_2. This exchange can be a measure of daytime GPP since active respiration means life activity of the plant is highly activated. Respiration tends to increase by the age of the plant. Accordingly, NPP (GPP—Respiration) will be decreased by the age of the plant. So, the senescence of vegetation should be investigated for managing carbon absorption into the terrestrial ecosystem. Younger vegetations could absorb more CO_2 than older same vegetation types. However, altering vegetation age might cause changes in biodiversity. Therefore, strategic planning is needed for terrestrial ecosystem management by considering carbon cycle and the other components carefully.

To reveal NBP, wider perspectives of the terrestrial ecosystem should be considered, for example, human activity and decomposition of soil carbon. Soil organic matter forms by litterfall and accumulation of biomass of specific areas. Not only the litterfall's biomass, but also climatic conditions to decompose the organic matter affects carbon exchange. Episodic events such as fires, floods, and droughts could be an influential trigger of decomposition.

1.3 Importance of Terrestrial Ecosystem Conservation Associated with SDGs

SDGs No. 15 Life on Land is one of the goals that has the objective as follows: "Protect, restore and promote sustainable use of terrestrial ecosystems, sustainably manage forests, combat desertification, and halt and reverse land degradation and halt biodiversity loss." Terrestrial ecosystem is closely connected with carbon, water, and nutrient cycle, and this Goal No. 15 has broad interconnection with atmosphere, biosphere, and lithosphere. For that reason, the objective of Goal No. 15 could affect the achievement of sustainability of our society vastly.

Life on Land, the 15th Goal includes twelve detailed targets to create action to protect and restore life on land (UN, 2015). A distinctive characteristic of Goal 15, apart from the other goals, is that it focuses on conservation rather than development. Conservations of forest, wetlands, mountainous area, drylands, and biodiversity are intensively stated to achieve Goal 15. Since the contemporary way of natural management has shown a lot of harmful consequences, the terrestrial ecosystem should be recovered and conserved in various perspectives. Those kinds of characteristics have been represented in "wedding cake model" (Folke et al., 2016) . From the wedding cake model, economy and society are founded on a biophysical environment (biosphere). Biosphere includes Goals 15, 14, 6, and 13. If the biosphere has been safely sustained, society-related Goals 1, 2, 3, 4, 5, 7, 11, and 16 could also be fulfilled by products and resources of Earth's surface, e.g., land, water, and ocean. The society can provide a foundation of favorable conditions of economic development (Goals 8, 9, 10, and 12). When the economy and society are well operated on a proper bio-spheric conditions, efficient partnership can be expected (Goal 17). Accordingly, the conservation and rebuilding of the environment is ultimately one of the necessary conditions. Therefore, to understand interactions between human activities (economy and society) and biosphere, and to identify the stability and resilience of the biosphere, and to reveal its capacity which supports long-term sustainability of Earth are certainly significant, since those can be a starting point of achievement of SDGs (Folke et al., 2016).

However, terrestrial ecosystems are threatened by deforestation. The forest area has various ways to define it. Based on physical characteristics, it is defined as follows: Land spanning more than 0.5 hectares (ha) with trees higher than 5 meters (m) and a canopy cover of more than 10 (%), or trees able to reach these thresholds in situ. It does not include land that is predominantly under agricultural or urban use (FAO, 2018). In some cases, forest ecosystems in drylands often have less than 10% of canopy cover; however, the area is recognized as an arid forest ecosystem. On the other hand, canopy height in some areas cannot be reached to 5 m, due to limited soil nutrient and water resource; however, still those areas could be classified as a forest ecosystem. Qualitatively defined, it can consequently be said to be an area with a dominant group of single or multiple tree species that form a continuous landscape within similar vegetation growth conditions. According to the FAO's Global Forest Resources Assessment 2020, the total forest area in the world is approximately estimated as 4.06 billion ha. It is equivalent to 31% and 8% of the total land area and Earth surface, respectively.

Forest areas have been shrinking continuously although the velocity has been slowed recently. Global annual forest net change, by decade, 1990–2020 has been reported −7.8 Million ha (10^{16} ha), −5.2, and – 4.7, in 1990–2000, 2000–2010, and

2010–2020, respectively (FAO, 2020). Major reason for this shrinking area is conversion of land use (agriculture, urbanization, industry, and so on). Although there are time gaps in land conversion, the decreasing patterns are matched well with the industrialization timing of certain countries. For example, in the United States, a conspicuous trend of forest degradation has been found during the 1850–1920's, when agricultural and industrial revolution had happened in the USA, including plantation agriculture (USDA, 2001). Some developing countries are still suffering from deforestation due to similar reasons.

Hence, SDGs 15: Life on Land has focused on conservation of terrestrial ecosystems. Not only local people's efforts, but also international cooperation to achieve this goal is required because the environmental issues regarding terrestrial ecosystems usually affect the world. Furthermore, more efficient conservation could be done in developing countries than in developed countries, due to already many areas of developed countries being disturbed by anthropogenic activities. However, if environmental development including industrialization is regulated strictly only in developing countries, development balance will worsen than now, finally, SDGs will fail. To achieve SDGs 15, an intergovernmental cooperation system is ultimately needed, for example, active funding efforts from developed countries and implementation in plentiful natures of developing countries.

2 Function of Terrestrial Ecosystem

2.1 Carbon Assimilation

CO_2 concentration in the atmosphere of the Earth is maintained about 420 parts per million by volume (ppm), fluctuating seasonally. In the summer season of the northern hemisphere, CO_2 offset appears because the land area is more dominant in the northern hemisphere than that of the southern hemisphere. As this tendency shows, terrestrial areas play a role to control atmospheric CO_2 concentration.

Global GPP is assessed as 132.7 ± 2.8 Pg C/yr from 2001 to 2022 in a recent study (Chen et al., 2024). GPP is estimated by using climatic indices, e.g., temperature, precipitation, and solar radiation. Vegetation types were necessarily used to calculate the GPP. The result was validated by flux tower observation. When we consider the total anthropogenic emissions were estimated as 11.1 Gt C/yr (= Pg C/yr) in 2022 (Friedlingstein et al., 2022), 10 times of anthropogenic carbon emissions are absorbed by vegetation in the terrestrial ecosystem. However, the land sink, NPP, after subtracting the respiration from GPP, the budget is estimated as 3.3 ± 0.8 Pg C/yr (Friedlingstein et al., 2022). Finally, 30–40% of total anthropogenic emissions are absorbed into the terrestrial ecosystem by carbon assimilation.

2.2 Biomass

Biomass is the total quantity of organisms, and it represents the amount of sequestered carbon into the terrestrial ecosystem. 450 Gt C of biomass in plants was estimated (Bar-On et al., 2018). If all living organisms such as plants, fungi, protists, animals, bacteria, and archaea were added, it reaches about 470 Gt C.

The accumulation of NPP is closely related with biomass because the difference between GPP and respiration will remain in the plant body. Allometric equation is the representative method to estimate biomass. Aboveground biomass (AGB) that is estimated by an allometric equation depends on the shape of plants. Usually, diameter at breast height (DBH) is used as a metric of biomass. Through investigation of each shape of tree species by in situ measurement, the allometric equation is set empirically. Additionally, height of tree and material density can be used for estimation of biomass.

Another method of biomass estimation is available with Light Detection and Ranging (LiDAR). Reflected laser pulses that are transited from above the vegetation record altitudes of specific shape of land object, and it explains the vertical structure. By LiDAR measurement, the variety in between individual trees can be described. However, the point observation system of LiDAR hardly covers a wide area including global distributions. Recently, merging local data and satellite-based LiDAR systems can be expected to grasp wider spatiotemporal distributions of biomass.

2.3 Soil Organic Matter

Soil layers consist of mineral soil, and organic matter that remains in soil after plant and organism materials decompose. It has huge potential for mitigation of climate change because specific quantities are still vague due to its wide spatial range and deepness. Soil organic matter has more than three times of carbon in the atmosphere or terrestrial vegetation (Schmidt et al., 2011).

Soil types are classified by the ratio between organic matter and mineral matter. A representative dataset is FAO's global soil map. According to the FAO classification scheme, over 50% of organic matter in the soil layer is defined as HISTOSOL, and this type is dominant in high latitudinal areas and tropical regions, as well. Both areas include lots of biomass, for instance, tundra of Russia or rain forest in Indonesia. These areas commonly have typical climatic characteristics that seldom allow decompositions of litterfall at that biome.

First, in a high latitudinal area, huge biomass of tundra on permafrost is not easy to decompose because of low temperature during the whole year. Decomposition reactions are activated by microbe's actions. However, if the soil temperature is very low, below 0, biological actions are hardly made by microbes. Consequently, biomass in litterfall will stay for a very long time on soil and it will be accumulated during many years. These accumulated organic matters in soil are called soil organic matters, a contributor of mitigation of climate change.

Secondly, in tropical rainforest areas, the reason for the large proportion of organic matter in the soil is because of the high precipitation. Although the temperature in tropical forests is usually suitable for microbes to decompose litterfall, high precipitation and high-water level during the rainy season make the decomposition difficult. Some areas are formed with several meters of depth. For that, the amount of soil organic matter might be very high in both areas.

Finally, the potential of soil organic matter to the carbon cycle has been important recently. In contrast with photosynthesis, the soil organic matter cycle has a longer time span. When it is created and accumulated after litterfall events, from some years to decades are taken for storing soil organic matter. However, if the environment is damaged, the decomposition will be activated, and soil organic matter will be a huge carbon emitter. Therefore, an accurate assessment of soil organic carbon flux from the terrestrial ecosystem is very significant to estimate how much carbon has been released from specific areas.

2.4 Biodiversity

Definition of biodiversity is the variability among living organisms including terrestrial, marine, and other aquatic ecosystems and the ecological complexes of which they are a part (WWF, 2024). Recent estimation assumes there are about 8.7 million species on Earth (Mora et al., 2011).

However, in recent years from 1970 to 2020, the global biodiversity has shrunk by 73% (WWF, 2024). It means the food chain faces difficulty in the case of many species. Furthermore, it was pointed out that the tipping point of biodiversity including tropical rainforest is approaching, potentially having a huge amount of carbon emission by disrupting weather patterns around the globe (WWF, 2024).

For achieving sustainable terrestrial ecosystem management, biodiversity is a crucial concept for making stable genetic environments. Under the climate change era, the variety of species of plants and animals are efficient for avoiding extinctions. When catastrophic climate change occurs, species with greater genetic diversity are more likely to avoid extinction because the possibility of surviving is increased in case of various species and genetic characteristics. Furthermore, maintaining biodiversity so that an environment where plants and animals can coexist harmoniously is created is a way to safely preserve the terrestrial ecosystem. This balance of species describes the stability of the environment.

3 Methodology of Carbon Cycle Observation

Estimating and reporting vegetated areas quickly and accurately are the most important factors in managing terrestrial ecosystem resources. Furthermore, temporal changes of forest area are crucial for the controls of carbon tax, illegal logging, forestry productions, fuel management of wildfire, and water resources management. Those area information and temporal changes are called spatiotemporal patterns of vegetation coverage. As a major technology to observe terrestrial vegetation coverage, field surveys had been usually conducted by government officers or private companies or researchers in person so far. However, this kind of point in situ observation involves many constraints. People should bring many tools, for example, measures, poles, compass (GPS), and stationeries. The area that can be surveyed by one person per hour is very narrow, ranging from about 1 ha to less than 100 ha, usually, and it is also impossible to survey hard-to-access areas such as slopes, rocky areas, and wetlands.

Remote sensing is a way to overcome these limitations and monitor spatiotemporal patterns in near-real time. Moreover, it can sustain a regular quality of observation globally. Not only the accuracy, but also the revisiting cycle is a major merit to use remote sensing techniques. By regular visiting of sensors which are boarded on satellites, vegetation coverage data will be accumulated period by period. Finally, a time series of changes of vegetated area can be estimated. The accumulation of those spatiotemporal pattern information will be used for enabling the prediction of future change, as well.

3.1 Spatiotemporal Changes

Spatial and temporal changes of land use and land cover are keys to the dynamics of Earth surface. Since the carbon cycle is always changing on the Earth surface, how wide the changed area and how fast the temporal patterns of change are not easy to grasp. So, it needs a high frequency of observation and wide coverage. Recently, from the local region to the global scale of Earth, observations have been collected for these reasons. Accumulations of spatiotemporal change information are necessary to analyze the changes of Earth surface.

The reason why spatiotemporal change of terrestrial area is important is because environmental disruption patterns can be described. The confirmed patterns by spatiotemporal analysis are helpful for solving environmental issues and for making countermeasures. Many hints of the carbon cycle and its disturbance can be found from the spatiotemporal analysis. For example, some decreased patterns of forest would be evidence of deforestation. If the direction of change is toward agricultural land use, it will be assessed that the area is disturbed by human activity. However, if there is no information about temporal change afterward deforestation, we cannot conclude whether deforestation was due to anthropogenic impact or some other natural reasons. This may make climate change mitigation policy vague, and it will be difficult to achieve sustainable development. To solve environmental problems, analysis of direct and underlying causes of those phenomena is necessary, but it will be difficult to find a solution unless the cause is clearly identified.

Spatial change information can be a key to the solution of environmental issues. In the terrestrial ecosystem, carbon and water cycles are the major systems to maintain the environment. Carbon cycle is often explained by vegetation coverage. Each vegetation type represents their own light use efficiency in the mechanism of photosynthesis. Under the same climate type, a vegetation type that has high light use efficiency has higher carbon absorption amount than other plants that have lower light use efficiency. Based on the difference between vegetation cover types, specific areas can be assessed in terms of carbon absorption function. To decide how much carbon credit or carbon tax is accepted in that area, we would need the spatial distribution of vegetation type.

Information on temporal changes is essential to determine whether land cover changes are caused by humans or not. Curtis et al. (2018) distinguished land cover changes and made several groups of change patterns. After the forest fire, if that area is recovered soon, it was classified as a natural hazard. However, if some area was not recovered after the fire event, it was recognized as deforestation by humans. Furthermore, different tendencies of land development were also revealed through image analysis.

The importance of confirming land destruction patterns whether it is anthropogenic or natural is because we can use that information for making decisions by policy makers. If some anthropogenic impact to the disappearance of vegetation was confirmed, specific people should take responsibility for CO_2 emission, the loss of carbon and environmental problems.

3.2 *In Situ Observation*

Flux tower is a representative system of direct observation of CO_2 around the vegetated regions. It uses infrared sensors and gas analyzers to catch the flux (inward and outward exchange) of CO_2 in the atmosphere. Usually, those sensors are established on a tall structure (about 40 m above the ground) for taking gases from stable air because the phenomenon of convection mixing due to friction is common near the surface. The direct observation in the field is a powerful tool to assess carbon assimilation functions of the local plants. The sensor records CO_2 and wind speed frequently, and finally, every 5 to 30 min of CO_2 exchange is provided to users. Near-real time observation is helpful to understand temporal change of CO_2 exchanges of relatively small area but encompassing from daily to yearly patterns. On the other hand, it is difficult to grasp CO_2 flux over a wider area than the tower's footprint. Based on the wind speed and wind direction at that moment of observation, footprints are changeable. For that, tower-based CO_2 flux still has some limitations, although it is almost the only way to reveal CO_2 in the atmosphere.

The most importantly observed CO_2 exchange by flux tower is nighttime respiration. Some studies that focus on CO_2 absorption into a biome aim to get daytime GPP. However, daytime CO_2 exchange is not used directly because air convection during daytime is strong, and a precise observation is hardly available. Relationships between daytime GPP and nighttime respiration are proportional to each other. Accordingly, if we have stable nighttime CO_2 emission data from the plant group (respiration), daytime vitality of that plant group for photosynthesis (GPP) is also able to be assumed. Stable air is essential to assume CO_2 flux correctly. So, the observation result on a rainy day has more uncertainties.

FLUXNET is the network system of local observations (Baldocchi, 2001). It collects many local flux tower data and is available online. Although the flux tower observation covers only a small area and it is not a seamless result, globally merged data show us spatiotemporal distributions of CO_2 on terrestrial areas. Many research projects regarding CO_2 flux observation are usually designed to cover specific biomes at a specific region to magnify efficiency when local data will be merged globally.

3.3 Satellite-Based Observation

Based on the observation equipment (vehicle), sensors of remote sensing would be classified into satellite-, airplane-, and drone-based systems. How to observe Earth's surface by remote sensing technology is by using differently designed sensors according to its wavelength. Wavelengths from gamma to radio are existing on Earth. The source of light is provided from the Sun. Since the sunlight has a peak around the visible wavelengths (0.4–0.7 μm), we can distinguish colors of many objects on Earth's surface. On the other hand, the Earth has a peak of longwave radiation because the temperature of Earth (about 288 K) is lower than the Sun (about 6000 K). More energy is needed to emit shorter wavelengths than longer wavelengths, according to Wien's displacement law (Eq. (1)).

$$\lambda_m = \frac{b}{T} \tag{1}$$

where $\lambda_m$is wavelength of maximum spectral radiant exitance (μm), b is Wien's displacement constant ($b \approx 2898$ μm • K), and T is temperature K (Lillesand et al., 2015).

On the Earth's surface, every object reflects sunlight in their own wavelength (radiometric) characteristics. As an example, there are colors and features of the target object. Based on this characteristic, land cover is classified. Among wavelength characteristics, vegetation has very clear peaks in NIR and red. This is distinguished from the other land cover types. A vegetation index is normalized difference vegetation index (NDVI) (Rouse, 1974).

$$\text{NDVI} = \frac{\text{SR}-1}{\text{SR}+1} = \left(\rho_{\text{NIR}} - \rho_{\text{red}}\right) / \left(\rho_{\text{NIR}} + \rho_{\text{red}}\right) \tag{2}$$

where SR is a simple ratio (SR = NIR/red) (Huete et al., 2002). SR is originally proposed to classify vegetation on land surface. However, there were some limitations that the difference of sunlight intensity causes daily difference in SR. Finally, a normalized ratio between NIR and red became used to classify vegetation cover.

NDVI enabled simple and accurate classification of vegetation; however, saturation is a limitation of it. For example, background (atmosphere and soil color) affects NDVI (blue and bright reflectance, respectively). Saturation of NDVI makes a spatial comparison difficult since the contrast in vegetation index will not be conspicuous. To overcome this problem, Enhanced Vegetation Index (EVI) was proposed (Huete et al., 2002).

$$\mathrm{EVI} = G * \frac{\mathrm{NIR} - \mathrm{RED}}{\mathrm{NIR} + C_1 * \mathrm{RED} - C_2 * \mathrm{BLUE} + L} \quad (3)$$

where $C_1 = 6$ (scattering calibration related to aerosol) ,$C_2 = 7.5$ (blue of atmosphere impact to red wavelength), $L = 1$ (including soil and biome impact), and $G = 2.5$ (EVI calibration parameter). This calibration improved vegetation index quality by decreasing atmosphere-induced error and saturation issues. However, blue bands were not usually captured by optical satellite sensors which were developed earlier. For that, some satellites have no blue band, accordingly, EVI is not available in this case. That is the limitation to use EVI.

When a research target is the carbon cycle of the terrestrial ecosystem, vegetation index should be calculated by remotely sensed data. The vegetation index mainly employs near-infrared (NIR) wavelength and visible (red) wavelength. Chlorophyll, exists in plant leaves, is an essential substance for photosynthesis using sunlight and CO_2 and actively absorbs red wavelengths. For that reason, a spectrometer recording very low red wavelength during photosynthesis is active in daytime when it observes daylight reflectance on plant leaves. On the other hand, if NIR wavelength was used for observation of plants, the reflectance signal will be very high (bright), because the mesophyll reflects NIR wavelength, and it exists in leaf tissue. Chlorophyll and mesophyll are found in healthy leaves, usually. Therefore, the difference between red and NIR wavelengths can be an indicator about vitality of photosynthesis.

Not only the spectral characteristics, but also the satellite orbit system can be considered when terrestrial ecosystem studies are conducted. Satellites around the Earth round along the defined orbit that eccentricity of a vehicle and gravity toward Earth are balanced. Briefly, we can classify orbits based on the altitude of the satellite: low, medium, and high Earth orbit. Low Earth Orbit (LEO) is a system below 2000 km from Earth surface in space. From 2000 km to 35,786 km is the range of Medium Earth Orbit (MEO). At 35,786 km or higher altitude is a route of High Earth Orbit (HEO) satellites. Spatial resolution is higher in case of a lower orbit system because the distance from Earth's surface is closer. Usually, LEO satellites have a purpose to observe the global area. Global observation by LEO is obtained by scanning with passing from northern to southern poles. MEO is often used for global positioning systems (GPS). HEO is occupied by geostationary satellites.

Geostationary orbit is designed as it is located 35,786 km above the Earth. Far distance from the Earth enables full disk observation which allows for a view of the entire circle of the Earth. Geostationary satellites are generally designed to be placed at specific points found from the Earth surface. It is available since the velocity of the satellite is equal to the rotation velocity of the Earth. Hence, high frequency of observation from several seconds to minutes is possible by geostationary satellite imagery. Those real time observation data can be usually used for meteorological research and as disaster countermeasures. However, geostationary satellites are not suitable to observe global observation due to the orbit always facing only one aspect of the Earth. Opposite side of the Earth is never detected by one system.

To overcome those limitations that geostationary can observe only one aspect of the Earth, a polar orbit system can be used. It is not matched to the velocity of the Earth rotation, but it is matched to the Sun movement that is found from the Earth. Running speed to pass the poles is designed to revisit the same point of space. While the satellite is moving, the Earth is also rotating. Finally, local time to be visited by satellite is always the same and the sun location from the Earth surface is not changed. It is possible to obtain one global map by one day if granule images taken by revisiting through a polar orbit were merged.

Consequently, satellite-based remote sensing that observes the Earth surface is helpful tool for terrestrial ecosystem research because surface information can be recorded widely without direct visiting, and furthermore its regular recording makes spatiotemporal changes be grasped.

3.4 Process-Based Modeling

Process-based modeling is a method to estimate GPP by using its physical relationship between environmental condition and plant's functions. Several specific models are the Dynamic Global Vegetation Model (DGVM), Lund-Potsdam-Jena Model (LPJ) (Sitch et al., 2003), ORganizing Carbon and Hydrology in Dynamic EcosystEms (ORCHIDEE) (Krinner, 2005), and Top-down Representation of Interactive Foliage and Flora Including Dynamics (TRIFFID) (Cox, 2001). Process-based modeling uses plant functional type and enzyme kinetic activation model to demonstrate the photosynthesis process that occurs inside of plants. Enzyme kinetic activations are closely related with temperature. Furthermore, water stress factor with intrinsic quantum efficiency is considered in DGVM. In ORCHIDEE, above/below ground biomass and agricultural soil carbon dynamics are applied. TRIFFID considers terrestrial carbon uptake models for the land-atmosphere interactions with simulated canopy height of given PFTs and leaf area index (LAI).

As models are specifically considering biological phenomena between plant body and environmental conditions, process-based models enable step-by-step analysis of CO_2 assimilation in detail. However, computing cost is a risk to use process-based models to areas that have wider and more complex biome types. In many cases of the modeling GPP by using process-based model, lots of input variables are needed; however, it is not easy to obtain global in situ observation data. It makes uncertainty of process-based models high. Process-based model is known as more complex than the empirical remote sensing-based model. However, details of daily dynamics of photosynthesis and relationships between plants and environments can be simulated by using process-based models.

3.5 Prediction

There are two kinds of predictions. One is spatial prediction toward neighboring regions when we have point-based observation data. The other one is a time-series prediction such as future prediction.

First, flux tower data as an in situ observation is used for spatial prediction with remote sensing data by extrapolation. In many cases, statistical empirical methods such as machine learning modeling are used for the extrapolation (Jung et al., 2009). In this spatial extrapolation, remote sensing data are useful because it provides even data quality between in situ observed pixel and not observed pixel. Remote sensing enables us to make links between the productivity of specific biomes which was obtained by flux towers and signals that were recorded by remotely sensed images.

However, there are some limitations. If the number of flux tower data are not sufficient, training of machine learning modeling will be failed or overfitted. As most of global extrapolation data suffer from, usually in situ observation sites are not easy to establish in every country. Difficulties in topographical condition, weather condition, and human accessibility make uneven distributions of observation sites. When the observation data of the site is not obtained, a model training is not available. In case of small numbers of training data were prepared, overfitting on that region will happen, and extrapolation will not be correctly operated. The small number of flux tower sites is the ultimate reason regarding low quality of calibration and validation. Therefore, many countries and institutes make efforts to establish more flux tower sites, recently. It is expected to improve the validation quality.

Gap filling of CO_2 flux under rainy or windy days is also a problematic issue of spatial prediction by using flux tower observation data. Very active kinetic characteristics of air convection in nature are not easy to grasp the CO_2 absorption or emission on specific biomes. Unfortunately, biomes in rainforest and temperate forest have many rainy and windy days, and it makes low quality of gap-filled CO_2 flux data as long as no-data days.

A future prediction of terrestrial productivity is available by accumulated historical climatic conditions and predicted climatic data. By combining climatic conditions and biome type, terrestrial productivities can be estimated. Physical meaning in the relationship between climatic variables (temperature and precipitation) and vegetation productivity (GPP) is used for this future prediction. However, still the spatial distribution of future land cover type is difficult to predict. This might be a large uncertainty for prediction of terrestrial productivity because land cover type is essential to estimate carbon flux.

4 Management of Terrestrial Ecosystem

In chapters 1–3, functions and study methodologies of the terrestrial ecosystem were introduced. In this chapter, practical applications are analyzed. Remote sensing data sets that are provided in recent days are described. As some representative providers of carbon cycle observing satellite data, National Aeronautics and Space Administration (NASA), European Space Agency (ESA), and Japan Aerospace Exploration Agency (JAXA) were considered. Carbon credit certification system was described, which is one of the practical methods to reduce global carbon emissions. As a direct campaign, which is associated with mitigation of deforestation, reducing emissions from deforestation and forest degradation in developing countries (REDD) was discussed. All these contents in chapter 4 are aiming to introduce global efforts for climate change mitigation by managing the carbon cycle on terrestrial ecosystems.

4.1 Satellite Missions of Each Country

NASA and JAXA operate CO_2 observation by satellite-based remote sensing sensors. As satellite missions of direct CO_2 observations from space, Greenhouse gases Observing SATellite (GOSAT) of JAXA, National Institute for Environmental Studies (NIES), and Ministry of the Environment (MOE) of Japan, and Orbiting Carbon Observatory (OCO) of NASA can be listed up.

Shared technology in common among these is to use infrared (IR) wavelength for detecting the absorption of CO_2 in the atmosphere. These CO_2 observation missions measure the Earth's surface radiance in the short-wave infrared (SWIR) and thermal infrared (TIR). Those wavelengths are reflecting solar scattered light by the surface of the Earth and atmosphere, and absorption characteristics are used for retrieval of CO_2 vertical profile. The key wavelength of CO_2 retrieval is 1.6 μm. Supplementally, 2.0 μm can be used.

The GOSAT of JAXA, NIES, and MOE that launched in 2009 was the earliest CO_2 observation mission in the world, and Thermal and Near-infrared Sensor for Carbon Observation Fourier Transform Spectrometer (TANSO-FTS) is on board. Every 3 days, data is obtained, and 56,000 points can be accumulated. Spatial representativeness is approximately 10.5 km diameter at nadir. Data with 2.5 degrees of grid size are obtained in case of monthly CO_2 column concentration, even spatial resolutions of data are varying in product levels. GOSAT-2, the next generation of GOSAT, was launched in 2018, and the resolution was improved to a circle of 9.6 km diameter (Kuze et al., 2012).

The OCO-2 was launched in 2014. It also has a sun-synchronous orbit, and the revisiting cycle is 16 days. Satellite's observation footprint is approximately 1.29 km by 2.25 km (3 km^2) (Feldman, 2023). Level 3 monthly data has spatial resolution of about 1 degree. Approximately 10,000 to 20,000 high-quality data points can be accumulated daily.

Ultimate merit to use satellite-based CO_2 observation is the possibility of visualization of relative CO_2 differences globally and regularly. However, there are also limitations in satellite-based direct CO_2 column concentration observation. Above all, it is difficult to compute an inversion model to get column-averaged CO_2 mole fractions (X CO_2) from satellite observed spectral radiances (Houweling et al., 2015). Secondly, reference data for validation differ between sites and defining its representativeness are not easy. Furthermore, validation results among sensors are found various. Due to differences in their orbits, data frequency varies. Finally, the accuracy of CO_2 retrieval shows differing datasets between GOSAT and OCO-2 (Wang et al., 2019).

Although there are limitations in data use and accuracy, global data coverage and regular visiting provide helpful information to grasp the carbon cycle in terrestrial ecosystems (Byrne et al., 2019). Globally, even quality of carbon flux data enables fair comparison between countries and continents in specific time span. Since spatiotemporal resolution of those sensors have been continuously improved, global estimation of CO_2 fluxes will be expected to provide an accurate quantitative evidence of climate change.

4.2 Carbon Credit Certification

Global society agreed to reduce CO_2 levels for the mitigation of global warming. Under the United Nations Framework Convention on Climate Change (UNFCCC), the Paris Agreement has been proposed and 195 nations have agreed with that. To promote practical reduction by conservation of the terrestrial ecosystem, various efforts became necessary. For the assessment of carbon reduction effort, some standards must be prepared.

Clean Development Mechanism (CDM) is one of the standards to promote cooperation for reduction of CO_2 under UNFCCC. Assessment and monitoring of the baseline of CO_2 emissions from a company or country are mandatory to apply CDM to a project (UNFCCC, 2022). Projects range from energy efficiency improvement to reforestation or agricultural practices. Key factor is that this carbon market is attempting to make developed countries support their achievement financially with developing countries by bringing in foreign investment and technology transfer. It enables sustainable development because usually developed countries have limited capacity to reduce carbon; on the other hand, developing countries have low financial resources to maintain infra and to manage environments.

A carbon credit is a voluntary system based on a market-based certification method to determine price of carbon for trading it, and it means the right of carbon emissions. This method makes it possible to mitigate climate change without stopping or delaying industrial technology development. For applying this system, above all, it needs an accurate certification of emission offset assessment. Once a project type was decided, the baseline of carbon flux without that activity is mandatory to be estimated. How much carbon was reduced should be presented, as well. Further plans of monitoring and its predictability are necessary to be given a certification to guarantee transparency of trade in the market (UNFCCC, 2022). The monitoring and assessment of the carbon cycle in that ecosystem are essential, however, simultaneously, it can be a high huddle to join this system by companies because of that. Specific technology to assess carbon storage has sometimes not existed due to lack of theory about carbon cycle in terrestrial ecosystems. If an in situ station is established for grasping the whole amount of carbon storage, the operation takes much cost and it might be a risk of investment to the carbon credit certification system.

4.3 Campaigns Regarding Carbon Emission Reduction

Representative campaign for carbon emission reduction is reducing emissions from deforestation and forest degradation in developing countries (REDD). In particular, REDD+, a next generation of normal REDD as part of the Paris Agreement, additionally includes forest protection activities, too, for example, climate protection, sustainable management, and the conservation and enhancement of carbon stocks. The framework of REDD+ is to give economic incentive when developing countries keep their forest and manage terrestrial ecosystems to sustain intact status. This incentive is conceptualized to follow a trend of benefits by products from forest before protection has been done. If some farmers quit agricultural activities that involve deforestation, forest area will be kept, and agricultural benefit will be decreased. It can be said that this is an incentive that converts the profits from previous production into money in exchange for deciding to stop agricultural production and maintain the forest ecosystem. This applies not only to agriculture, but also to livestock and mining.

It was also found that there were limitations in clarifying protective activities and the financial resources were insufficient. Furthermore, wide coverage and fair evidence are necessary. Monitoring, Report, and Verification (MRV) is a basic methodology to assess REDD+. To describe protective activities, remote sensing data can be used because data quality is equal globally, and it is relatively lower cost than an operation of an in situ observation system. In addition, the vegetation vitality is easily grasped by detecting various vegetation indices. For that reason, in many cases, global remote sensing data has been recognized as helpful to achieve intergovernmental forest conservation activities including REDD+ activity.

For example, Japan International Cooperation Agency (JICA) actively participates as a supporter of REDD+ initiatives through technical and financial assistance. Indonesia-Japan Project for Development of REDD+ Implementation Mechanism (IJ-REDD+) is representative. Peatlands in Indonesia are a major carbon pool; however, anthropogenic disturbances such as rice agriculture, forestry, oil palm plantation development, and land clearance-induced fires have made tremendous CO_2 emissions from those regions. As a REDD+ project between Indonesia and Japan, IJ-REDD+ was initiated in 2013. Specifically, the outputs were (1) mapping land cover and land use and forest dynamics, (2) monitoring improvement, (3) peat soil distribution and CO_2 emission investigation, (4) training of decision makers, and (5) communication system improvement with local people (Takahara, 2016). Additionally, MRV with satellite remote sensing made progress to visualize deforested or afforested areas in peat forest. Those technologies can be used for quantifying carbon emission and photosynthesis amount in peatlands in Indonesia.

4.4 International Cooperation Examples

Wetlands are one of the largest carbon pools in the terrestrial ecosystem. Among these, tropical peatlands especially have high potential to be converted as carbon emitters these days.

Restoration of WETlands to minimize emissions and maximize carbon uptake (REWET) is an example of wetland conservation and sustainable restoration activity. The project has several objectives that are listed, for example, to develop methodology of wetland restoration implementation and monitoring, to create an inventory of European wetlands, to estimate carbon footprint in EU wetlands, to create opportunities for green jobs for all stakeholders, and so on.

Another example is the International Tropical Peatland Center (ITPC). The target areas include Southeast Asia, the Congo Basin, and Peru. Key word is sustainable management of peatlands (Reed et al., 2022). Roles of the ITPC are listed as to support collaborative international relationships and connecting different stakeholders, to conduct scientific research on tropical peatland management, to support policy development, and to provide capacity building and technical services.

Norway and Indonesia agreed to have a partnership to reduce emissions from deforestation and forest degradation, including peatlands. Norway has provided funding to the Indonesian Peatland Restoration Agency (BRG).

Under REDD+ programme, many countries and institutions for example, UNEP, FAO, United States Agency for International Development (USAID), JICA, and so on are cooperating with the Indonesian government for supporting conservation and restoration activities on tropical peatlands.

Environmental issues including deforestation are not anymore local problems. Global society should cooperate to combat destruction of terrestrial ecosystems by supporting each other. These international cooperation activities seem hopeful to achieve sustainable development goal #15, Life on Land.

5 Conclusion

To sum up this chapter, the terrestrial ecosystem, the largest carbon absorber in the Earth, was investigated with various perspectives. Most of all, the importance had been found in photosynthesis of vegetation covers, biomass storage, and providing habitats. Distributions of terrestrial ecosystem types can be an indicator of climate change since vegetation habitat is resultant of adaptation toward the surrounding environment. Biome types are often used for revealing climatic conditions at that location.

Mechanisms of carbon cycle in terrestrial ecosystems are related to meteorological conditions such as temperature and vapor pressure deficit. Based on the temporal scale, carbon absorptions and emissions are classified into gross primary productivity (GPP), net primary productivity (NPP) (net ecosystem CO_2 exchange, NEE), and net biome productivity (NBP) from shorter to longer phenomena, that are an assimilated carbon by photosynthesis (daily), a budget between photosynthesis and respiration (daily), and a budget with anthropogenic disturbances (yearly to decadal), respectively.

In perspectives of SDGs, the Life on Land (No. 15) includes terrestrial ecosystem associated contents. Wedding cake structure of SDGs is helpful to understand how terrestrial ecosystem conservation activities can support the other goals. The terrestrial ecosystem is a fundamental resource of the environment, and human society is built on this ecosystem particularly. For its characteristics that support other goals, conserving and recovering efforts are basically required in SDGs No. 15 although development is relatively not focused.

Carbon assimilation is the most important role that is expected to the terrestrial ecosystem for global warming mitigation. Biomass provides products, e.g., timber woods, fuels, pulps, etc., that are fundamental materials to maintain human society. Soil organic carbon plays a role as the largest carbon pool for long-term preservation on terrestrial ecosystems. Biodiversity is important to keep genetic variety under changeable climate. Accordingly, significant conservation of the terrestrial ecosystem was found in many points of view listed above.

In the methodology of carbon cycle observation, above all, the importance of spatiotemporal change detection was described. To analyze causes and impacts of environmental problems in climate change, a wide data coverage in both spatial and temporal patterns is important. To guarantee the spatiotemporal analysis, remote sensing method had been recognized as efficient.

Tools for observing terrestrial ecosystem conditions were described. Above all, in situ measurement of carbon flux is operated by a flux tower system. Direct observation methods of flux tower have a lot of merit, for example, very high frequency of the data and accumulation of site validation data. However, high establishment cost, hard workload to maintenance, complex computation of gap filling, and especially difficulties in accurate observation under windy or rainy weather conditions are still problematic. Despite the uncertainties, tower-observed carbon flux contributes to estimate a two-dimensional seamless global distribution of primary production.

Satellite-based remote sensing is a powerful tool to observe a wide area of land cover change regularly. It uses electromagnetic characteristics of the Earth's reflectance toward solar radiation and Earth radiation itself. Sensors on satellites are operated along the satellite's orbital path, and moreover, the detectable wavelength is designed based on the purpose. For example, if a target object was the vegetation cover, visible red and near-infrared bands are used for calculation of vegetation indices. Satellite-based vegetation indices are helpful to visualize plant vitality globally. It is directly related to the carbon absorption ratio in the terrestrial ecosystem.

The last observation type was process-based modeling. Process-based models for carbon cycle on terrestrial ecosystems are using physical relationships between plant and climatic environmental conditions. Not only above ground but also soil organic matters can be considered in a process-based model. This physically explained calculation processing enables detailed analysis of each component of the carbon cycle in the terrestrial ecosystem.

Prediction of the carbon cycle can be done by using accumulation of Earth observation data. By using flux tower, satellite-based data, and process-based model, predictions of spatial distributions and time-series of land cover and land use changes are available. Spatial prediction enables us to grasp the environments of regions where in situ data are lacking because the accessibility is not good. Future prediction provides helpful information for the adaptation to climate change.

Some practical examples of terrestrial ecosystem management, satellite missions of CO_2 observation, carbon credit certification, campaigns regarding carbon emission reduction, and international cooperation examples were described.

CO_2 observing satellite missions are usually designed by using shortwave infrared and thermal infrared wavelengths. Although CO_2 retrieval methods from satellite-based data still have room for improvement, the satellite missions will continuously provide a larger amount of CO_2 column concentration globally. At least, relative differences of CO_2 fluxes that are recorded by satellite sensors will contribute to reveal climate change tendency.

Carbon emission reduction efforts are needed to achieve global warming mitigation. As a more practical way, a clean development mechanism (CDM) is proposed by UNFCCC. Carbon credit is a voluntary system based on a market-based certification method to determine the price of carbon for trading it. To be involved in this market, monitoring, and management of the carbon cycle of their sites are necessary. The expansion of this market can be expected to lead to the implementation of practical climate change measures.

REDD+ is an effective system to reduce carbon emission induced by deforestation and degradation of the forest ecosystem. The forest is a representative carbon absorber in the terrestrial ecosystem, and developing countries have many forest areas. However, regulation of industrialization in developing countries has the potential to create a widening gap with developed countries. For this perspective, resources for conducting conservation of forest should be provided to developing countries. REDD+ suggests a feasible way to qualify potential benefits when some developments have been done in that area. Several examples of peatland conservation cases done by two or more countries' cooperation were described in this chapter, additionally. In REDD+ and international cooperation, satellite remote sensing data is critically utilized since its fair quality for everywhere.

In conclusion, terrestrial ecosystems are important in providing a foundation for sustainable development through the carbon cycle, and it is necessary to conserve it to cope with climate change. For the conservation of the terrestrial ecosystem, remote sensing technology is utilized in almost all fields encompassing scientific research to local community's activity. Based on the clear objectivity of data, cooperation between stakeholders is essential. The SDGs No. 15, Life on Land is expected to be achieved by continuous cooperative activities internationally.

References

Baldocchi, D. (2001). FLUXNET: A new tool to study the temporal and spatial variability of ecosystem-scale carbon dioxide, water vapor, and energy flux densities. *Bulletin of the American Meteorological Society, 82*(11), 2415–2434.

Bar-On, Y. M., et al. (2018). The biomass distribution on earth. *Proceedings of the National Academy of Sciences, 115*(25), 6506–6511.

Byrne, B., et al. (2019). On what scales can GOSAT flux inversions constrain anomalies in terrestrial ecosystems? *Atmospheric Chemistry and Physics, 19*(20), 13017–13035.

Chapin, F. S., et al. (2002). *Principles of terrestrial ecosystem ecology*. Springer. https://doi.org/10.1007/978-1-4419-9504-9

Chen, X., et al. (2024). A 2001–2022 global gross primary productivity dataset using an ensemble model based on the random forest method. *Biogeosciences, 21*(19), 4285–4300.

Cox. (2001). Description of the "TRIFFID" dynamic global vegetation model. Hadley Centre technical note 24.

Curtis, P. G., et al. (2018). Classifying drivers of global forest loss. *Science, 361*(6407), 1108–1111.

FAO. (2018). FRA 2020 terms and definitions. In *Forest resources assessment working paper no. 188*. FAO.

FAO. (2020). *Global Forest Resources Assessment 2020: Main report*. FAO.

Feldman, A. F. (2023). Using orbiting carbon Observatory-2 (OCO-2) column CO 2 retrievals to rapidly detect and estimate biospheric surface carbon flux anomalies. *Atmospheric Chemistry and Physics, 23*(2), 1545–1563.

Folke, C., et al. (2016). Social-ecological resilience and biosphere-based sustainability science. *Ecology and Society, 21*(3), 41.

Friedlingstein, P., et al. (2022). Global carbon budget 2021. *Earth System Science Data, 14*(4), 1917–2005.

Houweling, S., et al. (2015). An intercomparison of inverse models for estimating sources and sinks of CO_2 using GOSAT measurements. *Journal of Geophysical Research-Atmospheres, 120*, 5253–5266. https://doi.org/10.1002/2014JD022962

Huete, A., et al. (2002). Overview of the radiometric and biophysical performance of the MODIS vegetation indices. *Remote Sensing of Environment, 83*(1–2), 195–213.

Jongman, R. H., et al. (2017). Global terrestrial ecosystem observations: Why, where, what and how? In *The GEO handbook on biodiversity observation networks* (pp. 19–38).

Jung, M., et al. (2009). Towards global empirical upscaling of FLUXNET eddy covariance observations: Validation of a model tree ensemble approach using a biosphere model. *Biogeosciences, 6*(10), 2001–2013.

Köppen, W. (1936). *Das geographische System der Klimate* (pp. 1–44). Gebrüder Borntraeger.

Krinner, G. (2005). A dynamic global vegetation model for studies of the coupled atmosphere-biosphere system. *Global Biogeochemical Cycles, 19*(1), 1–44.

Kuze, A., et al. (2012). Level 1 algorithms for TANSO on GOSAT: Processing and on-orbit calibrations. *Atmospheric Measurement Techniques, 5*(10), 2447–2467.

Lillesand, T., et al. (2015). *Remote sensing and image interpretation*. John Wiley & Sons.

Mora, C., et al. (2011). How many species are there on earth and in the ocean? *PLoS Biology, 9*(8), e1001127.

Reed, M., et al. (2022). Policy and governance options for peatlands conservation, restoration and sustainable management. In *Global peatlands assessment: The state of the world's peatlands: Evidence for action toward the conservation, restoration, and sustainable Management of Peatlands: Main report* (pp. 252–271).

Rouse, J. W. (1974). *Monitoring vegetation systems in the Great Plains with ERTS* (Vol. 351, p. 309). NASA Spec. Publ.

Running, S. W., et al. (2004). A continuous satellite-derived measure of global terrestrial primary production. *BioScience, 54*(6), 547–560.

Schmidt, M. W., et al. (2011). Persistence of soil organic matter as an ecosystem property. *Nature, 478*(7367), 49–56.

Sitch, S., et al. (2003). Evaluation of ecosystem dynamics, plant geography and terrestrial carbon cycling in the LPJ dynamic global vegetation model. *Global Change Biology, 9*(2), 161–185.

Takahara, S. (2016). Efforts to implement REDD-plus in Kalimantan, Indonesia: Activities of IJ-REDD+ project. *Japanese Journal of International Forest and Forestry, 95*(16).

UNFCCC. (2022). *CDM methodology booklet*. UNFCCC.

United Nations. (2015). *Transforming our world: The 2030 agenda for sustainable development (a/RES/70/1)*. United Nations. Retrieved from https://undocs.org/A/RES/70/1

USDA. (2001). *U.S. Forest resource facts and historical trends*. USDA.

Wang, H., et al. (2019). Terrestrial ecosystem carbon flux estimated using GOSAT and OCO-2 XCO 2 retrievals. *Atmospheric Chemistry and Physics, 19*(18), 12067–12082.

WWF. (2024). 2024 Living Planet Report.

Prevention of Biodiversity Loss

M. Savithri, V. Shobana, K. Thenmozhi, and R. Gunavathi

1 Introduction to Biodiversity

1.1 Definition and Types of Biodiversity (Genetic, Species, and Ecosystem Diversity)

The various types of living beings in a particular area like plants that have been grown, animals found, fungi, and other kinds of microorganisms like bacteria are jointly represented as biodiversity. The word diversity means variety and in Greek bio means life, and the word biodiversity is used to represent the various kinds of life. The various organisms and the species in an area will make the ecosystem that described as an intricate web. It helps to lead a balanced life by supporting and helping to manage all the natural things that are most important for survival. Biodiversity can be defined as every living organism located within a single ecosystem. It also includes the other environmental factors like oxygen, carbon dioxide, temperature, climate, etc. The term biodiversity is important in nature and in the artificial ecosystem. It involves various organisms and their frequency in the ecosystem along with how they organize at various levels. It provides us with the ecological significance and other resources like cloth, fuel, house, etc.

Biodiversity is classified into three different types as genetic, species, and ecological biodiversity. Genetic diversity is classified based on the genetic resources in the organisms (Klaus et al., 2020). Each member in a plant or animal species will

M. Savithri (✉) · R. Gunavathi
Christ University, Bengaluru, India
e-mail: savithri.m@christuniversity.in; gunavathi.r@christuniversity.in

V. Shobana · K. Thenmozhi
Dr. N.G.P. Arts and Science, Coimbatore, India
e-mail: shobana@drngpasc.ac.in; thenmozhi@drngpasc.ac.in

J. J. Puthenkalam S.J et al. (eds.), *Terra Thrive*,
https://doi.org/10.1007/978-3-032-02738-2_12

vary from other in its genetic constitution, and it also has some specific character based on its genetic. The genetic variability is due to the ability of the genes present in the organisms to combine with an infinite number of combinations. For example, if we consider rice, there are various varieties of rice each differ from it shape, size, colour, taste, etc., and this is due to the level of genetic variations. Gene pool is a term that has been used to represent the genetic diversity among the species including the wild species. As per the Biological Species Concepts (BSC) (Singh, 2024), species diversity is the basic classification used to group similar organisms. It refers to biodiversity in basic level including all the species living in the earth which includes plants, bacteria, fungi, algae, angiosperms, pteridophytes, and also the species of animals which include mammals, unicellular protozoans, etc. Regions with a balanced climate have a high degree of diversity. Hotspots of biodiversity are the regions which are rich in species diversity. Ecological biodiversity consists of organisms from various species which live together at a particular region and are connected with the stream of energy and nutrients. For example, the Sun is a source of energy for all living things. Plants convert solar energy into chemical energy that transferred to different animals that have eaten the plants. We can say that an ecosystem is a combination of all living and nonliving components like plants, animals, fungi, climate, matter, energy, etc. which are connected with the help of energy flow.

1.2 Importance of Biodiversity for Ecosystem Functioning and Human Well-Being

Biodiversity plays a vital role in the ecosystem for the stability, productivity, and the sustainability of ecological and human well-being in various ways. To maintain all life on the Earth, it reinforces the process of the ecosystem (Ali, 2023). It provides a stability to the ecosystem to enhance the resilience and to enable it to recover from the various disasters due to natural and human impacts. Various organisms in the ecosystem help in the cycling of nutrients which helps to ensure that the basic elements that are required for life like nitrogen, phosphorus, and carbon are available for plants and animals. It helps to improve the ecosystem's production to produce energy from various plants with the help of photosynthesis. Pollination and seed dispersal are done by bees, birds, and butterflies which are required for the reproduction of plant and food production. It also plays an important role in purifying the water in wetlands and preventing soil erosion in forests. It also helps in maintaining the fertility of soil.

1.3 Current State of Global Biodiversity and Trends in Biodiversity Loss

The state of biodiversity at present is alarming due to the decline of various species, habitants, and ecosystems. Some of the key features of the concerns are widespread loss of biodiversity, degradation of ecosystem, regional variations, tipping points, and irreversibility. Decline of wildlife population and the freshwater species lead to the widespread loss of biodiversity; this is due to the destruction of habitation, pollution, climate change, overutilization, etc. Degradation of ecosystem is due to the unprecedented threads like coral reefs, deforestation, and changes in usage of land (Isbell et al., 2022). Coral reefs support more than 25% of species living the marine life which have been degraded due to the ocean warming, water pollution, and acidification. Integrity of the tropical forest has been eroded by deforestation. Developing nations like Asia, Africa, and Latin America face the declines due to the pressure from agriculture, urbanization, and industrial activity (Sonia Llorente-Culebras et al., 2023). The various approaches which are considered as the tipping points of the ecosystem that are irreversible are due to the collapse of the critical ecosystem which might lead to various consequences which include availability of freshwater, climate regulation, and food security. Even though some efforts have been taken to prevent the loss of biodiversity, it still requires transformative changes like shifting toward a sustainable environment, reducing pollution, adopting renewable energy sources, and enhancing conservation. It requires Global cooperation to ensure the flexibility of the ecosystem for the future generation.

2 Causes of Biodiversity Loss (nextias, 2024)

2.1 Habitat Destruction and Fragmentation

One of the leading causes for the worldwide biodiversity loss is the habitat destruction and fragmentation which disrupts the ecosystem, displaces the species, and also reduces the availability of the various resources that are required for the survival of life. Cutting of trees for timber, production of paper and fuelwood, and clearing the forest for infrastructure development also led to the removal of species extinction and also the decline of its population. Deforestation is a major threat for chimpanzees and tigers. Due to deforestation, there is an increase in atmospheric CO_2 which leads to global warming. It also disrupts the structure of the soil due to which there is a huge erosion of fertile soil. The increase in population also caused the increased need for house, infrastructure, and other industry due to which the cities have been expanded. Due to this urbanization, the natural habitats have been divided into smaller and also isolated due to which the species find it difficult to migrate. The urbanization also increases the generation of waste which degrades the ecosystem and it also leads to the alteration in the local climate by creating urban heat islands due to which the biodiversity is being affected (Hogue & Breon, 2022). Due to the increasing demand for the food, biofuels, and cash crops, agricultural expansion is required for which the forests in the tropical regions are being burned and converted into farmland. Expansion of agricultural lands by destroying the forest has various impacts like loss of ecosystem services which reduce the pollination, quality of the water purification, and the stability of the soil. Genetic diversity and the resilience of the ecosystem have been reduced due to the monoculture by replacing diverse ecosystem by single-species crops. It also increases the conflict with human and wildlife due to the trespass on the wildlife habitats. Some of the consequences that are been faced due to the destruction and fragmentation of the habitat are reduced genetic diversity, disruption of migration patterns, and edge effect (Gustavo Ramirez, 2023). To prevent the disruption, we need to protect the animals by establishing national parks and wildlife reserves and need to promote the culture of agroforestry and generate reforestation and urban green spaces. It is also necessary to connect or group the fragmented habitats so that the movement of species will be improved. Policies like promoting the land usage regulations, and stopping deforestation should be strictly implemented. Efforts that have been taken to reduce the destruction of the habitat and fragmentation will help us to develop solutions for a sustainable environment and also to preserve biodiversity loss.

Climate change: Impact on ecosystems, species migration, and extinction risks.

Due to the change in climate, the ecosystem faces significant issues like disruption of species, behaviour of nature, and increased risk of destruction (Ahmed, 2022). It affects the global temperature, alters the pattern of rainfall leading to melting of ice, and increases the extreme weather. Due to the alteration in the habitat, oceans become warmer which might lead to coral bleaching and also threaten marine biodiversity. Due to the heatwaves and acidification, nearly 50% of the Great Barrier Reef has been lost during the past three decades. Due to the rising temperature, ecosystems like tundra's are decreased due to the terrestrial shifts. Forests and wetlands are the main source of the ecosystem that helps to regulate the carbon, and the water cycle has been degraded due to the disruption of the ecosystem. Deforestation of Amazon forest is aggravated by the climate driven droughts. The changes in the ocean, including rise in the sea level and increased CO_2, threaten the coastal and mangrove ecosystem, and it also leads to acidification of ocean which will affect the organisms like corals.

Pollution: Air, water, and soil pollution, including plastic pollution and its impact on marine life.

The various types of pollution like air, water, and soil have more impact on the ecosystem, affecting human health and biodiversity. Pollution due to plastic is more problematic due to the long-lasting effects and the pervasive nature, and its effects are more mainly to the marine ecosystem. Air pollution is caused due to the fossil fuels used in vehicles, industries, and power plants. Release of methane and ammonia from the various agricultural activities and the emission from various industries and wildfires also cause air pollution. The air pollution has various impacts on the human health, ecosystem, and climate change. Due to the polluted air, respiratory and cardiovascular diseases are increasing. PM2.5 is a fine particulate matter, and the nitrogen oxides that are found in the polluted air lead to premature deaths. Pollutants like sulphur dioxide cause various impacts to the ecosystem which contributes to the acid rain, which then damages forest, soil, and also the freshwater system. Emission of greenhouse gases like CO_2 and methane increases global warming.

Water pollution is caused due to the discharge of untreated waste from industry and agriculture into the rivers and oceans. Spills of oil and chemical runoff mixed in the water bodies. Plastic wastes are entering into the water bodies, causing water pollution. Due to the various chemicals, heavy metals, pesticides, and the plastic in the water bodies, it disrupts the food chain of aquatic life. It also affects human health due to the toxins like mercury and lead present in the water. Contaminated water is the main source of spreading various diseases like cholera. Polluted water also leads to an imbalanced ecosystem since the rivers and oceans will lose their ability to support marine life. Soil pollution is caused due to the improper disposal of waste materials from houses and industries, usage of fertilizers and pesticides in excess, and spilling of oil and various mining. The various chemical fertilizers that have been used will destroy the microbes in the soil and also reduce the capacity of the soil to grow the plants. Soil pollution also leads to risk in food security since it affects the safety of the food. Due to soil pollution, the groundwater is also contaminated. The increased usage of plastic and improper plastic waste management lead to plastic pollution which leads to various impacts like entanglement, ingestion, and habitat destruction. Global efforts should be taken to reduce the various pollutions like imitating some policy to control the air pollution and to address the climate changes. With the use of technology, wastewater should be treated to produce biodegradable plastics. Each individual and community should be aware about the various impacts due to the pollution by promoting the recycling, reducing the usage of plastic, and also participating in clean-up drives in rivers and oceans.

Over-exploitation of resources: Overfishing, hunting, logging, and unsustainable harvesting.

Due to the increased population and need for more resources, over-exploitation of natural resources leads to loss of biodiversity and also degrades the ecosystem. The demand for seafood has led to overfishing, the usage of advanced technology captures large quantities of fish. Some are doing illegal, unreported, and unregulated fishing that also lead to over-exploitation.

Due to the overfishing of some species like Atlantic cod, Bluefin tuna is in depletion. Due to the removal of key predators, there is an imbalance in the ecosystem which disrupts the food webs and also leads to overpopulation of certain species and collapse of the lives of other species. Due to the decline in the fish stock, the livelihoods of fishing communities are facing huge economic loss. Hunting of animals for poaching is like illegal hunting of elephants for their ivory, deer, and other wild animals. In some rural regions, people are hunting animals for food and also killing of wildlife for sport as a trophy hunting. Due to excessive hunting of animals, some species like dodo and passenger pigeon are in extinction stage. Due to ecosystem disruption, the loss of various species has reduced the spread of seed dispersal which also reduces the regeneration of the forest. It also leads to human–wildlife conflicts due to overhunting, and deforestation animals are closer to humans which increases the conflicts. Logging is another issue due to the demand for wood in the construction industry, paper industry and for furniture timber extraction is high and the increased need to grow crops also led to deforestation for agriculture. Due to deforestation, various habitats like orangutans and tigers have been destroyed. Deforestation also leads to 15% of global greenhouse gas emission which contributes to climate change. The loss of trees also led to soil erosion and also reduced the fertility of the soil. Over-cultivation in the agricultural land leads to soil exhaustion, and over-pumping of groundwater extraction for agricultural use and also for urban use leads to depletion in the under-groundwater level. Due to water scarcity, over-harvesting of plants and animals also leads to loss of biodiversity and diminishes agricultural productivity leading to insecurity of food.

Invasive species: Introduction of non-native species that outcompete local flora and fauna.

The species that are introduced to a new habitat accidentally or intentionally to outcompete the native flora and fauna are termed as invasive species. These kinds of species will thrive because of the natural predators and diseases that might affect their population in their native ecosystems. These species are introduced by global trade while transporting the seeds, microorganisms and agricultural products. through the ship unintentionally. Sometimes tourists will carry the seeds and insects while they travel to a country. Some species are introduced intentionally for the agriculture and aquaculture improvement while some are introduced to control pests. The invasive species also cause some impacts by competing with the native species for food, water, and sunlight. The non-native predators also destroy the population of the native species. It also changes the physical characteristics of the ecosystem and damages the agriculture, fishery, and forestry leading to economic cost. Some of the species might even spread diseases or allergies that threaten human health. Few strategies that can be used to manage the effects caused by the invasive species are to adopt strict biosecurity and provide awareness about the various risks involved in introducing the invasive species. Regularly monitor the ecosystem to identify the invasion species early and take necessary action to prevent and control their growth. Biological and mechanical controls should be taken to remove the invasive plants and animals.

Land use changes: Conversion of natural landscapes for agriculture, industrialization, and infrastructure.

There was a significant change in the environment due to the change in the usage of land by altering nature to meet the needs of the human that leads to loss of biodiversity, destruction of habitat, and climate change (Liang et al., 2020). Farmlands have been expanded to increase the production of food products to meet the increasing demand for food and the biofuels. To expand the farmlands, most of the forests are cleared especially in Southeast Asia and in the Amazon which also leads to soil degradation. Due to deforestation, it also destroys the diverse habitats and threatens the existence of species like orangutans and jaguars. Due to the increase in the number of industries like manufacturing industries, mining industry and power plants to support the economic growth of the people, the lands are being used to establish various factories. Due to the various activities and processes that have been involved in the various industries, the air, water, and the soil are being polluted. The facilities that have been done to improve the production of industry like road facility lead to isolation of the ecosystem, and it also disrupts the wildlife corridors. The greenhouse gas emission is increased due to the increased use of fossil fuel and deforestation. Infrastructure development like construction of dams, highways, cities by replacing the agricultural lands leads to increase in temperature and also the risk of flood in urban areas (Liang et al., 2020). Expansion of urban area leads to loss of habitat and displacement of species or even extinction of some species that leads to biodiversity loss. There are also some other combined consequences like disruption in water cycle, increased carbon emission, and cultural and social impacts due to the changes in the usage of land.

3 Consequences of Biodiversity Loss

Due to the biodiversity loss, there are various consequences that have been faced like disrupting the ecosystem (Li, 2024) and impacts on living beings. Few of the consequences that are been faced are listed below

- Impact on ecosystem.
- Food security (Hogue & Breon, 2022).
- Health.
- Climate change.
- Economic impacts.
- Social and cultural impacts.

Impact on Ecosystem: The ecosystem has been disrupted due to the loss of biodiversity which affects the stability and functionality of the biosystem. Due to which there is a decline in the service of ecosystems like pollination, purified water, fertile soil etc. that are more essential for the survival of humans.

Food Security: Due to the loss of biodiversity, there is a loss of genetic crops due to which there is a reduction of growth of crops and livestock. It also reduces the flexibility to pests, diseases, and the climate change due to which there is a global threat on food security (Mirzabaev et al., 2023).

Health: There is an increase in occurrence of diseases due to loss of biodiversity. The spread of transferable diseases can be regulated by the diverse ecosystem but the loss of biodiversity is leading to new pathogens.

Climate Change: The biodiversity plays a major role in regulating the climate of the Earth. The loss of forests and other vegetation have reduced the ability of the planet to absorb the carbon dioxide which leads to climate change.

Economic Impacts: The loss of biodiversity leads to various economic consequences especially for the industries that are based on the various natural resources like agricultural industry, forestry, and fishing. Due to the declining ecosystem service, the cost involved in the product is increased and the yield is reduced.

Social and Cultural Impacts: The culture and spirit of various communities are tied with the biodiversity. Due to the loss of species and the natural habitats, the various cultural identities of the indigenous population and their cultural identities are also being eroded which affects the social well-being and the traditional practices of the population.

4 Conservation Strategies

It is essential to take effective conservation strategies that might preserve biodiversity and help to maintain a good and healthy ecosystem. Some of the key strategies (Reeta Kumari, 2021) that can be considered to prevent the biodiversity loss are:

Protected Areas: These are specially designed or designated regions like national parks, wildlife reserves, and marine protected areas which can be used to safeguard the various habitats and the native species of the area where the human activities should be restricted and should promote the conservation of the ecosystem.

Habitat Restoration: Various steps should be taken to repair the ecosystem that has been destroyed. For example, reforestation helps to rebuild the natural habitats and support biodiversity and ecological health. Trees will absorb the carbon dioxide helping to mitigate the climate change and also reduce the greenhouse gas concentration. It prevents soil erosion and also plays an important role in regulating the water cycle.

Species Recovery Program: Breeding programs can be implemented to increase the population of various endangered species and maintain the genetic diversity and also prevent the extinction of the species. It helps to reintroduce the species into their natural habitats and maintain the ecosystem.

Ex situ Conservation: It refers to the conservation of various species outside their natural habitats. It helps to protect the species that have been endangered and ensure the survival and maintain the genetic diversity of the species. Zoos, botanical gardens, and seed banks are some of the places where the genetic diversity of the species can be preserved from their extinction.

Conservation Genetics: By applying various genetic methods and principles, we can conserve and restore biodiversity. It mainly focuses on the preservation of genetic diversity of the spices. It helps to strengthen the species resilience to the environmental changes and various diseases and also supports a long-term survival of the species adapting to the environmental changes.

5 Innovative and Sustainable Solutions

Innovative and sustainable solutions are essential to address the loss of biodiversity and to promote the healthy ecosystem. The various strategies that can be followed to prevent the loss of biodiversity are:

Sustainable agriculture and forestry (Zhou et al., 2022): It is a practice that has been designed to meet the requirements of current food and the resources without compromising the ability to grow and meet the needs of future generations. The main aim of this is to balance environmental health, economic profitability, and also social equity. It also promotes the management of resources, conservation of biodiversity, and sustainable livelihoods. Sustainable agriculture involves various farming methods that will help us to protect the environment and also the health of the human and animal welfare. Some of the key practices that have been involved in sustainable agriculture are agroforestry, organic farming, conservation tillage, integrated pest management, and water conservation. Agroforestry is a method where the trees and shrubs are integrated into the agricultural landscapes that will help to improve the biodiversity, enhance the fertility of the soil, reduce the soil erosion, and also increase the crop yields. Organic farming is a method where synthetic chemicals are avoided in the farming by applying only natural fertilizers. Crop rotation technique can also be used to improve the soil health, and it also helps to increase the yield of crops by decreasing the diseases and controlling the pests that affect the yield of the crops. Conservation tillage is another method of agricultural practice that helps to minimize the disturbance of soil during the planting and the crop cultivation stages. It helps to preserve the structure of the soil and also improves the retention of water by reducing the soil erosion and it enhances the health of the soil. Biodiversity-friendly land-use practice can also be implemented to protect the habitats and support the wildlife.

Sustainable agriculture and forestry: It is a practice that helps to meet the current needs of food, fibre, and timber while ensuring the long-term health benefits of the environment, society, and the economy. It aims to balance the ecological and social aspects of agriculture and forestry by ensuring that the available resources are used in such a way that they are productive and also beneficial for the future generation. Sustainable agriculture involves various farming methods like Agroforestry which combines the agriculture and the forestry practices that will help to create a diverse and sustainable land-use to improve biodiversity. Organic farming is a method that helps to reduce the use of chemical inputs, enhance the health of the soil, improve the yield of the crop, and promote biodiversity on agricultural lands. Biodiversity-friendly land-use practice is used to implement farming and forestry practice that will help to protect the various habitats and it will also support the wildlife.

Ecological corridors and connectivity: It is a strategy that helps to link the fragmented habitats and the ecosystems, and the physical structure will help the species to move freely between the landscapes. It helps to maintain biodiversity by facilitating the migration of species and also supports the formation of a healthy ecosystem. Ecological corridors are managed strips of land that are used to connect the isolated patches of the natural habitats. It serves as a pathway that helps the wildlife to travel between various habitats and ensure genetic exchange by providing access to food, shelter, and water to the wildlife while travelling. Some of the key features of the ecological corridors are to link the fragmented habitats, promote biodiversity, support the species migration, and also enhance the ecosystem services. Connectivity refers to how the various habitats or ecosystems are connected to each other by allowing the species to move among themselves. It helps to ensure that the species can interact with each other and their environment. There are various types of connectivity like landscape connectivity, genetic connectivity, and functional connectivity. Ecological corridors and connectivity will help to maintain and enhance the biodiversity and also reduce the risks of inbreeding or extinction. It also helps to maintain the connectivity between the ecosystems and preserves the process of the ecosystem.

Ecosystem-based approaches: These are strategies that have been used to serve the functions of the ecosystem that will help people to adapt to the changing climate and reduce the risks that have been associated with the disasters (Scarano, 2017). It helps to integrate biodiversity conservation into climate adaptation and disaster risk reduction. Some of the key principles of the ecosystem-based approaches are harnessing ecosystem services, promoting ecosystem health, integrating and multi-scale planning, and engaging community participation and local knowledge. Few examples to illustrate the ecosystem-based approaches are coastal protection through mangrove restoration which will act as a natural barrier against storms and protect the coastal community from extreme weather. Flood mitigation via wetlands restoration is a technique where wetlands are used to absorb the excess water during the heavy rainfall to reduce the risk of flood. Ecosystem-based approaches help to create a sustainable and integrated way that will help to address the various challenges of climate adaptation and disaster risk reduction and help to improve the well-being of the human and environment.

Payment for Ecosystem Services (PES): It is a market-based approach where individuals, communities, or organizations are provided with incentives for enhancing the ecosystem service. This program aims to provide rewards to landowners, managers, and the stakeholders who adopt various practices that help to conserve or to restore the ecosystem. The key concepts of the PES are ecosystem services, economic incentives, and voluntary participation. There are various types of ecosystem services in PES like carbon sequestration, biodiversity conservation, water management, soil protection, and flood and storm protection. The carbon credit system is one of the best examples for the PES program under this system. It is used to reduce the greenhouse gas emission by providing financial incentives to organizations or companies that help to reduce the carbon footprints. It plays a vital role in the mitigation of climate change. Some of the benefits of the PES are conservation and environmental protection, sustainable use of land, helps in mitigation of climate change, local and economic development with cost-effectiveness.

Nature-based solutions: Various strategies that can be implemented by using the natural process that will help to address the various environmental and economic challenges and to restore the ecosystem. Some of the nature-based solutions like green infrastructure, urban biodiversity, and ecosystem restoration can be used to enhance the ecosystem. Green infrastructure refers to a network of natural and semi-natural features which use natural processes. It helps to emphasize the sustainable and eco-friendly approach. Some of the key components of the green infrastructure are vegetative systems like green roofs, green walls, urban forests, and street trees. Water management systems like rain gardens, permeable pavements and bioswales are good nature-based solution. Wetlands and natural water bodiesl can be used as restroed wetlands, urban ponds and lakes to prevent the biodiversity loss. Green spaces and corridors like parks and open spaces, ecological corridors etc can also be a nature-based solution to prevent the biodiversity loss. Urban biodiversity mainly focuses on maintaining the biodiversity in cities and towns. Due to the expansion of urban areas, it is necessary to integrate nature into the urban planning such that it should ensure the ecosystem services for the residents in the urban area. Some of the strategies that can be implemented to improve the biodiversity imbalance in the urban area are to develop a biodiverse garden by planting various native species that helps pollination and other wildlife. Wildlife corridors should be built such that the pathway will help the wildlife to move freely and connect the green spaces in the urban areas promoting the local communities to participate in various activities that are related to urban greening and the conservation of biodiversity. To mitigate the climate change and to improve the biodiversity, ecosystem restoration can be implemented to repair and restore the natural functions and services like reforestation, wetland restoration, and coral reef restoration.

6 Role of Policy and Governance

It is most important to create an effective policy and governance to address the loss of biodiversity and to promote sustainable development. The policies should be made at international, national, regional, and local levels such that it ensures the conservation of biodiversity. Some of the key international frameworks that have been established to preserve biodiversity loss are Convention on Biological Diversity (CBD), Aichi Biodiversity Targets, and Sustainable Development Goals (SDGs). CBD is an international treaty with the main aim of conserving biodiversity, promoting the use of sustainable components, and ensuring the reasonable distribution of benefits from hereditary resources. It was established in the year 1992 during the Earth Summit in Rio de Janeiro and was approved by 196 parties as one of the universally supported treaties on international environment protection. Aichi Biodiversity Target was established in the year 2010 by setting 20 global targets that should be achieved by 2020 mainly aiming to reduce biodiversity loss and to enhance the resilience ecosystem by integrating the biodiversity into the broader policy and strategy. Sustainable Development Goals (SDGs) is a collection of 17 goals that have been set by the United Nation that are mostly related to biodiversity which include Goal 14 based on life below water and Goal 15 based on life on land to emphasize the importance of preserving the marine and terrestrial ecosystem.

National and regional policies: Apart from the international policies, national level policies are also built like biodiversity action plans, environmental protection laws, and green economy initiatives. Various strategies have been outlined by national strategies of various countries to conserve biodiversity and to improve the use of sustainable natural resources such that it also aligns with the international frameworks and also address the various challenges based on the local biodiversity (Whitehorn et al., 2019). Various acts have been passed in legislation of various countries to protect the endangered species, conserve the habitats, regulate the pollution, and also manage the available natural resources. For example, the USA has passed the Endangered Species Act, and Australia has passed the Environment Protection and Biodiversity Conservation Act. Various policies and initiatives are been taken in various countries to promote the Green Economy Initiatives for the sustainable environment growth which include renewable energy, sustainable agriculture, conservation of efforts, and preservation of the ecosystem (Bawa et al., 2021).

Corporate Responsibility: Corporate companies have more responsibility in preventing biodiversity loss. Biodiversity impact assessments are done by the companies to evaluate the understanding of the potential impact of the biodiversity loss and to identify the various mitigation measures that should be taken to protect the loss of biodiversity. Corporations should enhance the practice of sourcing sustainable products so that they can minimize the negative impacts on the ecosystem and also preserve the loss of biodiversity. Corporations can take voluntary pledges such that their business will reduce the ecological footprint and conserve the natural resources. They can also invest a part of their profit in various conservation projects such that it will align with the global standards and frameworks like SDGs and CBD.

Indigenous and community-led conservation: Indigenous communities have more knowledge about the local ecosystem and its biodiversity which can be used for the conservation of the local biodiversity loss. Traditional knowledge from the indigenous community and integrating it with modern conservation practice can help us to enhance the effectiveness in protecting biodiversity loss. Community-managed conservation areas that have been protected by the local communities with the help of traditional practice and the governance structure will help to support the local livelihoods. Identifying and recognizing the rights of indigenous people will help us to manage their lands and resources to conserve biodiversity. It is most important that the policy and the governance structure should facilitate the cooperation among the various countries, regions, and the local community such that it ensures the unified approach to prevent the loss of biodiversity. It is also necessary to have an effective governance, financial mobilization and human resource to make necessary initiatives and implement it as a large-scale project.

7 Community-Based Conservation (CBC)

Initiatives have been taken to protect the available natural resources and loss of biodiversity with the help of the local community. The main aim of the CBC is to create a sustainable solution by integrating the knowledge of the local community about the culture and the structure of the governance that will benefit the people who depend on the environment. This can be achieved by engaging the local communities to participate in various planning and decision-making on various projects that helps to conserve the culture with social acceptability. The ownership of the conservation project to the community people will make them more responsible and committed which will lead to a sustainable outcome. Awareness can be given to the local community about the importance of biodiversity and how it is important for the ecosystem? Educational programs can be included to provide the communities understanding which will be a long-term solution. Various training programs and capacity-building programs can be conducted to enhance the skills of the people to effectively use the available natural resource and conserve it. By strengthening the local governance like cooperative society, community conservation committees can be used to ensure the sustainable effort for the conservation of natural resources.

Here are some of the success stories of various countries based on the Community-led conservation efforts (Kathleen Galvin, 2018). In Namibia, they established a community conservancy for managing wildlife and the natural resources which helped to increase the population of the wildlife and also generated income by attracting tourism which benefit both the local people and also conservation of resources. Same way in Nepal Chitwan National Park was built to manage the buffer zone which helped to reduce the human-wildlife conflicts and also strengthen the forest health. It also helped the local community's income through the tourism and use of sustainable resources (Fariss et al., 2022). In Philippines, various coastal communities established marine protected areas (MPAs) which have been managed by the local fishers. It helps to increase the fish population due to which the livelihood of the community has been improved and also, they strengthened the community resilience to the environmental changes. In Mexico, a community-managed forest has been developed which helped to preserve biodiversity and also to improve sustainable products using timber which helped to increase the economic growth of both the people and the governance. A community-based conservation is one of the most useful approaches for the prevention of biodiversity loss. It is a win-win scenario where the local people and also the ecosystem get benefited even though there are various challenges that should be faced to implement. The various success stories that have been seen help us to significantly contribute to the CBC approach to prevent biodiversity loss.

8 Biodiversity and Climate Change

There is a strong interconnection between biodiversity and climate change. The change in climate will have a significant impact on biodiversity which is one of the major challenges faced globally. But on the other side, a healthy ecosystem will help to mitigate the changes in the climate and help to enhance the resilience.

Climate change impacts on biodiversity: Due to the change of climate, there is a huge change in the behaviour, distribution, and survival of various species that are habitats of the ecosystem. Due to the rising temperature, various species are being forced to migrate to regions that are cooler like high altitudes or latitude which will disrupt the ecosystem. Due to the thermal stress, few species face increased mortality and also reduction in reproduction. The change in the precipitation pattern like altered rainfall led to severe drought or flood which has more impact on freshwater ecosystems and also affects the species that are mostly dependent on the stable water source. Due to the change in the precipitation, the growth of plants has been affected, thereby affecting the availability of the food for the habitat. Due to the increase in the frequency of storms, most of the habitats in the coastal and marine ecosystem like coral reefs and mangroves are destroyed. The increase in temperature and prolonged dry periods has increased the risk of wildfires which also destroys the various species in the forest and its inhabitants.

Mitigation and adaptation strategies: It is necessary to take action to preserve the loss of biodiversity which plays a more important role in the climatic change. Various strategies like ecosystem-based adaptation (EbA), conservation of carbon-rich ecosystem, and climate-smart protected areas can be implemented to mitigate the changes (Gillingham, 2024). Restoration of the ecosystem can be done by restoring the forest and wetlands which helps the buffer community to protect themselves from the various impacts like flooding and erosion and also provide habitats for the wildlife.

Ecosystem services and climate regulation: It is the need of the hour to regulate the climate and mitigate the various effects of climate change by building a biodiversity rich ecosystem. The forest has more capacity to absorb the carbon dioxide in huge amount that are been present in the atmosphere which will help us to mitigate the global warming (Zafar et al., 2024). It also helps to influence the weather pattern of the local area like change in temperature which will help to contribute to a stable climate. Increasing wetlands will help to control the floods since it will act as a natural sponge that will absorb the excess water during the flood and the absorbed water will be released during the hot and dry period which will help to mitigate the impact of extreme hot weather. Peatlands also act as a carbon sink by storing more carbon per unit area which helps us to create a good ecosystem (Wang et al., 2022). Near the coastal areas, the marine ecosystem can be implemented by growing mangroves, seagrasses, and salt marshes which have the capacity to capture and store the carbon which plays an important role in the global carbon cycle (Chen et al., 2023). The absorption capacity of heat by ocean is also significant which also helps to moderate the variation in the climate and the global temperature.

9 Technological and Scientific Advances

The role of innovative technology is more important in the conservation of biodiversity, it also acts as a powerful tool to monitor and analyse climate change impacts on the various steps that have been taken to mitigate the challenges. With the help of the Geographic Information System (GIS), the detailed information about the ecosystem has been created and tracking the changes in the ecosystem during the time period is made easy. With the help of the GIS, we can also identify the areas where more importance should be given for ecology. Spatial data that are being available can be integrated with the ecological information which can be used to predict the distribution of various species and identify the habitats that are under critical situation which should be preserved. Remote sensing is made easier with the use of satellite images and aerial drones. The high-resolution images that have been received from the satellite help to monitor the large-scale environmental changes like deforestation, fragmentation of various habitats, and changes in the usage of lands. Aerial drones are also used to collect the real-time data in the remote areas and also used to monitor and assess the habitat in the dense forest.

DNA Barcoding and Genetic Analysis: DNA barcoding helps to identify species based on the short genetic sequence based on the standardized region quickly and accurately. It also helps in monitoring the biodiversity by cataloguing the various species in the different parts of the ecosystem and also in monitoring the illegal wildlife trade and to detect the invasive species movement. With the help of genetic analysis, we can identify the various strategies that can be used to assess the genetic diversity within the population and it can also be used to identify the risks in inbreeding. It is also necessary to understand the relationship between the spices (Phylogenetics) which helps us in the reconstruction of the ecosystem and prevent the loss of biodiversity.

Citizen Science: It is all about involving the general public in a particular region for the purpose of data collection and scientific research. For example, we can involve the public in collecting the data regarding the various species occurrences like monitoring of butterfly, bird counts and in mapping the biodiversity apps. They also provide valuable data which will help the researchers to track back on the various changes in the biodiversity during a period of time. It also helps to provide awareness about the changes in biodiversity and the need to conserve it by fostering the connection between nature and the people with the help of the native people.

Artificial intelligence and machine learning: The various AI and machine learning algorithms have been used to predict and analyse the environmental changes and predict the details about the climate that most suits for the various species. It also helps to predict the likelihood of the species based on their survival factors like food, climate, habitat, and impact on humans. Various AI powered tools are used to analyse the real-time images that have been received from the satellite that helps to monitor the wildlife. It also enables the integration of various datasets from different sources that will provide a complete insight about the pattern and trends of various species and habitats. Advances in technology have made a huge transformation in the conservation of biodiversity. The ability to monitor and analyse the changes in the ecosystem has been improved. The various tools that have been available for remote sensing, GIS, and AI- driven models are providing closer insight about nature which helps us to address the various challenges that we are facing due to the loss of biodiversity.

10 Education and Awareness

Providing education and awareness about biodiversity preservation are more crucial. It is necessary to incorporate and educate the public about the need to preserve biodiversity with the help of various sectors like social media and various media channels, and the public can be motivated to participate in various conservation programs in their region. Educating about biodiversity to every individual and community will help them to know the value of the biodiversity and the need to protect it, they also will know the various threats that they need to face due to the loss of biodiversity.

In schools, a separate topic can be integrated in the curriculum along with science, geography, and environmental studies which will help the students to know more about the ecosystem, various species in the ecosystem, and also the ways they can preserve them.

Schools can organize field trips to natural reserves and wildlife sanctuaries and give awareness to them how they are connected with nature. An experiential learning will have more impact with the students. They can also reinforce them in the classroom teaching so that the students can relate it easily.

Awareness can be given to the global audience with the help of various media which plays a vital role in the transformation of information. Documentary films that showcase the importance of "Mother Earth", its beauty, and importance about the prevention of loss of biodiversity that highlights the challenges that have been faced in the prevention of loss along with various success stories will have a great impact. Awareness can be provided by broadcasting these films in various social media platforms, broadcasting it in various television channels so that it will reach a diverse audience and make them aware about the various issues that they are facing due to the biodiversity loss. Social media campaigns can be conducted by sharing the awareness education about the ecosystem and the ways to preserve along with the ongoing initiatives. Interactive content like conducting biodiversity awareness quizzes, engaging people in social media about complex topics on prevention of biodiversity loss. To reach the local community, information about the various projects that have been implemented in the local region can be regularly published in newspaper and magazines and announced in various local channels and radio programs which will help to provide more awareness among the local community and motivate each individual to contribute to the prevention of biodiversity loss.

Non-governmental organizations and various other national and international organizations are also playing an important role in creating awareness about the conservation of biodiversity among the public. Grassroot is an NGO which mostly works with the local community people to provide awareness about the various available resources in the community and also educate them to conserve the resource. Global campaigns are also organized by various organizations like WWF and Conservation International to provide awareness about the loss of biodiversity and the changes in the policy to prevent the loss. International organizations like CBD and SDGs are more involved in promoting global awareness about biodiversity. The various educational materials for the programs that highlight the importance of biodiversity have been offered by UNESCO's Man and the Biosphere Programme (MAB) (Ishwaran, 2012). Educating the people about the importance of loss of biodiversity is more important, and it is necessary to protect nature by fostering the global culture in the conservation of natural resources.

11 Future Directions in Biodiversity Conservation

Due to the various challenges that the globe is facing due to the significant changes in biodiversity, the conservation efforts that have been taken are more innovative with collaborative ideas and integrating various organizations. The key focus area at present to prevent the loss of biodiversity is to align with the sustainable development goals with global partnership (Hossian et al., 2018). Various ongoing research and implementation strategies are mostly focused on the biodiversity framework. SDG 15 (life on land) mainly aims to protect and promote the sustainable use of the terrestrial ecosystem. It also helps to manage the sustainable forest and help to stop the biodiversity loss (Opoku, 2019). While SDG 14 (life below water) mainly focuses on the conservation of marine resources and the sustainable use of oceans, seas, and the marine resources, it is also necessary to incorporate cross-sectoral integration like agriculture, forestry, and urban development such that it ensures the sustainable development practices. It is also necessary to identify the economic values and its services and integrate it along with the national development planning. Governments, business organizations, and NGOs can collaborate in framing national level policies that can be incorporated to set the conservation policies and legal frameworks. Bilateral and Multilateral agreements can be made to protect the transboundary ecosystem. A global funding mechanism can be created which can provide financial support to various initiatives that have been taken in the developing country to protect the loss of biodiversity. Currently most of the researchers are focusing on identifying the sustainable goals and in monitoring the ecosystem. The goal that has been set at present is to stop the biodiversity loss by the year 2030 and also to recover the ecosystem by 2050. Most of the countries in the global are at present developing various implementation strategies based on sustainable development goals.

References

Ahmed, A. K. (2022). Climate change: An overview. In *Global agricultural production: Resilience to climate change*. Springer.

Ali, A. (2023). Biodiversity–ecosystem functioning research: Brief history, major trends and perspectives. *Biological Conservation, 285*, 110210.

Bawa, K. S., et al. (2021). Securing biodiversity, securing our future: A national mission on biodiversity and human well-being for India. *Biological Conservation, 253*(1), 108867.

Chen, F., et al. (2023). Cropland carbon stocks driven by soil characteristics, rainfall and elevation. *Science of the Total Environment, 862*, 1–11.
Fariss, B., et al. (2022). Catalyzing success in community-based conservation. *Conservation Biology, 37*(1), e13973.
Gillingham, P. K. (2024). Climate change adaptation for biodiversity in protected areas: An overview of actions. *Biological Conservation, 289*, 110375.
Gustavo Ramirez, K. R. (2023). study.com. Retrieved December 18, 2024, from https://study.com/academy/lesson/what-is-habitat-destruction-effects-definition-causes.html
Hogue, A. S., & Breon, K. (2022). The greatest threats to species. *Conservation Science and Practice, 4*(5), e12670.
Hossian, S., et al. (2018). Identifying future research directions for biodiversity, ecosystem services and sustainability: Perspectives from early-career researchers. *International Journal of Sustainable Development and World Ecology, 25*(3), 249–261.
Isbell, F. I., Balvanera, P., Mori, A. S., & He, J. S. (2022). Expert perspectives on global biodiversity loss and its drivers and impacts on people. *Frontiers in Ecology and the Environment, 21*(2), 94–103.
Ishwaran, N. (2012). Science in intergovernmental environmental relations: 40 years of UNESCO's man and the biosphere (MAB) Programme and its future. *Environmental Development, 1*(1), 91–101.
Kathleen Galvin, T. A. M. L. (2018). African community-based conservation: A systematic review of social and ecological outcomes. *Ecology and Society, 23*(3), 39.
Klaus, V. H., Whittingham, M. J., Baldi, A., et al. (2020). Do biodiversity-ecosystem functioning experiments inform stakeholders how to simultaneously conserve biodiversity and increase ecosystem service provisioning in grasslands? *Biological Conservation, 245*, 108552.
Li, J. (2024). "Earth.org". Retrieved December 20, 2024, from https://earth.org/what-are-the-consequence-of-biodiversity-loss/
Liang, X., Li, Y., Ran, C., et al. (2020). Study on the transformed farmland landscape in rural areas of Southwest China: A case study of Chongqing. *Journal of Rural Studies, 76*, 272–285.
Mirzabaev, A., Kerr, R. B., Hasegawa, T., et al. (2023). Severe climate change risks to food security and nutrition. *Climate Risk Management, 39*, 1–10.
nextias. (2024). Retrieved December 18, 2024, from https://www.nextias.com/blog/biodiversity-loss/
Opoku, A. (2019). Biodiversity and the built environment: Implications for the sustainable development goals (SDGs). *Resources, Conservation and Recycling, 141*, 1–7.
Reeta Kumari, D. A. S. B. (2021). Biodiversity loss: Threats and conservation strategies. *International Journal of Pharmaceutical Sciences Review and Research, 68*(1), 242–254.
Scarano, F. R. (2017). Ecosystem-based adaptation to climate change: Concept, scalability and a role for conservation science. *Perspectives in Ecology and Conservation, 15*(2), 65–73.
Singh, B. N. (2024). Species: Concepts and categories. *Journal of Scientific Research, 68*(2), 9–16.
Sonia Llorente-Culebras, S. L., Ladle, R. J., Santos, A. M. C., & Santos, E. (2023). Publication trends in global biodiversity research on protected areas. *Biological Conservation, 281*, 109988.
Wang, B., et al. (2022). Modelling and mapping soil organic carbon stocks under future climate change in South-Eastern Australia. *Geoderma, 405*, 115442.
Whitehorn, P. R., et al. (2019). Mainstreaming biodiversity: A review of national strategies. *Biological Conservation, 235*, 157–163.
Zafar, Z., et al. (2024). Predictive modeling of regional carbon storage dynamics in response to land use/land cover changes: An InVEST-based analysis. *Ecological Informatics, 82*, 102701.
Zhou, Z., et al. (2022). Agricultural transformation and its impact on ecosystem services and human well-being in peri-urban areas: The case of Xi'an, China. *Land, 11*(1), 110.

Sustainable Usage of Terrestrial Ecosystem

G. Manivasagam, C. Balaji, R. Nithya, and T. Genish

1 Introduction

SDGs are the goals set by the United Nations (UN) to address various challenges that affect human life globally and to achieve sustainable development. Across the world, 17 SDGs and 169 targets have been monitored and examined using some indicators. The leaders across the globe had made a decision to safeguard the rights and well-being of every individual in the world in terms of a healthy and poverty free world. SDG offers a multidimensional view of growth, and almost all the nations have SDG indicator data. There are 230 indicators mapped to the 17 SDGs by the Inter-Agency and Expert Group.

Forests and biodiversity play a vital role in achieving sustainable development. Among the 17 Sustainable Development Goals (SDGs), Goal 15 focuses on land conservation. When the United Nations adopted the SDGs in 2015, indicators revealed a decline in biodiversity (Laurance et al., 2013). SDG 15 calls for the restoration, protection, and sustainable use of ecosystems, along with responsible forest management. However, challenges such as population growth, economic expansion, and increasing consumption threaten the sustainability of life on land. This chapter explores the key opportunities and challenges that must be addressed to achieve SDG 15 and halt biodiversity loss. As an integral part of the SDG

G. Manivasagam
Jain (Deemed to be University), Bangalore, India

C. Balaji
SRM Institute of Science and Technology, Trichy, India

R. Nithya
KPR College of Arts Science and Research, Coimbatore, India

T. Genish (✉)
Arumugam Pillai Seethai Ammal College, Tiruppattur, India

J. J. Puthenkalam S.J et al. (eds.), *Terra Thrive*,
https://doi.org/10.1007/978-3-032-02738-2_13

framework, Goal 15 underscores the importance of incorporating conservation concerns into broader policy and decision-making processes.

The Agenda of SDG for the year 2030 underlines the critical role of sustainably managing natural resources in promoting social and economic development. It highlights the importance of conserving and sustainably using seas, oceans, forests, freshwater resources, mountains, and drylands, as well as protecting biodiversity, ecosystems, and wildlife.

Human activity has transformed nearly 75% of Earth's surface, confining wildlife and natural ecosystems to shrinking areas. Yet, nature remains vital to human survival—it generates oxygen, regulates weather, pollinates crops, and sustains food production, all of which rely on soil. As a finite resource, soil is under immense pressure: More than 2 billion people depend on agriculture, but around 50% of cultivated land is moderately to severely destroyed. According to the UN, agricultural land is being lost at a staggering rate of 230,000 square meters per minute, disproportionately impacting 74% of the world's poor.

Forests, which host 80% of all land-based animals and plants, are also crucial in combating climate change. Critics of biofuels argue that their production competes with the need to feed a growing global population and contributes to deforestation, as forests are often cleared or burned to expand agricultural land. Life-cycle analyses show that ethanol provides 25% more energy than is required for its production, while biodiesel from soybeans offers 93% energy return. Additionally, biodiesel reduces greenhouse gas emissions more than three times as effectively as bioethanol, compared to the fossil fuels they replace (MacDicken, 2015).

2 Sustainable Development Goals: An Overview and Needs

For millions of years, flora and fauna flourished and diversified on Earth, but this began to change with the rise of human dominance. The field of agriculture was focused on cultivating wild plant species and basic livestock management. Over a period, individuals added knowledge and practice in plants and animals. To make way for agriculture, humans started clearing forests. In the last 10,000 years, more than 50% of forest area has been gone in the world, with the most significant decline occurring in the recent Five decades, coinciding with a rapid explosion of population. Population growth is a major contributing factor, and agriculture is considered as a direct cause of deforestation. Since the onset of deforestation, nearly 46% of the world's forest cover has been destroyed (Gupta & Khandelwal, 2019).

Population explosion and the activities of industries have placed immense strain on the world's ecosystems. Industrialization shifted societies from agrarian lifestyles to those dominated by technological innovations. This transition spurred rapid population growth and urbanization. The count of World's population was found as One Billion in the year 1804. Then the population reached Two Billion in another 120 years. Since then, the rate of population has surged, and today, the Earth is home to over billions of people. Various Earth's species ranging from plants to microorganisms have been essential for maintaining healthy and thriving ecosystems. Biodiversity ensures the continuing sustainability of life on the earth. Normally, species evolve and go extinct at a natural pace. Research shows that over the last half-billion years, there have been several extinctions on earth. We are now experiencing a sixth mass extinction. However, unlike past events driven by natural catastrophes like asteroid impacts or volcanic eruptions, the current extinction crisis is because of human activity.

Population and an industrial improvement have contributed to destruction, over consumption of several natural resource deforestation, climate change, etc. While the extinction rates occur at about one to five species per year, studies suggest that species are now disappearing gradually, with several species extinct daily. If such a trend continues, one third or half of the species will be lost within mid-century. Human activities have dramatically disrupted the balance of life on Earth (Kakar et al., 2021).

Human life is deeply connected to both the Earth and the ocean for food and living. Plant life offers 80% of the food in agriculture and is a significant key factor for the development of an economy. One third of the surface of the earth is covered by forests, offering essential benefits to thousands of species and providing critical resources. They also play an important role in combating climate change. Today, modern agriculture depends on certain crops, which lack genetic variety, making them more susceptible to diseases, pests, and environmental fluctuations. The extinction of related wild species results in the loss of genetic material that could be crucial for future agricultural resilience and combating emerging diseases.

In light of the accelerating rate of extinction and the exhaustion of biological assets, the SDGs emphasize the conservation and restoration of global ecologies, such as forests, mountains etc., by 2020. Uncertain deforestation is essential for modifying environment variations. Serious action is required to protect biodiversity and natural resources, as they are essential heritage of a country. The rates of biodiversity harm, poaching, trafficking, and the spread of invasive species pose major challenges to achieving these objectives. The UN approved the 2030 Agenda in 2015 that comprises 17 Goals which aim to address global issues such as environmental degradation, poverty, climate change, prosperity, inequality, etc.

2.1 SDG 15: Targets

Goal 15 seeks to safeguard, restore, and promote the sustainable use of terrestrial ecosystems, confirm maintainable management of forests, fight desertification, and stop or land degradation. It stresses the importance of integrating ecosystems and biodiversity into national and local planning, development strategies, poverty reduction efforts, and national accounting. Additionally, the goal aims to guarantee the impartial and unbiased distribution of assistance from the resources, enable responsible contact to these assets, and curb the stealing and transferring of endangered plant and animal types. This goal is deeply interconnected with nearly all other

SDGs, including those addressing poverty alleviation, hunger eradication, inequality reduction, gender equality, and employment.

The targets and indicators which have been outlined to accomplish this goal by 2030 are given in Table 1.

2.2 *Significance of Goal 15 from an Indian Perspective*

Biodiversity in India has developed history over several billion years, with adequate resources deeply rooted in Indian culture. As a nation housing almost one-fifth of the global people, India is witnessing a rapid shift from an agrarian economy to an expanded one, causing increasing burdens on terrestrial and natural resources. This transition has triggered significant habitat loss and fragmentation, posing the greatest threat to biodiversity. The devastation and deprivation of ecosystems, coupled with over manipulation of types, have severely impacted India's natural wealth. Over recent decades, the country has lost a huge area of forests, made the water bodies polluted in large numbers, replaced much of its grasslands with urban or agricultural development, and degraded several coastal areas.

India is home to three of the world's 34 biodiversity hotspots, areas teeming with life yet facing urgent conservation challenges. These include tropical rainforests, which are among the most biodiversity-rich habitats but are being destroyed due to human activities. Such habitat destruction not only diminishes biodiversity but risks the extinction of species that cannot survive elsewhere. Currently, around 9% of wild flora documented in India, particularly plants, and a huge percentage of fauna, are under risk. The true extent of species nearing extinction remains unclear, highlighting the need for more robust data and conservation measures.

India has demonstrated global leadership in biodiversity conservation by being an early signatory to the Nagoya Protocol and committing to the Aichi Biodiversity Targets. Given the accelerating degradation of ecosystems, it is imperative to implement immediate and effective measures to protect India's unique biodiversity and ensure its sustainable future.

2.3 *The Status of India in "Life on Land"*

Of the approximately 190 countries in the world, only 17 are considered "mega diverse," collectively harboring 70% of global biodiversity. India is proudly one of these mega diverse nations. Although it occupies only 2.4% of the biosphere's terrestrial zone, it supports 7% of the plant's classes; containing around Ninety One thousands documented animal types and Forty Five thousands plant types, which constitute 13.8% of the world's flora. India's biodiversity spans ten biogeographic regions and includes over Seventeen thousands species of blossoming flowers, in which Four thousand Nine fifty are prevalent to the nation. Globally, the nation

Table 1 Targets of SDG-15

Goal 15	Targets of SDG 15	
15.1	By 2020, secure the conservation, restoration, and sustainable utilization of terrestrial and inland freshwater ecosystems and their services, in particular forests, wetlands, mountains and drylands, in line with obligations under international agreements	
15.2	By 2020, promote the implementation of sustainable management of all types of forests, halt deforestation, restore degraded forests, and substantially increase afforestation and reforestation globally	
15.3	By 2030, combat desertification, restore degraded land and soil, including land affected by desertification, drought and floods, and strive to achieve a land degradation-neutral world	
15.4	By 2030, ensure the conservation of mountain ecosystems, including their biodiversity, in order to enhance their capacity to provide essential benefits for sustainable development	
15.5	Take urgent and significant action to reduce the degradation of natural habitats, halt the loss of biodiversity and, by 2020, protect and prevent the extinction of threatened species	
15.6	Promote fair and equitable sharing of the benefits arising out of the utilization of genetic resources and promote appropriate access to such resources, as internationally agreed	
15.7	Take urgent action to end poaching and trafficking of protected species of flora and fauna, including by tackling both demand and supply of illegal wildlife products	
15.8	By 2020, introduce measures to prevent the introduction and significantly reduce the impact of invasive alien species on land and water ecosystems and control or eradicate the priority species	
15.9	By 2020, integrate ecosystem and biodiversity values into national and local planning, development processes, poverty reduction strategies and accounts	
15	15.1.1	The percentage of forest area in relation to total land area
15	15.1.2	The percentage of significant terrestrial and freshwater biodiversity sites that are covered by protected areas, broken down by ecosystem type
15	15.2.1	Progress in achieving sustainable forest management
15	15.3.1	The percentage of degraded land relative to the total land area
15	15.4.1	The extent of protected area coverage for key sites of mountain biodiversity
15	15.4.2	Mountain Green Cover Index
15	15.5.1	Red List Index
15	15.6.1	The number of countries that have established legislative, administrative, and policy frameworks to guarantee fair and equitable sharing of benefits
15	15.7.1	The proportion of wildlife trade involving poached or illicitly trafficked species

(continued)

Table 1 (continued)

Goal 15	Targets of SDG 15	
15	15.8.1	The proportion of countries that have implemented appropriate national legislation and provided adequate resources to prevent or control invasive alien species
15	15.9.1	(a) The number of countries that have established national targets in line with or similar to Aichi Biodiversity Target 2 of the Strategic Plan for Biodiversity 2013–2020 within their national biodiversity strategies and action plans, and the progress reported toward these targets; and (b) the incorporation of biodiversity into national accounting and reporting systems, defined by the implementation of the System of Environmental-Economic Accounting

Sources
https://www.csir.res.in/corporate-social-responsibility-csr/policy-and-guidelines-accepting-and-utilization
https://www.cbd.int/doc/meetings/cop/cop-13/official/cop-13-06-en.doc

ranks ninth in higher plant species richness. Its faunal diversity is equally remarkable, with 91,307 animal species, representing 7.46% of the world's fauna. This includes approximately Eight lakhs insects, Twenty one thousand fish, Two hundred and forty amphibians, Four hundred and sixty reptiles, Thousand two hundred bird species, and Three hundred nineteen seven mammals. Among these, invertebrates account for about 86,413 species.

The level of endemism in India's biodiversity is significant. Additionally, it is estimated that around 400,000 more species in India remain to be discovered and documented. India also boasts a rich avifauna, with several birds' species. However, 78 Indian bird species are currently threatened with extinction. Beyond visible biodiversity, India also hosts an impressive diversity of microscopic life, including at least 850 species of bacteria and viruses, and around 12,500 species of fungi (Kakar et al., 2021). This vast biodiversity is a result of India's diverse landforms, which support environments ranges from tropical to temperate forests, alpine ecosystems, and arid deserts. This immense variety not only highlights India's biological richness but also underscores its critical role in global conservation efforts. India is home to globally recognized biodiversity hotspots, including Himalayas which ranks among the most biologically rich regions in the world. Islands of India further enhance this richness with a remarkable variety of species and subspecies of animals and birds, many of which are endemic—found only in India.

The Andaman and Nicobar Islands alone host approximately 2000 types of flowering floras and 120 types of leaves in which of 135 land mammal genera in India, 85 (63%) are concentrated in the northeastern states, which also boast 1500 endemic plant species. The Western Ghats, another biodiversity treasure, are particularly notable for their amphibians, reptiles (especially snakes), and 1500 endemic plant species. India's coral reef ecosystems are equally significant. Found around the Lakshadweep Islands, and the Gulf regions of Tamil Nadu, these reefs are closely as species-rich as steamy forests and serve as critical habitats for marine biodiversity.

These hotspots and ecosystems are not only vital for their unique biodiversity but also for their ecological significance, making their conservation a top priority for India and the world.

2.4 Integration of Water and Land Resources to Combat Land Deprivation in India

With approximately 30% of land in India classified as corrupted, the Indian Government is prioritizing the integration of several terrestrial and water management plans to address land deprivation comprehensively within countrywide programs. The nation has set an ambitious goal to attain land deprivation impartiality in the next Five years, ensuring that any new land degradation is offset by equivalent restoration efforts, thereby maintaining the balance of land-based natural capital.

Although the combined budgets related to the schemes at the central and state levels are sufficient to tackle the issue, better convergence and coordination of these resources are essential for maximizing their effectiveness. The TERI (The Energy and Resources Institute) says that "Economics of Desertification, Land Degradation, and Drought in India" highlights that though the harm of prolific land across different environments is a pressing issue, the degradation of existing ecosystems poses an even greater concern.

For 2014–2015, the report says that the cost of land deprivation and use of land changes in India yearly were estimated at nearly Three Lakhs crores. The cost is as same as 2.5% of the Gross Domestic Product and approximately 15% of the gross value added from the fishing and agriculture sectors. Notably, 80% of these costs are attributed to land degradation, with the remaining 18% due to land-use changes. Under two projected scenarios, the land area likely to face degradation by 2030 is estimated at 94.53 million hectares and 106.15 million hectares, respectively. Both scenarios indicate an increase in the regions affected from erosion and a rise in forest cover. These trends underline the urgent need for India to intensify reclamation and conservation efforts, particularly in areas impacted by erosion and deforestation.

3 Sustainable Development Issues

Sustainability is not just about prioritizing environmental preservation but about maintaining a balanced relationship between ecological health, social, and political stability and essential human needs. True environmental sustainability must be considered within the broader context of societal actions and their environmental consequences. A well-rounded discussion on sustainable development and urban sustainability should incorporate four critical aspects: population dynamics and urbanization, the role of economic production systems, awareness of global environmental challenges, and a thorough evaluation of development strategies and future objectives. Addressing these factors collectively is essential for achieving long-term sustainability.

3.1 Effects of Population Growth

China and India together make up more than 40% of the global population. While this demographic strength has played a crucial role in their economic growth, both nations remain aware of the long-term challenges associated with rapid population expansion. The neo-Malthusian viewpoint, which argues that unchecked population growth can place excessive pressure on environmental resources, is particularly relevant in this scenario. Between 1950 and 2050, China's population is estimated to grow by 160%, while India's is projected to increase by 400%. As their populations continue to expand, both countries face mounting environmental challenges due to increasing demands on resources.

Studies indicate that food shortages are often not solely due to insufficient supply but are influenced by disparities in distribution, economic inequalities, and restricted access in affected regions. Ensuring food security requires not only advancements in agricultural production but also a comprehensive approach that considers political and economic structures. Since the 1960s, concerns over rapid population growth have led many developing nations to implement family planning policies. China enforced the one-child policy as a strict population control measure, while India introduced sterilization programs in an effort to curb population growth.

3.2 The Capitalist Economic Model

As society continues to progress with scientific and technological advancements, it has also established a production system that heavily exploits natural resources and a social structure that, at times, disregards human well-being. A common assumption is that economic growth alone will resolve societal challenges. Capitalism, as the dominant global economic system, has been a major driver of development but has also contributed to significant environmental concerns. Understanding its full impact requires examining its historical development and worldwide expansion. While capitalism takes on different forms in various cultural contexts—thriving in countries like India and China—globalization and rapid technological advancements have made it nearly impossible for any nation to remain unaffected by the broader economic framework.

There is growing awareness of the environmental link between industrialized and developing nations, as ecological challenges in less developed regions are often shaped by global economic forces. One of the key factors is the large-scale extraction of natural resources to meet the consumption demands of wealthier countries, leaving developing regions to bear the environmental costs. Additionally, these nations frequently function as raw material suppliers, supporting industrial and service economies in more developed nations while dealing with the long-term consequences of resource depletion.

3.3 *The Nature of Development*

Development is commonly associated with modernization, yet its success depends on a country's political leadership and administrative efficiency. In many developing nations, however, development is often influenced by nationalistic narratives and political motives, sometimes prioritizing symbolic projects over tangible progress. To achieve meaningful and lasting development, policies must be rooted in realistic objectives, well-structured planning, and effective execution. Moreover, they should be designed to align with the cultural and social fabric of society to ensure sustainable and impactful change.

3.3.1 Urban Sustainability

The ongoing expansion of urban areas and the continuous migration of people from rural regions present complex challenges that require well-planned governmental intervention. The emergence of megacities, alongside the growth of smaller urban hubs, has intensified demographic shifts, straining infrastructure, increasing unemployment, and raising concerns about social stability. The economic opportunities in these cities continue to attract both domestic and international migrants, but prolonged economic instability could reduce their appeal.

Even as overall population growth slows, urban migration remains a significant factor contributing to resource shortages and inadequate infrastructure. To foster sustainable urban development, authorities and urban planners must prioritize two key objectives: gaining a deeper understanding of urbanization trends and implementing strategies to mitigate the environmental footprint of expanding metropolitan areas.

3.4 *Asian Sustainable Land Management*

Forests, which support 80% of animal and plant species, also play a vital role in mitigating climate change. However, the production and use of bioethanol present both environmental and ethical challenges. Critics argue that biofuels compete with food production, raising concerns about global food security, and contributing to deforestation as forests are cleared for agricultural expansion. Life-cycle assessments indicate that ethanol produces 25% more energy than is used in its production, while soybean-based biodiesel offers a 93% energy gain. Furthermore, biodiesel reduces greenhouse gas emissions over three times more effectively than bioethanol when compared to fossil fuels. Nevertheless, even if all U.S. corn and soybean crops were dedicated to biofuel production, they would meet the demands for 12% of gasoline and 6% of diesel, highlighting the limitations of biofuels as a large-scale energy solution.

Asia is expected to undergo a period of population growth followed by a decline, bringing about complex socio-economic changes that will challenge land management practices. Effective and sustainable management of land and natural resources will be essential in navigating these shifts. A key aspect of this approach is recognizing the diverse functions that land serves to support long-term regional sustainability. Across the globe, land degradation is causing a decline in vital ecosystem services, raising serious concerns. Without prompt intervention, ongoing climate change and increasing consumption—driven by population growth, economic shifts, and evolving consumer demands—will further accelerate this issue. In many rural areas of Asia, land-use changes have been implemented to enhance food production. However, these efforts have largely prioritized efficiency and yield improvement, often neglecting the broader multifunctional value of land in contributing to the SDGs.

Sustainable land management aims to balance the use of land-based resources—including soil, water, plants, and animals—to meet human needs while safeguarding long-term productivity and ecological integrity. Tackling land degradation and advancing sustainable development require strategic policy measures. Regional land-use planning plays a crucial role in reducing disaster risks and enhancing climate resilience, ultimately minimizing the social and economic consequences of extreme weather events. This policy brief explores methods for evaluating the diverse roles of land use and management and provides recommendations for promoting sustainable land practices.

3.5 Ecosystem Services in the Asia-Pacific Region

Several countries in the Asia-Pacific region are experiencing development trends that resemble the historical changes seen in Europe and the United States. This often involves the transformation of natural resources into urban infrastructure, leading to the disruption of ecosystem services and contributing to environmental harm. Moreover, such patterns of growth can increase economic inequality and weaken social unity. To ensure sustainable development, it is vital to adopt effective land-use policies and cultivate a deeper recognition of the importance of ecosystem services.

3.5.1 Market Forces

Global development is driven by the gradual evolution of institutions, the growth of globalization, and the slow integration of less affluent regions into the economic frameworks of wealthier nations. Despite the continued progress in international economic integration, significant income disparities remain both across and within countries. At the same time, the worsening state of the environment is emerging as a key factor in addressing global challenges.

3.5.2 Policy Reform

The Policy Reform scenario imagines a future where committed political leadership leads to coordinated, swift actions for a fairer and more environmentally sustainable world. Rather than predicting future trends, this approach utilizes backcasting—starting with a desired sustainable future and working backward to identify feasible paths to reach it. The focus is on finding ways to meet social and environmental objectives while maintaining strong economic growth, similar to the path envisioned in the Market Forces scenario.

3.5.3 Fortress World

The Fortress World scenario is a particular interpretation of the broader Barbarization themes created by the Global Scenario Group. These scenarios envision a future where societal, economic, and ethical systems collapse due to crises that surpass the ability of markets and policy reforms to manage. In the Fortress World variant, authoritarian rule emerges as a response to the looming breakdown. Wealthy elites isolate themselves in secure enclaves, consolidating their control over critical resources, while the rest of society suffers from repression, environmental decay, and widespread hardship.

4 Health and Terrestrial Ecosystems

The connections between biodiversity, ecosystem services, and human health are complex, with human-driven environmental changes influencing well-being through multiple channels. The region is expected to face major environmental challenges, including biodiversity loss, climate change effects on agriculture and coastal areas, worsening air pollution and chemical exposure, and growing scarcity of water in Central Asia.

Efforts to protect biodiversity and sustainably manage land have had varying degrees of success, underscoring the immense challenge ahead. In the past, environmental degradation was often visibly linked to health problems, making it easier to address. However, today's large-scale environmental shifts—many of which remain insufficiently understood—present new and complex policy challenges.

Achieving Sustainable Development Goal (SDG) 15 requires strong political commitment, as well as collaboration across different sectors and policy areas. Strengthening policy coherence can lead to better long-term outcomes. Building strong alliances between governments, businesses, and civil society is essential for tackling interconnected social and environmental risks. Effective coordination between national and local authorities will also be key in managing critical system changes. Since environmental conditions play a significant role in public health, achieving SDG 15 is crucial to ensuring access to essential resources such as clean water, energy, etc.

The principle of "leaving no one behind" is at the core of the SDGs. In the context of SDG, it means improving conditions for those who rely most on natural resources for their survival, particularly marginalized and vulnerable communities. Restoring ecosystems and preserving biodiversity can help mitigate climate change while enhancing resilience to environmental shocks. Additionally, ecosystem services in both urban and rural areas contribute to pollution reduction and climate adaptation, creating healthier and more sustainable environments.

4.1 Health Benefits of Green Spaces

A well-designed urban environment with interconnected and visually appealing green spaces provides safe opportunities for residents to engage in physical activity, recreation, and social interaction while also supporting relaxation and stress relief. Such spaces contribute to urban resilience against extreme environmental challenges. Studies emphasize the necessity of both small, easily accessible green areas near residential and workspaces and larger parks that allow for meaningful interactions with nature.

The presence of green spaces in cities offers several health benefits, including:

- Improved mental well-being.
- Enhanced cognitive function.
- Greater social and cultural engagement.
- Better pregnancy health outcomes.
- Lower mortality rates.

Despite these benefits, any negative impacts are typically linked to inadequate maintenance. The need for green spaces and their positive impact on public health are universally recognized. However, urban planners and designers must consider local geographic and cultural factors when developing these areas. Important aspects influencing health outcomes include the perceived quality and accessibility of green spaces, the variety of available facilities, overall size, tree coverage, and canopy density.

To aid in urban planning and policy-making, a geographic information system (GIS)-based toolkit has been created to assess green space distribution, accessibility, and usage through objective and standardized indicators. This tool helps maximize the health advantages of urban greenery while ensuring that these spaces remain beneficial to all communities. Standardized metrics also allow for better evaluation of policy interventions aimed at improving green environments.

4.2 Strengthening Health Systems

Countries that are part of the WHO have made a commitment to promote universal health coverage, guaranteeing that everyone has access to essential healthcare services—such as prevention, treatment, rehabilitation, and palliative care—without incurring financial hardship. The growing risks to public health and safety posed by natural disasters and extreme weather events are becoming more severe due to climate change and the increasing strain humans place on ecosystems, disrupting their ability to maintain environmental stability. To address these concerns, health systems must focus on:

- Promoting environmentally sustainable behaviors that also support human well-being.
- Strengthening national capabilities for early warning systems and global health risk management.

The pursuit of sustainable solutions that align with ecological limits must be intensified, along with improving strategies for their effective implementation to prevent severe health consequences. Although sustainable development efforts offer many advantages, they can sometimes pose risks to both ecosystems and human health. For example, while wind farms provide a renewable energy source, their construction and operation may negatively impact the environment through noise pollution and habitat disruption, affecting species such as marine mammals and seabirds. Continuous assessment and monitoring are necessary to understand and mitigate these effects.

Additionally, certain industrial activities, including fracking and bioprospecting, raise concerns about environmental damage, community displacement, and the exploitation of indigenous knowledge for commercial purposes. A thorough evaluation of these practices is essential to ensure that sustainability efforts do not come at the cost of ecological integrity or social justice.

5 SDG 15 and National Context

The implementation of SDG 15 varies by country, depending on national priorities, environmental conditions, and the policy frameworks that guide land and biodiversity conservation efforts. Many countries have integrated SDG 15 into their national development agendas, adopting specific environmental policies aimed at safeguarding ecosystems. Nations with extensive forest resources, such as India and Brazil, have implemented strategies that seek to reconcile economic growth with ecological preservation, promoting responsible land use and sustainable forestry practices to maintain ecological integrity.

To address the challenge of land degradation, several countries have committed to global initiatives, including the UN Convention to Combat Desertification (UNCCD). Moreover, many governments recognize the vital role of indigenous and local communities in managing land and resources, incorporating their traditional knowledge into policies for conservation and sustainable land use.

However, challenges remain, such as deforestation driven by agricultural expansion and insufficient financial support for conservation projects. Large-scale reforestation initiatives are helping to restore degraded lands and increase carbon storage. Agroforestry, which integrates trees into agricultural practices, plays a significant role in preserving biodiversity and supporting rural communities. Indigenous and local populations continue to be key players in forest conservation, combining their traditional knowledge with contemporary sustainable practices.

5.1 Degradation of Land Resource

It is the reduction, either temporary or permanent, in the land's productive capacity, affecting its key uses such as rain-fed agriculture, irrigation, and forestry. It also impacts farming systems like smallholder subsistence and diminishes the land's economic value. This issue is considered one of the most urgent global challenges in the present day. Over the past 200 years, human activities have converted 70% of grasslands, 50% of savannahs, 27% of tropical forests, and 45% of temperate deciduous forests into agricultural land.

The geographical area of India covers around Three hundred million hectares, with 228.3 mha designated as drylands. The important natural resource is land, water cycle regulation, recycling process of nutrients, carbon storage, and other vital ecosystem services. Currently, 105.4 mha, or about 32% of land has undergone degradation. The erosion caused by water and wind is the leading contributor, accounting for over 71% of the total degradation. Water erosion, the most widespread form, impacts 33.56 mha across all agroclimatic zones, while wind erosion, prevalent in the western regions, leads to topsoil loss and sand dune shifting (Bai et al., 2008).

5.2 *Desertification*

Land humiliation is prevailing in sub-humid, semi-arid, and dry regions, where limited water availability hampers productivity and contributes to desertification. Around 81.45 million hectares (mha) or 25% of the nation's total geographical zone are ruined by desertification. With 69% of India's land comprising drylands, this degradation has severe consequences for livelihoods and food security. Desertification deteriorates the productivity of fragile ecosystems and deepens poverty (SRO, 2007). To address these challenges, India must focus not only on sustainable land management but also on restoring and regenerating degraded lands to enhance productivity and resilience.

5.3 *Ecosystem: Mountain*

The 1992 Rio Meet recognized the vital part of mountain environments, noting that around 10% of the global population relies on resources like forests, water, agricultural products, etc., for their livelihoods. These regions face significant environmental pressures from population growth, tourism, infrastructure expansion, mining, and other activities, making them particularly susceptible to degradation.

6 India's SDG Index: Interventions for Goal 15

Land is a crucial and irreplaceable global resource, supporting a huge part of various species such as plants, animals, and insects. However, unsustainable practices like deforestation have affected 21% of India's land area, while protected areas make up nearly 5% of the country's total land. With a large portion of the population reliant on land resources for livelihoods, the country is committed to incorporating the environment.

The performance of India in "Life on Land" is evaluated through national indicators, covering 3 of the 12 SDG targets. These indicators were selected based on the availability of data at the national level and to maintain comparability across states and Union Territories (UTs). Table 2 represents the national indicators used for SDG 15.

Table 2 National indicators of SDG 15

SDG target	Indicator for SDG India Index	National target value for 2030
15.1	1. Percentage of total land area covered by forests	33
15.1	2. Percentage change in the extent of water bodies within forests from 2005 to 2015	0
15.2	3. Variation in forest area from 2015 to 2017	0
15.7	4. Five-year percentage change in the estimated wild elephant population	0

6.1 National Mission for Green India

With a total budget of Rs 46,000 crores, the Green India Mission (GIM) focuses on increasing the coverage of forest and thus enhancing the value of present forestry. In the financial year 2018–2019, Rs 160 crores were allocated to the Green India Mission (National Afforestation Programme), marking a 48.8% increase from the previous allocation. The mission, planned for a 10-year period, has the following objectives:

- To expand forest cover by 5 mha.
- To boost livelihood who depend on forests.

It adopts a general approach to greening, emphasizing different environmental services such as water resources, biomass production, biodiversity, and the preservation of habitats. It promotes the advanced recent technology to monitor both inputs and outcomes effectively. It aims to integrate efforts with other national missions, programs, and schemes. To this end, convergence guidelines have been issued for coordination.

The mission outlines an ambitious plan to plant 1000 crore saplings across the country over a decade (2021–2030) under initiatives like CAMPA and GRM. Despite its potential, GRM has progressed slowly and, even in its fifth year, remains in the initial phases. The Ministry is currently understaffed and lacks the necessary expertise to implement the mission effectively. Although the task emphasizes an active meeting as the central body for distributed governance, no significant steps have been taken to build capacity at the grassroots level. The envisioned decentralized and participatory governance remains largely theoretical, with limited on-ground realization. At present, the mission appears more focused on being a plantation initiative rather than addressing broader objectives like carbon sequestration and forest diversion. These critical aspects have been largely overlooked, reducing the mission's effectiveness and scope.

6.2 Action Programme to Fight Desertification

Approximately 25% of area, equivalent to 82 million hectares (mha) is currently undertaking desertification, whereas 32% (105 mha) is experiencing deprivation, leading to low efficiency. This has significantly impacted the lives and nourishment safety of thousands and thousands of a common man in the country. Addressing this problem involves the interconnected management of three critical resources like water, land, and biodiversity. Those resources are co-dependent in one inevitably distresses the other. The hydrological cycle sustains life, while soil plays a vital role in regulating water movement. These three elements are inseparable in maintaining ecological balance.

ISRO, Ahmedabad, in collaboration with 19 partner institutes, conducted a comprehensive inventory and monitoring of desertification across India. The findings were compiled into an Atlas, providing valuable insights for policymakers and planners. According to the report, 69.6% of the nation falls below the category of dry land. Area affected by land degradation amounts to 105 mha, representing 32.07% of the country's area, while 81.45 million hectares are undergoing desertification. It aims to encourage a community-driven method to growth, plan actions that enhance the excellence of life for local groups, elevation of awareness, strengthen deficiency awareness, and implement locally appropriate R&D initiatives. Additionally, the focus is on empowering local communities through improved self-governance.

India is currently preparing an Action Programme to fight Desertification, aligned with the UNCCD 10-Year Strategy. This program will be finalized through regional consultations and workshops with stakeholders, keeping national circumstances and development priorities in view. India's current efforts prioritize sustainable land and resource management to support community livelihoods, restore land productivity, and make a healthier, more supportable environment for the people. The country is firmly committed to achieving Land Degradation Neutrality by 2030.

6.3 Integrated Development of Wildlife Habitats (IDWH)

Key activities in this plan have included the initiatives such as promoting ecotourism, anti-poaching initiatives, wildlife research, capacity building, etc. The scheme likewise provides financial assistance to situations for the repositioning of people from the conserved areas. This program supports States through the below components to enhance wildlife conservation efforts.

- Wildlife Conservation Beyond Protected Areas: Numerous wildlife habitats exist outside designated protected areas. This component provides funding based on the schemes developed by the corresponding states for regions adjacent to threatened zones.

- Restoration Schemes for Endangered Species and Habitats: A total of 16 species such as Nicobar megapode, Dugongs, Nilgiri tahr, Asian wild buffalo, etc., have been identified for recovery under this initiative.

In addition to offering substantial environmental benefits, these initiatives strengthen the execution of tiger protection efforts within the reservation of tigers by Project Tiger. They also support wildlife protection in safe areas and nearer parts via the Habitats program, along with the program of Project Elephant. Together, these programs contribute to the broader goal of enhancing and consolidating wildlife conservation.

These efforts will generate employment opportunities, driving economic growth for communities around tiger reserves and protected areas. By promoting the usage of clean energy, they will also decrease reliance on natural resources. Indirect benefits will extend to surrounding communities, offering jobs in areas like guiding, driving, hospitality, and other support services. Additionally, the schemes will focus on skill development through eco-development projects, helping individuals achieve self-employment and greater self-sufficiency.

7 Conservation of Resources and Environment in India

The growing population is driving an increased consumption of natural resources. Coupled with rapid industrialization and urbanization, the need to conserve these resources has become critical, as their depletion could disrupt the ecological balance. Conservation comprises the natural resources to avoid their destruction, degradation, and exploitation. It encompasses a range of activities aimed at utilizing these resources sustainably, ensuring that benefits are derived while avoiding overuse or harm that could lead to their depletion or degradation.

7.1 The Importance of Conserving Natural Resources

Natural resources are fundamental to human life, providing necessities such as food, water, air, and materials for shelter. However, many of these resources are not renewable. Conserving them is essential to protect the health of the planet and secure the livelihood of humanity. The significance of natural resources highlights the necessity of their conservation, as outlined below:

- Water is a renewable natural resource vital for drinking, power generation, irrigation, and various industrial and everyday uses. Its scarcity could lead to vegetation loss, harm to flora and fauna, soil erosion, and other environmental issues.
- Plants and animals offer various biological and industrial resources, playing a critical part in pharmaceutical production and other essential applications.

- Fossil fuels, including oil, natural gas, and coal, are crucial for energy production and powering industries worldwide.
- Forests are invaluable natural resources that drive economic growth. They provide raw materials such as timber, paper, medicines, gum, and more, while also preserving ecological balance, protecting wildlife, and reducing soil erosion.
- Land is a fundamental resource that supports natural vegetation, wildlife, and transportation. It provides the foundation for food production, housing, clothing, and other essential human needs.

7.2 Traditions in India to Conserve Natural Resources

India's forefathers understood the importance of conserving natural resources, laying the foundation for a rich heritage of respecting and protecting nature. This legacy is reflected in the establishment of springs, lakes, sacred groves, and revered species, such as the sacred river Ganges. The conservation of natural resources like forests has been a longstanding practice in India, dating back to the reign of King Ashoka. Sacred forests, maintained by tribal communities, were dedicated to deities and ancestral spirits. Activities such as tree felling and other disturbances were banned in such places, contributing to the protection of numerous species.

Similarly, certain water bodies, like Khecheopalri Lake, were considered holy by local communities, to ensure safeguarding marine ecosystems. Worshipping specific plants such as banyan, peepal, and tulsi has also fostered their preservation and growth. India's history is replete with instances where people have made great sacrifices to protect trees. A modern example is the Chipko Movement, initiated by women in Gopeshwar village, Chamoli district, Uttarakhand. They prevented deforestation by embracing trees and halting the lumbermen's efforts to cut them. This movement successfully protected approximately Twelve thousand km^2 of a fragile water basin and inspired same initiatives across the country.

7.3 Schemes in India to Conserve Environment

The protection of natural assets and ecosystems is achieved by various action plans designed to protect corals, mangroves, biosphere reserves, wetlands, and lakes.

The Environment Ministry is taking care of executing several key programs focused on environmental conservation and the sustainable development of ecosystems. These include the National River Conservation Programme, National Mission on Himalayan Studies, etc., all of which are part of the Climate Change Program driven by the Indian Government. Such programs are integral to tackling environmental challenges and safeguarding the environment over a long time. Several initiatives protect vital natural resources such as corals, mangroves, biosphere reserves, wetlands, and lakes. The Green India scheme aims to regenerate and restore forest and its surrounding areas. The National Programme works to improve the water quality of polluted rivers by reducing pollutant inflow. The National Coastal Management Programme aims to conserve coastal ecosystems while supporting the livelihoods of fishing communities. The National Mission on Himalayan Studies is focusing on tackling desertification and land degradation. It works to strengthen scientific collaborations, enhance data collection, and improve reporting mechanisms related to desertification. Through the Environmental Information System (ENVIS) program, the Ministry compiles and stores environmental data to inform future policies and practices aimed at ensuring a sustainable value for generations to come.

The Government of India fully funds Central Sector Schemes. The Government has allocated 50% funding in the year 2015–2016 for most states, 80% for the North Eastern states, and 90% for Jammu & Kashmir, Himachal Pradesh, and Uttarakhand. For forestry and wildlife schemes, the Government of India contributes 60% of funds. The National Scheme is focused on safeguarding aquatic ecosystems, including lakes and wetlands. The National River Protection scheme is promoting the sustainable development of coastal areas, based on technical principles. Additionally, the study underscores the importance of protecting the Himalayan ecology and supporting its sustainable development.

8 Progress on the SDGs in the Asia-Pacific Region

Progress toward the SDGs in the Asia-Pacific region remains uneven, with many countries, especially developing states, significantly falling behind. As a region marked by diverse levels of economic development, efforts to implement an Agenda of 2030 have been unreliable.

The World Bank highlights that rising the rate of interest has increased debt burdens, constraining investments in vital sectors like education, healthcare, and infrastructure, which are critical for social advancement. Moreover, resource shortages are threatening the energy transition, putting climate targets in jeopardy. Developing nations face an annual financing gap of $1.7 trillion for the SDGs, while the global funding shortfall for achieving these goals exceeds $4 trillion each year. These gaps underscore the urgent need for enhanced national and global efforts to improve data collection and sharing, ensuring more effective progress toward achieving the SDGs in the Asia-Pacific region.

8.1 SDG 15 Progress in India

India has made notable strides in advancing SDG 15, which focuses on life on land. Forest cover, as a percentage of total geographic area, grew from 21.35% in 2015 to 21.71% in 2021. The area designated as Ramsar sites, as a percentage of total wetland area, saw a significant increase, rising from 4.15% in 2015 to 7.14% in 2021. Additionally, the number of cases filed under the Wildlife Protection Act, 1972, decreased from 698 in 2015 to 632 in 2021, indicating progress in wildlife conservation.

The Indian government has initiated several legislative measures, schemes, and programs to achieve the objectives of SDG 15. The Indian Forest Act, 1927 highlights laws to ensure the safety of birds, plants, and wild animals. The act was introduced to conserve the nation's forests and focuses on safeguarding and improving the environment. The scheme called umbrella for the Protection of Resources and Ecosystems includes various sub-schemes aimed at protecting lakes, biosphere reserves, corals, wetlands, and mangroves, thereby preserving the country's resources and ecosystems. The Integrated Development of Wildlife Habitats (IDWH) supports state and union territory governments in wildlife protection and conservation, along with recovery programs for critically endangered species.

Additionally, the "Environmental Knowledge and Capacity Building" umbrella scheme addresses the training and exposure needs of forestry personnel, as well as ecological restoration efforts. The "Environmental Protection, Management and Sustainable Development" umbrella scheme focuses on promoting clean technologies and reducing pollution from industrial plants.

8.2 *India's Contribution to Regional Development*

India has been developed as a significant player in the development of sustainability, thanks to its strategic geographic location, robust economic growth, and dynamic foreign policy. The nation has achieved notable progress toward the Sustainable Development Goals, mainly in poverty eradication, promoting work and development in the economy, addressing changes in climate, and conserving terrestrial ecosystems. Transformative government initiatives such as PM-Mudra, Ayushman Bharat-PMJAY, Pradhan Mantri Awas Yojana, Saubhagya, Ujjwala, Ayushman Arogya Mandir, Start-up India, etc. have been instrumental in driving rapid and inclusive advancements. SSDN—Sustainable Development Solutions Network has given a Report 2024 as given in Table 3. In the table, the SDG score of India is compared with the other countries in the World.

Table 3 Sustainable development report

Year	India		World	
	Score	Change in percentage	Score	Change in percentage
2019	61.57	1.72	65.30	0.81
2020	62.28	1.15	65.69	0.60
2021	62.86	0.93	65.87	0.26
2022	63.57	1.13	66.15	0.44
2023	63.99	0.65	66.30	0.22

India has embraced a comprehensive approach to regional challenges by engaging with key forums. India has prioritized strengthening economic integration through active engagement in bilateral and multilateral trade agreements across a region of Asia-Pacific region. Advancing internal abilities, particularly by innovations in IT Sector and digital solutions, remains a central focus. Furthermore, the country's strategic maritime initiatives are vital to address challenges in the environment, promote sustainable practices, and accelerate the development of the economy

Several nations in the Asia-Pacific region remain dedicated to the attainment of targets, but their growth depicts the critical need for higher quality data and more efficient data management systems to improve policy outcomes. Addressing these challenges calls for aligning objectives in the nation's specific context and embracing a holistic, coordinated method. Prioritizing personalized solutions, fostering regional collaboration, and leveraging technological innovations will be vital to driving sustainable progress and achieving long-term objectives

8.3 Role of Educational Institutions to Improve India SDG Score

Higher Education Institutions (HEIs) are vital drivers of the SDGs, serving as hubs of innovation and critical thinking. HEIs show a vital part in preparing upcoming generations of experts and leaders to address global challenges through research, dialogue, and actionable solutions. They also foster a culture of sustainability, equipping their communities with the skills and mindsets necessary to develop sustainable campuses and, by extension, sustainable societies. For India, it is imperative to learn from global examples and leverage insights from SDG-based rankings to enhance the role of its HEIs in contributing to sustainable development.

9 Sustainable Forestry

Sustainable forestry is key to addressing climatic problems and safeguarding that the people can enjoy the countless benefits forests and trees provide in future. While the global forest cover is steadily shrinking, a sustainable approach to forest management can help reverse this trend, delivering environmental, social, and economic advantages to communities around the world. Sustainable forestry is a forest management approach designed to balance social, economic, and environmental goals.

Sustainable forestry recognizes that forests are valuable assets that, when managed responsibly, can deliver long-term benefits. Unfortunately, many forests are mismanaged in unsustainable ways, leading to significant deforestation. In some instances, poor management practices can even cause forests to vanish entirely.

9.1 Role of Sustainable Forestry in Climate Change

People committed to sustainability recognize the complexity of our global challenges. However, one crucial yet often overlooked factor is the vital role forests play in mitigating climate change. When managed sustainably, working forests contribute to slowing climate change by capturing and storing carbon over the long term. The United Nations has even identified sustainably managed forests as a key strategy in limiting global warming to levels that ensure the planet's health. Forests are the second-largest carbon sink on the planet, only surpassed by oceans in their ability to absorb carbon dioxide. This crucial function makes forests a key element in our efforts to tackle global warming and climate change.

To make meaningful progress, we need more forests and a significant reduction in deforestation. Sustainable forestry provides a way to utilize forest resources while ensuring they are not depleted. By protecting and expanding forested areas, we increase our chances of mitigating the harmful effects of climate change. Trees are particularly efficient at capturing carbon through photosynthesis, and even when harvested responsibly, the carbon stored in wood remains sequestered for decades, making sustainable forestry a vital approach in the fight against climate change.

9.2 Advantages of Sustainable Forestry

When we consider sustainable forestry, its environmental importance and contribution to combating climate change usually come to mind first. While tackling climatic problems is certainly an important global issue, sustainable forestry also offers a variety of other environmental benefits. These are highlighted below:

9.2.1 Addressing Climate Change

Forests are key players in combating climate change, as they capture carbon dioxide from the atmosphere through photosynthesis. Sustainable forestry practices enhance this by fostering healthy, resilient forests that continue to absorb and store carbon over time. As natural carbon sinks, trees and forests help reduce greenhouse gas levels, playing a vital role in mitigating climate change.

9.2.2 Protection of Forest Ecosystems

Sustainable forestry practices work to preserve and enhance the productivity, diversity, and resilience of forest ecosystems. By maintaining this balance, forests can stay healthy and continue to provide vital ecological services, such as clean air, water, and safe habitats for wildlife.

9.2.3 Reduced Erosion

Unsustainable work activities can result in soil erosion, negatively affecting soil quality and reducing nutrient availability, which in turn decreases forest productivity. Conversely, sustainable forestry practices help prevent soil erosion by strategically managing tree harvesting—considering factors like timing, method, and the number of trees removed. This careful approach protects the soil, keeping it largely intact and fertile.

9.3 Economic Gains from Sustainable Forestry

Forests offer a broad array of economic and social benefits to humanity. These include their contributions to the economy, such as through employment, the processing and trade of forest products, and energy production, as well as investments in the forestry sector. Additionally, forests host and protect areas of cultural, spiritual, and recreational significance. Ensuring the preservation and improvement of these functions is a key component of sustainable forest management. Tracking socio-economic trends and their status is essential for assessing progress in sustainable forest management, alongside the environmental data typically used in other areas.

Economic benefits are generally expressed in monetary terms and may include income from employment in forestry, the value of goods and services given by forests, and contribution to the economy of the nation, energy production, and international trade. Additionally, an economic sustainability of the sector can be gauged by factors such as the profitability of forest-based enterprises and the level of investment in the sector.

The social functions of forests are often more challenging to measure and can differ greatly depending on a country's level of development and cultural practices. For example, in industrialized societies, the social value of forests may be tied to their recreational and amenity benefits or their role in preserving rural traditions. In contrast, in developing countries, the availability of forests for subsistence uses or the people who are employed in the sector may better reflect their social importance.

9.3.1 Economic Benefits

Job Creation and Income: Forests are a source of employment for millions of people worldwide, supporting industries such as timber, paper, non-timber forest products, and ecotourism.

Forest Products: Forests provide a variety of goods, including timber, fuelwood, medicinal plants, and edible products like fruits and nuts, contributing significantly to livelihoods and local economies.

Sector Contribution: The forestry sector is an important contributor to national economies through trade, GDP, and energy production, particularly from renewable resources like wood and biomass.

Energy Resource: Forests serve as a renewable source of energy, with wood and biomass used for heating, cooking, and electricity generation, especially in rural and developing areas.

9.3.2 Social Benefits

Cultural Significance: Forests are integral to the cultural identity of many communities, often serving as sites for religious rituals, traditions, and cultural practices.

Recreation and Well-being: Forests provide spaces for outdoor recreation such as hiking, camping, and wildlife observation, promoting physical health and mental well-being.

Subsistence and Livelihoods: In many developing countries, forests are essential for subsistence, providing food, water, and materials for shelter, as well as sources of income for local populations.

Community and Rural Development: Forests support rural communities by offering resources for sustainable livelihoods, maintaining local economies, and helping preserve traditional ways of life.

10 Challenges and Opportunities

India faces several pressing environmental challenges, including the pollution of rivers and water bodies, desertification, biodiversity loss, land degradation, depletion of freshwater and groundwater resources, habitat fragmentation, and the unsustainable exploitation of natural resources. Even though forest cover has shown a slight increase, this is partly due to the inclusion of monoculture plantations alongside biodiverse natural forests, without distinguishing between the two. Regular evaluations are essential to track the loss of biodiversity in non-forest ecosystems, areas outside protected zones, and undisturbed forest regions (Oza, 2007).

With the world's geographical area of 2.4% in India, assists 18% of the livestock population and 17% of the global population, creating significant pressure on its land resources. In 2013–2012, 29.5% of the population lived in poverty (Planning Commission, Government of India, 2014), and 200 million people depended on forests for their livelihoods. Forests supply 30% of livestock fodder and 40% of domestic fuel wood needs. Ensuring food security, a key pillar of human development, poses substantial risks to biodiversity. These include the introduction of habitat loss, chemical runoff, non-native species, excessive water use, overreliance on chemical fertilizers and pesticides, soil, contamination of water, and decline of soil fertility from unsustainable farming practices. Additionally, extensive ecological degradation occurs by soil erosion, salinity, micronutrient deficiencies, waterlogging, alkalinity, and the rapid depletion and contamination of groundwater, with irrigation being a primary driver. As the population is likely to attain Nine Billion by 2050 globally, guaranteeing access to energy, water, and food will become a significant one. India, which is facing a population where there are about high undernourishment, will need an additional 100 million tonnes of food grains to attain nourishment. This increased demand places greater burden on dryland regions, leading to the overexploitation of resources.

Irrigated agriculture is expected to contribute 64 million tonnes to food grain production by 2020 through area expansion and yield improvements, leaving a gap of 36 million tonnes. This gap will require enhanced productivity from drylands. However, the report projects that, at the current productivity, the country's domestic food production will meet only 59% of its food demand by 2030. Establishing clear linkages between goals and specific, important indexes is essential for effective observation and assessment. SDG 15, with its interconnections to targets in multiple other goals, often leads to issue fragmentation across various departments, resulting in coordination challenges and inefficient use of resources. In a diverse nation like India, the formulation of the SDG framework must address the complexities of a system involving multiple institutions and stakeholders with differing objectives and capabilities.

To overcome these challenges, it is vital to ensure the open and free access to diverse data and establish ability mechanisms among scientific institutions. Strengthening the science-policy interface will further enhance collaboration and drive cohesive progress toward achieving the goals (FAO, 2014). However, challenges persist in effectively mainstreaming biodiversity conservation. These include the need for better coordination across institutions, both horizontally and vertically, enhanced human resources and capacity within sector ministries, the collection and sharing of policy-relevant data, and mobilizing financial resources for biodiversity, especially within national budgets. While the CBD and governments acknowledge the critical role of mainstreaming and financial mobilization, progress has been limited, and finding practical solutions remains a challenge.

11 Conclusion

Experts warn that India is nearing a critical point where lack of resource management and poverty could lead to substantial and possibly irretrievable decline in environmental resilience. Strong ecosystems—such as soils, forests, oceans, and freshwater systems—and the essential services they offer are crucial to addressing these challenges. Safeguarding terrestrial ecosystems requires protection of inland waters, restoration of an environment, and protection of biodiversity—all of them are aligned with the goals of SDG 15. While India has made notable progress in achieving certain MDG targets, sustainable development for its 1.27 billion people will require clear and measurable targets and indicators.

An integrated approach is essential to protect terrestrial ecosystems, linking land use, food security, and biodiversity conservation with ways to renewable energy, clean water, climate change mitigation, and sanitation. Centering people in the sustainable development agenda, the success of SDG implementation will depend on how effectively these priorities are embedded within a decentralized governance framework.

The Great Transition (GT) scenario aligns with the recently adopted SDGs, which all countries have pledged to achieve. This scenario assesses the potential benefits to ecosystem services if these goals are successfully implemented. In contrast, the Market Forces (MF) and Fortress World (FW) scenarios predict a considerable decline in ecosystem services and overall human well-being. It is essential to move beyond the traditional focus on GDP growth that has guided past development strategies and instead embrace a more balanced approach—such as the SDGs—that supports ecosystem restoration and long-term sustainable well-being.

References

Bai, Z. G., Dent, D. L., Olsson, L., & Schaepman, M. E. (2008). Global assessment of land degradation and improvement, 1. In *Identification by remote sensing. Report 2008/01*. ISRIC World Soil Information.

FAO. (2014). *The state of food insecurity in the world*. UN.

Gupta, H., & Khandelwal, N. (2019). Protecting the diversity of life on land sustainable development goal 15: Life on land. In *Sustainable development goals in Indian context: Interventions and schemes*.

Kakar, N., Popovski, V., & Robinson, N. A. (Eds.). (2021). *Fulfilling the sustainable development goals: On a quest for a sustainable world* (1st ed.). Routledge.

Laurance, W. F., Sayer, J., & Cassman, K. G. (2013). Agricultural expansion and its impacts on tropical nature. *Trends in Ecology & Evolution, 29*(2), 107–116.

MacDicken, K. G. (2015). Global forest resources assessment 2015: What, why and how? *Forest Ecology and Management, 352*, 3–8.

Oza, A. (2007). *Irrigation and water resources*. India Infrastructure Report.

Planning Commission, Government of India. (2014). Report, Government of India of the Expert Group to Review the Methodology for Measurement of Poverty.

SRO. (2007). *Desertification and land degradation atlas of India*. Space Application Centre, India Space Research Organisation (ISRO), Government of India. Retrieved from www.moef.nic.in/sites/default/files/desert_atlas

Role of Humans Toward Biodiversity Protection

Nilima Gaikwad and S. Vijayalakshmi

1 Introduction

Biodiversity variety of life forms on Earth is very important for ecosystem constancy, human well-being, and economic prosperity. It encompasses genetic diversity, species diversity, and ecosystem diversity, all of which contribute to the balance and functioning of the environment. However, human acts such as deforestation, pollution, and climate change have led to a decline in biodiversity (Convention on Biological Diversity (CBD), 2020). The deprivation of biodiversity has far-reaching effects including the disruption of ecosystems, food security, and less demand for potential medicinal resources. It explores the role of humans in biodiversity protection, emphasizing sustainable practices, conservation efforts, and policy initiatives that can help mitigate biodiversity loss and promote ecological balance (Tilman et al., 1996).

N. Gaikwad (✉)
Department of Computer Science, School of Sciences, Christ University, Bengaluru, Karnataka, India
e-mail: nilima.gaikwad@res.christuniversity.in; nilima.gaikwad@christcollegepune.org

S. Vijayalakshmi
Department of Data Science, School of Sciences, Christ University, Bengaluru, Karnataka, India
e-mail: s.vijayalakshmi@christuniversity.in

J. J. Puthenkalam S.J et al. (eds.), *Terra Thrive*,
https://doi.org/10.1007/978-3-032-02738-2_14

1.1 What Exactly Does Biodiversity Mean?

Biodiversity encompasses the vast diversity of life within a particular region, including animals, plants, fungi, and microorganisms like bacteria, all of which are essential to the natural world. These organisms interact within ecosystems, forming complex relationships that sustain life and preserve ecological balance. Biodiversity is crucial for human survival, supplying vital resources such as food, clean water, medicine, and shelter (Tilman et al., 1996).

Human activities now stress the planet beyond all previous measures leading to quick resource loss and damaging ecosystem equilibrium. WWF reported in their 2024 Living Planet Report that mammal, fish, bird, reptile, and amphibian community populations fell by 73% between 1970 and present day. The 2019 Global Assessment Report from the Intergovernmental Platform on Biodiversity and Ecosystem Services predicts the highest number of extinction in recorded history due to one million species of animals and plants facing disappearance (Tilman et al., 1997).

Human actions have significantly altered 75% of terrestrial habitats and around 66% of marine environments. Additionally, more than one-third of the planet's land area and nearly 75% of its freshwater resources are now used for agriculture and livestock. Climate change worsens these environmental pressures, further endangering biodiversity and human well-being. Overfishing, deforestation, water pollution, and the ongoing climate crisis have severely disrupted ecosystems worldwide, from remote natural landscapes to urban centers (IPBES, 2019).

1.2 What Are the Different Types of Biodiversity?

Biodiversity is of three types (Fig. 1):

1. Species diversity.
2. Genetic diversity.
3. Ecological diversity.

1.2.1 Species Diversity

The Biological Species Concept (BSC) identifies a species as the primary unit of biological classification, defined by a group of organisms that can successfully interbreed and produce fertile offspring, all descending from a shared evolutionary lineage. Species diversity, considered the most fundamental aspect of biodiversity, refers to the range and relative abundance of various species inhabiting a specific geographic area. This includes the entire spectrum of living organisms—from microscopic entities such as bacteria and viruses to more complex life forms like

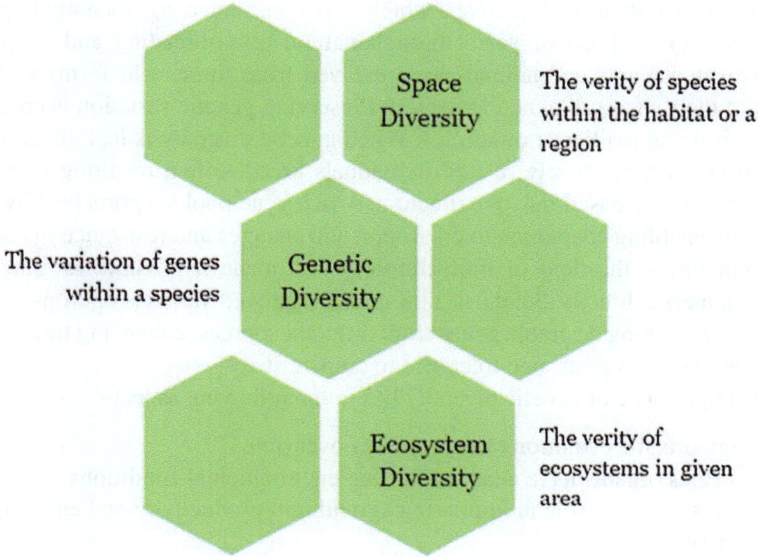

Fig. 1 Types of biodiversity

fungi, algae, mosses, ferns, gymnosperms, flowering plants, and animals, including both single-celled protozoans and multicellular mammals.

The distribution of species across the planet is uneven, influenced by ecological and environmental factors. Areas with fertile soil, balanced temperatures, adequate rainfall, and sufficient sunlight tend to support richer biological communities. Tropical regions, in particular, exhibit significantly greater biodiversity than arid deserts or icy polar zones. For example, natural tropical forests often contain a wider variety of species compared to monoculture plantations. Locations that harbor exceptionally high levels of species richness are known as biodiversity hotspots, and they play a crucial role in global ecological balance.

1.2.2 Genetic Diversity

It refers to the range of genetic characteristics found within and among populations of organisms. Each organism, whether a plant or an animal, has a unique genetic blueprint, which is responsible for its distinct traits. These variations arise due to the numerous ways genes can combine and express themselves, resulting in differences even among individuals of the same species. For example, all humans belong to the species *Homo sapiens*, yet no two individuals (except identical twins) are genetically identical. Likewise, different varieties of crops like rice, wheat, apples, and mangoes show diversity in traits such as size, color, flavor, and yield, all because of underlying genetic differences. The total collection of genes and their variations within a species is referred to as the **gene pool**. This includes not only cultivated species but also their wild relatives, which have contributed significantly to genetic diversity over millions of years through natural cross-breeding and adaptation. Domesticated plants and animals have evolved from these wild forms and carry traces of their ancestral genetic material. Preserving genetic variation is crucial for the survival and resilience of species. When genetic diversity is lost, it can lead to inbreeding—where closely related individuals breed—often resulting in reduced fitness and an increased risk of extinction. A rich gene pool supports healthy populations by enabling adaptation to environmental changes and resistance to diseases. In recent times, the field of **biotechnology** has made it possible to manipulate genetic material directly. Scientists now create improved varieties of crops and livestock by combining desirable genes from different sources, enhancing traits such as yield, resistance to pests, and tolerance to climate stress.

The importance of genetic diversity lies in the following aspects:

(a) It supports the evolution of new species over time.
(b) It enables organisms to adapt to shifting environmental conditions.
(c) It plays a critical role in improving agricultural productivity and ensuring food security.

1.2.3 Ecological/Ecosystem Diversity

An ecosystem is made up of a wide variety of species that coexist in a shared environment, interacting through the continuous exchange of energy and nutrients. The primary energy source for all ecosystems is the Sun. Plants capture solar energy and transform it into chemical energy through photosynthesis. This stored energy then moves through the ecosystem when herbivores consume plants, and these herbivores are, in turn, eaten by carnivores. Meanwhile, decomposers like fungi and bacteria break down dead organisms, obtaining energy and returning essential nutrients to the soil, thus sustaining the cycle. In this context, an ecosystem includes both living elements—such as microbes, plants, animals, and fungi—and nonliving factors like sunlight, temperature, water, and minerals, all of which are interconnected through energy flow and nutrient cycling.

Ecological diversity refers to the variation and abundance of life forms—both plant and animal species—interacting and coexisting within these ecosystems. It encompasses diversity not only among different species but also within the same species and includes microorganisms. This kind of diversity reflects the richness and variety of biological communities within ecosystems or habitats. The immense variety of life on Earth is a result of the diverse ecosystems present across the planet. These include environments such as forests, grasslands, deserts, wetlands, marine zones, freshwater bodies, swamps, and marshes. Each ecosystem supports its own unique combination of plant, animal, and microbial species. Ecological diversity captures the complex web of species and the dynamic relationships they maintain within their habitats. This diversity has evolved over millions of years and plays a crucial role in maintaining the balance and functionality of life on Earth.

1.3 Why Is It So Important?

Biodiversity plays a vital role in sustaining life on Earth, including human health and survival. A diverse range of animals, plants, and microorganisms is crucial for maintaining balanced ecosystems that provide essential resources like clean air and food. Additionally, nature holds intrinsic value for people (Maestre et al., 2012).

While some aspects of biodiversity are well known, scientific research continues to highlight the importance of all species, even those that are microscopic, hidden, or often disregarded, such as insects and bacteria. Humans depend on biodiversity in many ways, making its conservation essential. For instance, pollinators like birds, bees, and other insects contribute to nearly one-third of global crop production. Without them, key foods such as apples, cherries, blueberries, and almonds would become scarce. Likewise, invertebrates play a crucial role in sustaining soil health, which is vital for agriculture (Venail et al., 2010).

Soil is teeming with microorganisms that release essential nutrients necessary for plant growth, which eventually pass on to humans through the food chain. Furthermore, marine life serves as the primary source of animal protein for a significant portion of the world's population.

The combination of trees, bushes, and wetlands along with wild grasslands naturally produces water deceleration and promotes soil moisture absorption during rainfall. The removal of these vegetation types creates conditions which result in more flood occurrences. Plants along with trees breathe clean air and fight climate change by collecting carbon dioxide from the environment. Coral reefs together with mangrove forests perform as ecological shields which shield shoreline areas from ocean waves and storm events (Vila et al., 2011).

Plants provide the foundation for producing our medicine inventory and the production of rubber latex as well as other sophisticated chemical substances we depend on daily. Research shows that spending time in natural settings enhances both physical wellness and mental well-being of humans. Research indicates that urban tree presence alongside green areas reduces hospital visits and stress while it also lowers blood pressure rates (WWF, 2022).

1.4 How Do Humans Affect Biodiversity?

Humans control biodiversity decline through direct intentional acts alongside unintentional natural causes. Modern human development methods such as land conversions into agriculture and resource extraction activities have caused the greatest biodiversity decline to date but climate change will expand this negative impact on ecosystems.

Biodiversity loss stems directly from land-use change pertaining to large-scale food production systems and results in about 30% ecosystem depletions worldwide. Overharvesting procedures relating to fishing and hunting for human uses and timber harvesting contribute to roughly 20% of biodiversity decline worldwide. Climate change emerges as the third primary direct factor responsible for biodiversity decline since it combines with pollution to create a loss of 14% of biodiversity. Invasive alien species account for 11% (National Geographic Society, 2021).

Multiple scientific studies indicate that climate change might eclipse all other causes of biodiversity decline throughout the next few decades. All primary contributors to biodiversity decline are now speeding up which results in rising biodiversity reduction rates.

Since human population size grew and people started increasing their resource use and switching their consumption approaches, there has been greater conversion of natural spaces into agricultural land along with mining sites and industrial structures and urban areas (Power et al., 1996).

Primary human activities contributing to biodiversity loss include:

- **Deforestation**, particularly in tropical rainforests, which are among the most biodiverse ecosystems and are being rapidly cleared.
- **Habitat destruction**, caused by gradual human encroachment, such as urban expansion, leading to wildlife displacement.
- **Pollution**, especially the overuse of pesticides and fertilizers, which have increased sixfold and twelvefold, respectively, since 1961.
- **Agricultural expansion**, which endangers nearly half of all at-risk species.
- **Water resource depletion**, where large dams and irrigation systems in major river basins drastically reduce water availability.
- **Overhunting and excessive exploitation**, including unsustainable fishing practices and illegal wildlife trade.
- **Introduction of invasive species and disease spread**, often intensified by international trade and travel.
- **Climate change**, which alters species distribution, disrupts rainfall patterns, and impacts vital water and chemical cycles.
- **Plastic waste pollution**, with long-term effects on biodiversity still unclear.

1.5 How to Assess the Impact?

Although there is no more definitive way to measure the full extenuating impact on biodiversity, it is clear that human action plays a significant part in its decline. Assessing the overall effect of human actions on an environment requires considering factors such as the amount of land and water utilized for resource production and the necessity of waste management. The degree of this impact also depends on prevailing management and construction practices.

Biodiversity loss is driven by both direct and indirect human influences. There are factors categorized into primary and secondary drivers. Direct drivers include land-use changes, species introduction or removal, climate fluctuations, resource exploitation, water pollution, and global warming. Indirect drivers encompass economic, commercial, sociopolitical, technological, scientific, cultural, and religious factors (Hector et al., 1999).

2 Threats to Biodiversity

As previously mentioned, biodiversity is essential for human survival. However, it is currently undergoing a crisis, with ecosystems worldwide facing severe threats due to human activities. From green rainforests and open grasslands to pine forests and coral reefs, our planet is home to many different types of natural environments. The rapid deterioration of ecosystems endangers the very species that depend on them, largely due to human intervention. This extensive biodiversity loss is increasingly impacting livelihoods, economies, and food security.

2.1 The Current Mass Extinction Crisis

Scientists predict that Earth is approaching its sixth mass extinction process because biodiversity shows rapid depletion. The definition of mass extinction involves the disappearance of 75% of all species during a fairly short geological period. The scientific definition of a brief geological period amounts to any span shorter than 2.8 million years because Earth formed 4.5 billion years ago and life emerged after 3.5 billion years (Paquette & Messier, 2011).

To date, Earth has experienced five mass extinction events, with the earliest occurring roughly 540 million years ago and the most recent around 60 million years ago. Each of these extinctions was caused by drastic yet entirely natural environmental changes. However, the current sixth mass extinction stands apart due to two key factors: (a) the extinction rate is occurring at an unprecedented pace and (b) human activities are the primary cause.

Researchers forecast the disappearance of more than one million species in the next few generations. Two out of five amphibian species face vulnerability alongside a third of reef-building corals and 10% of insects as well as 30% of various marine species. This present biodiversity loss exceeds all recorded levels because human population growth combined with excessive resource consumption leads to this state. Earth needs millions of years to restore its current biodiversity loss that results from human behavior patterns (Fox & Kerr, 2012).

To combat this crisis, it is essential to recognize and address the major threats posed by human actions. These include habitat destruction, illegal wildlife trade, and overexploitation of resources, invasive species, and human–wildlife conflicts. Raising awareness and implementing conservation strategies are crucial steps toward preserving Earth's biodiversity.

2.2 *Habitat Destruction*

Habitat destruction is one of the most significant threats to biodiversity and a primary cause of species extinction. A habitat encompasses the environmental conditions necessary for a species to survive, including access to food, shelter, protection, and breeding grounds. Human activities have dramatically accelerated habitat degradation, particularly since the twentieth century.

Deforestation for agriculture, plantations, urbanization, mining, and infrastructure development, along with wetland drainage and dam construction, has contributed to large-scale habitat loss. Natural and human-induced forest fires further exacerbate this destruction. It is estimated that approximately 18 million acres of forest are lost annually due to human activities, severely disrupting ecosystems that sustain countless species.

Tropical rainforests face the most severe destruction because they contain major proportions of worldwide biodiversity. Some notable cases include:

- Deforestation throughout Brazil and other South American countries including Peru, Colombia, Bolivia, Venezuela, Suriname, and Guyana together with French Guiana has resulted in the permanent loss of 750,000 square kilometers of Amazon rainforest.
- West Africa has experienced the total dismantling of 90% of its coastal rainforest terrain.
- Forests within the Congo Basin stretching across Cameroon and six other nations experienced a double increase of deforestation since 1990.
- Research shows the Western Ghats forests in India experienced a 35.3% loss of forested area amounting to 33,579 square kilometers during the last nine decades.
- The destruction of 80 million hectares of forest occurred across Southeast Asia throughout the years 2005 to 2015.

Tropical forest destruction exists only as one type of habitat destruction phenomenon. The destructive actions of humans have led to severe damage throughout temperate forests, grasslands, wetlands, coral reefs which are known as marine rainforests, and marine ecosystems.

Freshwater ecosystems, such as rivers and streams, which host unique biodiversity, have also suffered due to habitat alterations. The construction of dams disrupts water flow, affecting species adapted to seasonal changes in water levels. Furthermore, industrial waste, sewage discharge, mining runoff, and oil spills have significantly degraded aquatic habitats, putting countless freshwater and marine species at risk.

3 Impacts of Habitat Fragmentation

Habitat fragmentation is a major human-induced threat to biodiversity, leading to the division of continuous ecosystems into smaller, disconnected patches. This disruption results from activities such as road and railway construction, deforestation for agriculture and plantations, urban expansion, and livestock grazing. Such alterations create physical barriers that hinder wildlife movement, limiting access to essential resources like food and mates. Consequently, species in fragmented habitats struggle with survival and reproduction, making them more vulnerable to population decline and extinction (Finke & Snyder, 2008).

4 Poaching: A Severe Threat to Wildlife

The illegal hunting and capture of wildlife, known as poaching, has surged in recent years, posing a serious risk to global biodiversity. It has significantly contributed to the extinction of numerous plant and animal species while pushing many others toward endangerment. Despite stringent laws, poaching remains prevalent due to the high demand for animal body parts, the exotic pet trade, and the use of wildlife as a food source. Conservationists warn that poaching not only threatens biodiversity within protected areas but also disrupts ecosystems beyond them. The loss of key species destabilizes ecological balance, weakening ecosystems and causing long-term environmental consequences (Wilson, 2016).

4.1 Consequences of Poaching on Biodiversity

Poaching has widespread negative effects on biodiversity, threatening wildlife populations, ecosystems, and even human health. Some of its major consequences include:

4.1.1 Declining Wildlife Populations

Many species of mammals, reptiles, and birds are already classified as Critically Endangered, Endangered, or Vulnerable under IUCN criteria. Poaching remains a major threat, continuously reducing their numbers. Combined with habitat destruction and human-wildlife conflicts, this decline increases the risk of local extinctions, further jeopardizing species and pushing some toward total extinction (Chivian & Bernstein, 2008).

4.1.2 Ecological Disruption

Poaching upsets the delicate balance of ecosystems by depleting populations of key species, particularly apex predators and keystone species. As top predators, apex species play a crucial role in controlling prey populations and maintaining ecological stability. Their removal can lead to uncontrolled prey population growth, triggering a domino effect that disrupts the entire ecosystem. Such imbalances can have long-lasting consequences for biodiversity and environmental health.

4.1.3 Rising Cases of Zoonotic Diseases

Zoonotic diseases, or zoonoses, are infections caused by viruses, bacteria, parasites, or fungi that spread between animals and humans. These diseases can range from mild to severe, sometimes leading to deadly outbreaks. Once transmitted to humans, zoonotic diseases may result in epidemics (affecting large populations in specific regions) or even pandemics (spreading across multiple countries and impacting global populations). Examples include SARS, Ebola, Swine Flu, and COVID-19 (Jones et al., 1994).

Scientists attribute the increasing frequency of zoonotic diseases to human intrusion into wildlife habitats, poaching, and the illegal wildlife trade. These activities heighten human-animal interactions, increasing the risk of disease transmission and posing serious global health threats.

4.2 Pollution

Pollution is a severe threat to biodiversity, disrupting ecosystems and endangering species across the planet. The different forms of pollution, including air, water, soil, noise, and light pollution, have led to irreversible damage to the natural environment. As human activities continue to expand, the negative impacts of pollution on biodiversity intensify, leading to habitat loss, species extinction, and ecosystem imbalances.

4.2.1 Air Pollution and Its Impact on Biodiversity

Air pollution arises from the emission of harmful substances, including carbon dioxide, sulfur dioxide, nitrogen oxides, and particulate matter, into the atmosphere. These pollutants primarily come from industrial emissions, vehicle exhaust, and the combustion of fossil fuels.

- Acid Rain Formation: When sulfur dioxide and nitrogen oxides mix with atmospheric moisture, they form acid rain. Acid rain damages forests, acidifies water bodies, and reduces soil fertility, affecting plant and animal life.
- Ozone Depletion: The depletion of the ozone layer due to pollutants like chlorofluorocarbons (CFCs) increases ultraviolet (UV) radiation exposure, harming plants, animals, and microorganisms.
- Respiratory Issues in Animals: Airborne pollutants can cause respiratory diseases in wildlife, leading to declining populations of sensitive species.
- Disruption of Photosynthesis: Particulate matter and smog block sunlight, reducing the rate of photosynthesis in plants and affecting entire food chains.

4.2.2 Water Pollution and Its Consequences

Water pollution occurs due to the discharge of industrial waste, plastic debris, agricultural runoff, and untreated sewage into rivers, lakes, and oceans. It severely affects aquatic and terrestrial ecosystems.

- Oil Spills: Large-scale oil spills coat marine animals, leading to suffocation, poisoning, and destruction of marine habitats.
- Plastic Pollution: Plastics and microplastics enter marine ecosystems, causing entanglement and ingestion by fish, birds, and marine mammals, often leading to fatal consequences.
- Eutrophication: Excess nutrients from fertilizers cause algal blooms, which deplete oxygen in water bodies and create dead zones, killing fish and other aquatic organisms.
- Heavy Metal Contamination: Mercury, lead, and arsenic released from industrial waste accumulate in the food chain, causing toxicity in aquatic and terrestrial species.
- Decline in Coral Reefs: Pollutants such as sunscreen chemicals and sewage discharge cause coral bleaching, destroying marine biodiversity.

4.2.3 Soil Pollution and Its Effects

Soil pollution results from excessive use of pesticides, chemical fertilizers, and dumping of industrial and household waste. This leads to contamination of the land, reducing its ability to support plant and animal life.

- Loss of Soil Fertility: Harmful chemicals alter soil composition, reducing its ability to support vegetation growth.
- Harm to Microbial Communities: Microorganisms play a crucial role in nutrient cycling; pollution disrupts these communities, affecting plant health and food production.
- Bioaccumulation of Toxins: Toxic substances accumulate in plants and move up the food chain, affecting herbivores, carnivores, and eventually humans.

- Deforestation and Land Degradation: Polluted soil leads to poor plant growth, contributing to desertification and habitat loss for wildlife.

4.2.4 Noise Pollution and Its Disturbance to Biodiversity

Noise pollution, mainly caused by urbanization, industrialization, and transportation, disrupts animal behaviors and communication patterns.

- Interference with Animal Communication: Many species rely on vocalization for mating, navigation, and hunting; excessive noise interferes with these natural behaviors.
- Disruption of Migration Patterns: Animals like birds and marine mammals rely on sound for navigation. High noise levels from human activities disorient them, leading to population declines.
- Increased Stress in Wildlife: Prolonged exposure to loud sounds increases stress hormone levels in animals, affecting their reproductive success.
- Impact on Predatory and Prey Relationships: Predators rely on sound cues to locate prey; noise pollution alters their ability to hunt effectively.

4.2.5 Light Pollution and Its Ecological Impact

Artificial lighting from urban areas, vehicles, and industrial sites affects nocturnal wildlife and ecosystem dynamics.

- Disruption of Circadian Rhythms: Many species depend on natural light cycles for survival; artificial lighting disturbs their sleep-wake patterns.
- Threat to Nocturnal Animals: Animals like bats, owls, and fireflies suffer from excessive lighting, reducing their ability to hunt and reproduce.
- Impact on Bird Migration: Birds navigate using natural light cues; excessive artificial lighting causes them to become disoriented, leading to fatal collisions with buildings.
- Disruption of Marine Life: Hatchling sea turtles use moonlight to find their way to the ocean. Artificial lighting confuses them, leading to increased mortality rates.

4.2.6 Plastic Pollution and Its Long-Term Effects

Plastic pollution is one of the most pervasive threats to biodiversity, affecting both terrestrial and marine ecosystems.

- Entanglement and Suffocation: Marine animals such as turtles, seals, and seabirds get trapped in plastic waste, leading to injuries and death.

- Microplastic Ingestion: Tiny plastic particles enter the food chain, accumulating in organisms and causing long-term health issues.
- Chemical Leaching: Plastics release toxic substances into soil and water, harming organisms at all trophic levels.
- Disruption of Ecosystem Balance: The presence of plastic in ecosystems disrupts nutrient cycles, water flow, and habitat structures.

4.2.7 Consequences of Pollution on Ecosystem Stability

The combined effects of pollution result in long-term damage to ecosystems, leading to biodiversity loss and ecosystem collapse.

- Decline in Keystone Species: Pollution affects keystone species that play a vital role in maintaining ecosystem balance, leading to cascading effects on biodiversity.
- Disruption of Food Chains: Pollutants accumulate in organisms at various trophic levels, leading to weakened populations and altered predator-prey dynamics.
- Loss of Genetic Diversity: Pollution-induced stress reduces reproductive success, leading to genetic bottlenecks and increased extinction risk.
- Reduced Ecosystem Services: Healthy ecosystems provide essential services like air purification, water filtration, and climate regulation; pollution weakens these functions.

4.3 Overexploitation

Overexploitation refers to the excessive and unsustainable use of natural resources, leading to a significant decline in species populations and ecosystem stability. Human activities such as overfishing, deforestation, hunting, and mining have drastically altered natural habitats, making it difficult for species to survive and reproduce. When species are harvested at a rate faster than their natural replenishment, their populations dwindle, leading to extinction risks. Overexploitation disrupts food chains, reduces genetic diversity, and affects ecosystem services such as pollination, carbon sequestration, and water purification. The excessive extraction of resources not only threatens biodiversity but also undermines the long-term sustainability of ecosystems that support human life (Power et al., 1996).

One of the most severe examples of overexploitation is overfishing, which has led to drastic declines in marine species such as tuna, sharks, and cod. Similarly, illegal wildlife trade threatens endangered species like tigers, elephants, and rhinos, pushing them closer to extinction. Deforestation for timber and agriculture destroys critical habitats, affecting countless plant and animal species. Mining and resource extraction contribute to land degradation and pollution, further exacerbating biodiversity loss. Addressing overexploitation requires urgent measures such as enforcing strict conservation laws, promoting sustainable resource management, and raising awareness about the consequences of unsustainable practices on biodiversity and ecosystems (Tilman et al., 1996).

4.3.1 Invasive Species

Invasive species are non-native organisms that are introduced into new environments, often by human activities, and they outcompete native species for resources. These species can rapidly reproduce and spread, leading to habitat destruction, altered food webs, and the extinction of indigenous species. Invasive species disrupt ecosystems by preying on native species, introducing new diseases, and modifying habitats in ways that make them unsuitable for local wildlife. This phenomenon has significantly contributed to biodiversity loss worldwide, particularly in island and freshwater ecosystems where native species have not evolved defenses against these new competitors (Letourneau et al., 2009).

Examples of invasive species include the Burmese python in the Florida Everglades, which preys on native mammals, and the zebra mussel in North American lakes, which clogs water systems and disrupts aquatic ecosystems. Invasive plants like kudzu in the USA overgrow native vegetation, suffocating biodiversity. Controlling invasive species requires early detection, habitat restoration, and the implementation of biosecurity measures to prevent further introductions. Raising awareness and enforcing strict regulations on trade and transportation of species can help mitigate the impact of invasive species on biodiversity (Folke et al., 1996).

4.3.2 Role of Humans in Biodiversity Protection

Despite the negative impact, humans play a crucial role in biodiversity protection through various conservation strategies, sustainable practices, and policy initiatives. Governments, organizations, communities, and individuals must collaborate to mitigate biodiversity loss and ensure a sustainable future.

4.3.3 Conservation Strategies

1. Protected Areas and Wildlife Sanctuaries—Establishing different parks and biosphere reserves, and marine protected areas help conserve ecosystems and species. Countries around the world have designated conservation areas to safeguard biodiversity.
2. Reforestation and Afforestation—Planting trees and restoring degraded ecosystems help in maintaining biodiversity by improving habitat quality and carbon sequestration.

3. Natural practices in sustainable farming together with fishing methods employing organic methods and crop rotations combined with minimized pesticide usage and responsible fishing techniques prevent environmental destruction. Agroforestry and permaculture methods promote biodiversity-friendly agriculture.
4. Community Involvement—Indigenous communities and local populations play a vital role in conservation through traditional knowledge, sustainable resource use, and ecotourism initiatives. Community-led conservation projects have proven successful in protecting endangered species.
5. Legislation and Policies—Governments implement and enforce laws such as the Convention on Biological Diversity (CBD), the Endangered Species Act, and the Paris Agreement to regulate human activities that affect biodiversity.

4.3.4 Technological and Scientific Contributions

1. Biotechnology—Genetic research and biotechnology help in species conservation, habitat restoration, and breeding programs for endangered species.
2. Remote Sensing and GIS—Monitoring biodiversity through satellite imagery and geographical information systems (GIS) enables scientists and conservationists to track habitat changes and plan conservation efforts effectively.
3. Environmental Education—Raising awareness about biodiversity conservation through school curricula, media campaigns, and community outreach fosters responsible behavior among individuals and organizations.

4.3.5 Sustainable Practices for Individuals

Every individual can contribute to biodiversity protection through sustainable lifestyle choices. Simple actions can have a significant impact on conservation efforts.

- Reducing single-use plastics to prevent pollution.
- Supporting sustainable products and eco-friendly brands.
- Planting native trees and maintaining home gardens to support pollinators.
- Participating in conservation programs and volunteering for environmental initiatives.
- Advocating for biodiversity-friendly policies and supporting environmental organizations.

5 Case Studies

1. Amazon Rainforest Conservation—Global efforts to curb deforestation and promote sustainable land use include policies to reduce illegal logging and support indigenous land rights (Vila et al., 2011).
2. Coral Reef Restoration in Australia—Scientists and conservationists employ artificial reef structures, coral farming, and marine protected areas to restore damaged coral reefs and protect marine biodiversity.
3. Project Tiger in India—Launched by the Government of India, Project Tiger focuses on habitat conservation, anti-poaching measures, and community involvement to protect the endangered Bengal tiger population.
4. Costa Rica's Ecotourism Model—Costa Rica has successfully integrated biodiversity conservation with ecotourism, generating economic benefits while preserving natural ecosystems.

6 Future Directions in Biodiversity Conservation

The future of biodiversity conservation depends on global cooperation, policy implementation, technological advancements, and community engagement. Some key areas that require further attention include:

- Strengthening international collaboration for biodiversity protection.
- Enhancing climate adaptation strategies to reduce biodiversity loss.
- Promoting sustainable land-use practices to prevent habitat destruction.
- Encouraging investments in conservation projects and green technologies.
- Expanding educational programs to create biodiversity-conscious societies.

7 Conclusion

Biodiversity faces significant threats due to pollution, overexploitation, and invasive species, each contributing to habitat destruction, species extinction, and ecological imbalance. Pollution disrupts ecosystems through air, water, and soil contamination, affecting the health of both flora and fauna. Overexploitation depletes natural resources faster than they can replenish, endangering species and altering food chains. Meanwhile, invasive species outcompete native organisms, leading to biodiversity loss and ecosystem instability. To combat these challenges, urgent conservation measures, sustainable resource management, and public awareness are essential. By adopting eco-friendly practices, enforcing strict environmental policies, and promoting habitat restoration, we can work toward preserving biodiversity and ensuring the stability of ecosystems for future generations.

References

Chivian, E., & Bernstein, A. (2008). *Sustaining life: How human health depends on biodiversity*. Oxford University Press.

Convention on Biological Diversity (CBD). (2020). *Global Biodiversity Outlook 5*. Retrieved from https://www.cbd.int/gbo5

Finke, D. L., & Snyder, W. E. (2008). Niche partitioning increases resource exploitation by diverse communities. *Science, 321*, 1488–1490.

Folke, C., Holling, C. S., & Perrings, C. (1996). Biological diversity, ecosystems, and the human scale. *Ecological Applications, 6*, 1018–1024.

Fox, J. W., & Kerr, B. (2012). Analyzing the effects of species gain and loss on ecosystem function using the extended Price equation partition. *Oikos, 121*, 290–298.

Hector, A., et al. (1999). Plant diversity and productivity experiments in European grasslands. *Science, 286*, 1123–1127.

IPBES. (2019). *Global assessment report on biodiversity and ecosystem services*. Intergovernmental Science-Policy Platform on Biodiversity and Ecosystem Services.

Jones, C. G., Lawton, J. H., & Shachak, M. (1994). Organisms as ecosystem engineers. *Oikos, 69*, 373–386.

Letourneau, D. K., Jedlicka, J. A., Bothwell, S. G., & Moreno, C. R. (2009). Effects of natural enemy biodiversity on the suppression of arthropod herbivores in terrestrial ecosystems. *Annual Review of Ecology, Evolution, and Systematics, 40*, 573–592.

Maestre, F. T., et al. (2012). (ref 21), exemplify how to quantify biodiversity effects on ecosystem functions at large scales in real ecosystems.

National Geographic Society. (2021). *Biodiversity conservation: Challenges and strategies*. National Geographic Publications.

Paquette, A., & Messier, C. (2011). The effect of biodiversity on tree productivity: From temperate to boreal forests. *Global Ecology and Biogeography, 20*, 170–180. This paper, along with Maestre et al.

Power, M. E., et al. (1996). Challenges in the quest for keystones. *BioScience, 46*, 609–620.

Tilman, D., Wedin, D., & Knops, J. (1996). Productivity and sustainability influenced by biodiversity in grassland ecosystems. *Nature, 379*, 718–720.

Tilman, D., Lehman, D., & Thompson, K. (1997). Plant diversity and ecosystem productivity: Theoretical considerations. *Proceedings of the National Academy of Sciences, 94*, 1857–1861.

Venail, P. A., Maclean, R. C., Meynard, C. N., & Mouquet, N. (2010). Dispersal scales up the biodiversityproductivity relationship in an experimental source-sink metacommunity. *Proceedings of the Royal Society B: Biological Sciences, 277*, 2339–2345.

Vila, M., et al. (2011). Ecological impacts of invasive alien plants: A meta-analysis of their effects on species, communities and ecosystems. *Ecology Letters, 14*, 702–708.

Wilson, E. O. (2016). *Half-earth: Our planet's fight for life*. Liveright Publishing Corporation.

WWF. (2022). *Living planet report 2022*. World Wildlife Fund. Retrieved from https://www.worldwildlife.org/publications/living-planet-report-2022

Integrating Indigenous Knowledge and Practices into Land Management Systems

S. Vijayalakshmi and Vandana R. Bais

1 Introduction

Indigenous knowledge acts as a fundamental tool for land management through sustainable conservation practices which indigenous communities learned and evolved over time. ANCs employ proven cultural techniques which prioritize sustainable land practices while maintaining biodiversity and adapting to both climate changes and managing water availability. Native communities sustain soil fertility through three methods including crop rotation together with agroforestry and floodwater farming which also helps stop land degradation. Ecological protection through sacred grove conservation in India along with other preservation projects demonstrates how communities should retain authority over environmental care (Roué & Molnar, 2017). Rainwater harvesting along with waru waru farming allows effective water resource use and controlled burning methods, and seasonal migratory practices protect the environment from climate challenges. The indigenous land management system implements community-based governance through collective decision-making to stop resources from getting overused while maintaining environmental stability. Traditional knowledge shows its sustainability and practicality through low-cost solutions which include Himalayan terracing as one example (Sukartiko et al., 2018). Land sustainability receives a boost through IK's collaborative work with current scientific methods which include traditional fire management combined with satellite technology. The integration of Indigenous Knowledge in

S. Vijayalakshmi
Department of Data Science, Christ (Deemed to be University),
Pune Lavasa, Maharashtra, India

V. R. Bais (✉)
Department of Computer Science, Christ College, Pune, Maharashtra, India
e-mail: baisvandana.rajendrasingh@res.christuniversity.in

J. J. Puthenkalam S.J et al. (eds.), *Terra Thrive*,
https://doi.org/10.1007/978-3-032-02738-2_15

modern land management strategies yields efficient environmentally conscious and cultural-aware solutions to worldwide ecological issues (Miralles-Wilhelm, 2021).

1.1 Definition and Importance of Indigenous Knowledge

The indigenous communities have developed Indigenous Knowledge (IK) through numerous generations by interacting directly with their environment to create this accumulated wisdom of customs and practices. The traditional values extend from cultural practices to oral history along with spiritual practices across generations. Study approaches from the modern era differ from Indigenous Knowledge because it views ecosystems through a comprehensive lens which includes practices for sustainable land management alongside biodiversity protection and climate resilience (Ajayi & Mafongoya, 2017; Karki et al., 2017; Roué & Molnar, 2017).

1.2 Historical Perspective

For thousands of years, indigenous communities have demonstrated successful ecosystem management strategies including rotational agricultural systems, conservation practices, and water resource distribution. Resource extraction serves as less important to their methods than their objective of natural harmony. The Maya employed permaculture techniques while the Zuni people used irrigation systems together with the Kayapo people who practiced forest management in the Amazon. The sustainable use of resources as a result of these practices has allowed natural ecosystems to remain resilient (Food and Agriculture Organization of the United Nations, 2019).

1.3 Need for Integrating Indigenous Practices

Modern land management needs indigenous knowledge integration at this time because of growing threats to deforestation and land degradation and climate change.

Indigenous practices offer:

- Sustainable agricultural methods should be used to fight soil erosion.
- Forest management techniques that prevent wildfire outbreaks form a part of the prevention strategy.
- Protecting biodiversity through community leadership-based conservation strategies forms part of their approach.

Such integration into official policies will provide better environmental sustainability along with elevated climate resilience (Miralles-Wilhelm, 2021; Sillitoe, 2017).

Figure 1 illustrates how traditional knowledge contributes to agroforestry, water conservation, and wildlife protection through addressing challenges corresponding to deforestation and policy gaps. These factors integrate into sustainable land-use strategies, leading to biodiversity conservation and climate resilience.

1.4 Overview of Traditional Land-Use Practices

For centuries, traditional indigenous land-use concepts have been built through cultural knowledge systems to sustainably manage natural resources. Such practices employ local ecological characteristics to maintain natural-human-environment harmony which creates lasting ecological stability with accessible resources.

1. Traditional agricultural practices.
2. Traditional water management techniques.
3. Traditional forest and biodiversity management.

1. **Traditional Agricultural Practices:** Rural communities use traditional farming techniques like agroforestry combined with intercropping as well as shifting cultivation (Jhum farming) to maximize land use and preserve soil fertility together with biodiversity. Sustainable agricultural techniques implemented by both Gond and Khasi communities have produced more harvests while preserving their environmental elements (Miralles-Wilhelm, 2021).
2. **Traditional Water Management Techniques:** Stepwells or Baolis and tank irrigation along with Zabo system from Nagaland and bamboo drip irrigation from Meghalaya are among ancient water conservation systems that effectively manage water resources by enhancing groundwater recharge while supporting sustainable irrigation (Eslamian & Eslamian, 2021).

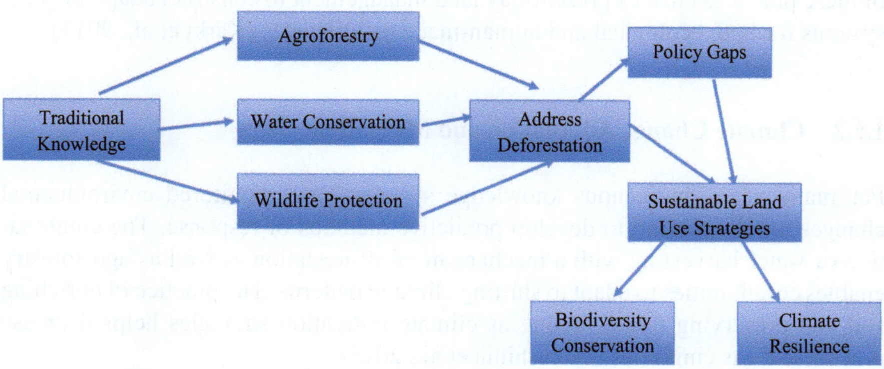

Fig. 1 Flowchart on Indigenous Land Management Systems

3. **Traditional Forest and Biodiversity Management:** The preservation of biodiversity and ecosystem health in Northeast India is supported by sacred grove conservation and community-managed forests and carefully managed controlled burning techniques. The inhabitants of Rajasthan identify the Bishnoi community as being dedicated to protecting wildlife in their region (Singh et al., 2019).

1.5 Significance of Integrating Indigenous Knowledge with Land Management Strategies

For sustainable environmental conservation and climate resilience with equitable resource management strategies, you need to integrate Indigenous Knowledge (IK) with modern land management systems. Native community members have perfected sustainable land management techniques which function alongside scientific procedures. A united land management strategy produces sustainable solutions which benefit human communities and nature ecosystems (Nakashima et al., 2018).

1. Enhancing sustainability and ecosystem resilience.
2. Climate change adaptation and mitigation.
3. Biodiversity conservation and ecosystem management.
4. Sustainable agriculture and food security.
5. Water conservation and management.

1.5.1 Enhancing Sustainability and Ecosystem Resilience

The management practices of indigenous peoples originate from sustainability principles because they handle natural resources with the aim of sustaining ecosystem stability. Soil fertility as well as land degradation prevention and environmental stress resistance improvement come from traditional rotational farming alongside controlled land burning approaches and land regeneration practices. The integration of these practices enables present-day land management to construct adaptable ecosystems for both ecological and human-made disturbances (Karki et al., 2017).

1.5.2 Climate Change Adaptation and Mitigation

For many years, indigenous knowledge systems have monitored environmental changes by observation to develop predictive methods of response. The combination of water harvesting with a mechanism of afforestation as well as agroforestry enables communities to adapt to shifting climatic patterns. The practice of enriching soil and preserving forests acting as climate mitigation strategies helps decrease greenhouse gas emissions (Nakashima et al., 2018).

1.5.3 Biodiversity Conservation and Ecosystem Management

The sustainable hunting together with fishing and harvesting practices of indigenous communities has protected biodiversity in the past. The preservation of native species and their genetic diversity depends on sacred groves in addition to protected areas and rotational land-use methods. Local ecosystem knowledge permits indigenous communities to create conservation solutions that defend habitats while building migration areas and maintaining ecosystem wellness (Karki et al., 2017).

1.5.4 Sustainable Agriculture and Food Security

Through traditional agricultural practices, polyculture and intercropping combined with organic fertilization increase soil productivity along with agricultural diversity and reduce requirement for chemicals. The methods protect agricultural landscapes from three main threats: soil erosion, pest infestations, and extreme weather conditions at the same time as improving food security. Modern agronomic research together with indigenous techniques helps develop food production systems which demonstrate enhanced sustainability and resilience (Hunter et al., 2020).

1.5.5 Water Conservation and Management

Rainwater harvesting together with step-well systems and floodplain farming practices enable effective use of existing water resources according to indigenous management approaches. Through these techniques, soil becomes protected from salinization while groundwater stays viable and both droughts and floods receive lower intensity. People can learn effective contemporary water resource management for arid and semi-arid areas through studying traditional water conservation techniques (WWAP, 2018).

1.6 Objectives of Integrating Indigenous Knowledge and Practices into Land Management Strategies

The integration of Indigenous Knowledge (IK) and practices into modern land management strategies aims to create sustainable, resilient, and equitable land-use systems. By combining traditional ecological wisdom with scientific advancements, policymakers, researchers, and conservationists can develop more effective environmental management solutions.

1. Promote sustainable land use and resource conservation.
2. Enhance climate resilience and adaptation.
3. Improve water management and soil fertility.

1.6.1 Promote Sustainable Land Use and Resource Conservation

Indigenous knowledge systems place special importance on sustainable land and resource use. Cultural land management methods, including shifting cultivation, agroforestry, and regulated burning, support soil fertility, avoid overutilization, and promote the long-term sustainability of ecosystems. If we combine these practices with modern land management strategies, we can create environmentally supportive approaches that accommodate human needs without compromising ecological protection (Food and Agriculture Organization of the United Nations, 2018a).

1.6.2 Enhance Climate Resilience and Adaptation

Indigenous societies have learned to cope with environmental fluctuations over centuries by evolving land-use practices that reduce the effects of climate variability. Water harvesting, terracing, and drought-resistant agriculture are some of the traditional practices that enhance resilience to extreme weather events. Integrating these practices into contemporary land management can enhance climate adaptation measures and minimize vulnerabilities (Filho et al., 2020).

1.6.3 Improve Water Management and Soil Fertility

The traditional methods of water conservation include step wells, rainwater harvesting, and floodplain farming, which have been effective at maintaining groundwater levels and preventing soil erosion. Similarly, indigenous practices for soil management like organic compost, crop rotation, and agroforestry maintain soil fertility. Incorporation of such methods should enhance water efficiency and soil health and lessen the dependency on chemical fertilizers and artificial irrigation (Food and Agriculture Organization of the United Nations, 2018a).

2 Theoretical Framework

2.1 Core Principles of Indigenous Ecological Knowledge (IEK)

Indigenous Ecological Knowledge (IEK) is based on fundamental principles that guide sustainable environmental interactions and resource management. These principles have enabled indigenous communities to maintain ecological balance for generations. Indigenous Ecological Knowledge (IEK) is based on several key principles:

2.1.1 Holistic Approach

There is an interconnectedness between land, water, flora, and fauna with human communities recognized within IK. According to this perspective, nature is an interconnected system of humankind with water, land, flora, and fauna. The indigenous people in their practices emphasize respect for all living species and sustainable resource use. For example, most indigenous groups view rivers as sacred to them, which help them exercise a level of conservation ethics over water sources, obviating overuse and pollution (Saito et al., 2020).

2.1.2 Intergenerational Knowledge Transfer

Oral customs, myths, and legends are the main vehicles for the transfer of knowledge. IEK is transferred mainly by oral customs, myths, folklore, rituals, and legends, not by documentation. The elders are important figures in teaching new generations about biodiversity, seasonal cycles, and ancient conservation methods (Saito et al., 2020).

2.1.3 Adaptive Strategies

By shifting cultivation, seasonal migration, and rotational farming, dynamic responses to changes in the environment are established. Flexible, site-specific responses to environmental change are developed by indigenous communities. It includes practices such as shifting cultivation, seasonal migration, and controlled burning to maintain soil fertility and prevent overexploitation of resources (Kumar et al., 2020).

2.1.4 Sustainable Resource Use

Practices of selective harvesting and natural resource conservation facilitate long-term ecological balance. IEK favors appropriate harvesting methods to achieve long-term availability of resources. Traditional fishing populations of indigenous communities tend to practice selective fishing and not catch juvenile fish to facilitate population replenishment. Traditional agroforestry systems also incorporate a variety of crops, trees, and livestock to improve soil health and biodiversity (Tietenberg & Lewis, 2018).

2.1.5 Sacred Relationship with Nature

Environmental stewardship is augmented by spiritual and cultural beliefs. Many indigenous cultural groups consider nature to be sacred and, accordingly, they combine spiritual beliefs with their conservation practices. Several wildlife species and landscapes are considered sacred and hence are preserved accordingly by means of cultural taboos and rituals. For instance, this is followed by the Bishnoi community in India, which has protected forests and wildlife because of a spiritual devotion to nature for centuries (Tietenberg & Lewis, 2018; Verschuuren & Brown, 2018).

3 Indigenous Land Management Practices in India

3.1 Agroforestry and Sustainable Agriculture

The agroforestry system (Fig. 2) represents a traditional indigenous practice of land management in India, combining trees, crops, and livestock in order to ensure biodiversity and sustainable agriculture. The Gond, Bhil, and Khasi communities implement techniques like intercropping, mixed cropping, and silvopasture to maintain soil fertility and check against soil erosion at maximum productivity levels. They also enhance water retention, carbon sequestration, and food security and reduce dependence on chemical inputs through organic fertilization. As an example, in Jharkhand and Odisha, the Taungya system involves the plantation of sal (Shorea robusta) and teak (Tectona grandis) trees along with seasonal crops. Through timber and non-timber forest products, the system provides for economic sustainability and improves soil moisture, thereby preventing degradation of land. Agroforestry combines indigenous knowledge with modern agricultural advances, thus enhancing climate resilience and facilitating sustainable rural livelihoods (Hegde, 2020).

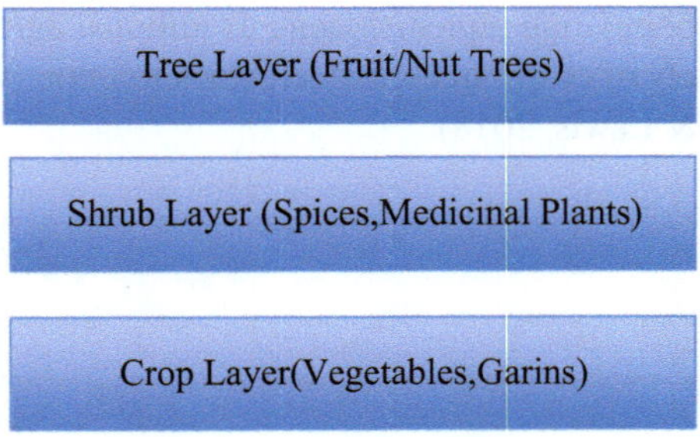

Fig. 2 Agroforestry system—multi-layered farming approach

3.2 *Water Resource Management*

Water conservation has been an integral part of traditional land management in India (Fig. 3), where indigenous people have ensured sustainable utilization and conservation of water resources through their customary practices. Indigenous societies have created effective water management mechanisms in accordance with their localized environments. The Zabo system in Nagaland is a distinctive rainwater harvesting system that incorporates forestry, livestock husbandry, and paddy cultivation to ensure efficient water conservation and retention of soil moisture. Likewise, holy water bodies, including the Ganges and Pushkar Lake, are conserved by cultural and religious practices that avoid pollution and overexploitation. In dry lands like Rajasthan and Gujarat, ancient stepwells act as important water storage tanks, facilitating groundwater recharge and increasing drought resistance. These tried-and-tested traditional approaches still contribute meaningfully to sustainable water management, maintaining ecological harmony and long-term resource availability (Kumar & Nair, 2007; Wester et al., 2019).

3.3 *Forest and Biodiversity Conservation*

Indigenous communities have often played a vital role in maintaining the ecological balance in the worlds' forests and biodiversity through traditional practices (Fig. 4). Sacred groves, in the Western Ghats, Meghalaya, and Rajasthan, are small patches of forests protected due to religious and cultural beliefs, as they are some of the most important hotspots for biodiversity. Similarly, in Odisha, Jharkhand, and Chhattisgarh, community forest management systems embody indigenous governance models for regulating the use of forests to ensure conservation and sustainable management of resources. In Northeast India, controlled burning is practiced to test fire-dependent ecosystems, clear physiognomic underbrush, enhance soil fertility, and discourage large-scale deforestation. These time-honored practices manifest long-standing principles of environmental stewardship providing traditional forest management with ecological and biodiversity premises (Saito et al., 2020).

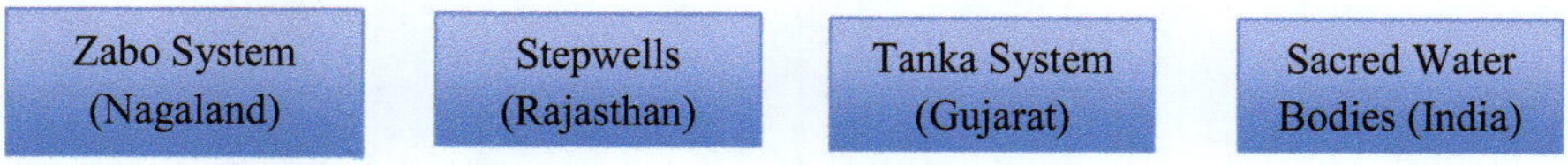

Fig. 3 Water conservation techniques in India

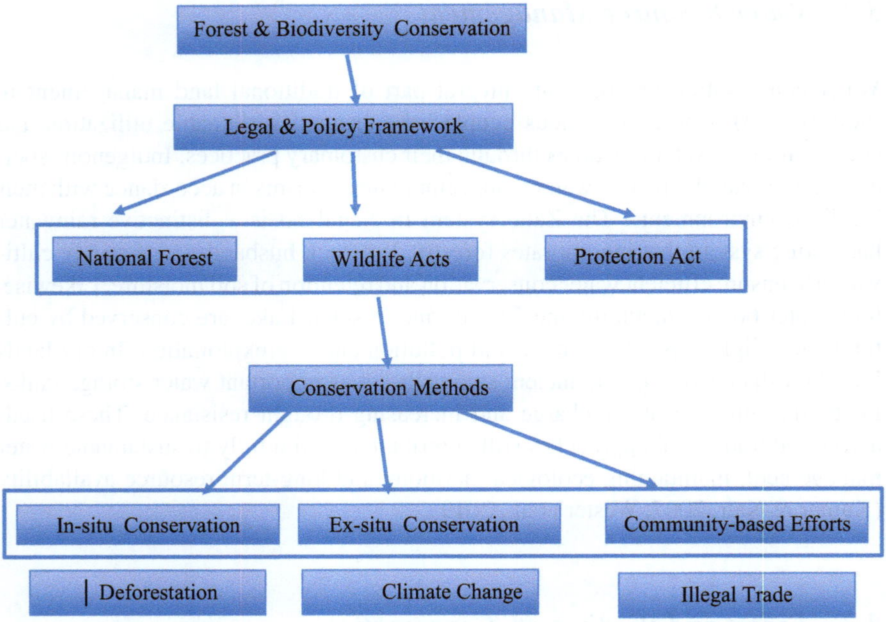

Fig. 4 Flowchart for forest and biodiversity conservation

3.4 *Wildlife and Habitat Protection*

Traditional communities in India have worked out with great skill plans and techniques for wildlife conservation to ensure the proper balance of nature and protect endangered species. In the Andaman & Nicobar Islands, tribal groups apply sustainable hunting practices, introducing seasonal restrictions on their harvest to prevent overexploitation and allow for species regeneration. In Central India, the Adivasi people support coexistence with wildlife by maintaining traditional wildlife corridors, enabling animal migration, and decreasing conflict with surrounding inhabitants. The Bishnoi community in Rajasthan is famous for their devotion to keystone species, particularly blackbucks, whom these people protect for religious and moral reasons. These traditional conservation practices demonstrate the significance of traditional knowledge in the maintenance of biodiversity as well as the protection of wildlife habitats (Bijoy et al., 2010).

4 Integrating Indigenous Knowledge into Land Management

4.1 Policy Recognition and Legal Frameworks

The October 2023 update pointed to the various governmental policies and legislation aimed at ending local communities with the requisite power while promoting environmental conservation in India that recognize the importance of traditional territorial practices related to land management. In compliance with the Forest Rights Act (2006), indigenous communities were granted legal entitlements to forest resources and assured of actively managing and utilizing forest resources while conserving them. The Biodiversity Act (2002) recognizes the equal importance of traditional knowledge of local communities in biodiversity conservation and ensures local communities to be involved in this conserving respectable use of biological resources. Lastly, Joint Forest Management (JFM) programs allow collaboration between forest departments and indigenous communities for participatory governance in sustainable forest management. Such policies encapsulate a confluence between indigenous wisdom and national conservation efforts and further substantiate the significance of community-led environmental stewardship.

4.2 Challenges in Integration

Despite their ecological significance, indigenous land management practices face several challenges:

- **Land Encroachment and Deforestation:** Commercial farming, urbanization, and industrial developments have rapidly increased at the expense of traditional land management systems. The activities result in wide-scale deforestation, destruction of habitats, and loss of biodiversity, upsetting the balance of nature as sustained by indigenous people. The loss of local population through land acquisitions also compromises the continuity of sustainable indigenous practices (Bijoy et al., 2010; Sukartiko et al., 2018).
- **Loss of Traditional Knowledge:** Modernization, migration, and emerging socio-economic conditions are leading the younger generations to very often depart from indigenous land management practices. This gradual loss of ancestral knowledge and sustainable techniques consists of agroforestry, water conservation, and biodiversity protection. However, the decline in oral traditions and the reduced transmission of knowledge, combined with a lack of formal vehicle for its documentation, will further speed up this descent into oblivion, which is slowly growing as a concern regarding the values of indigenous environmental stewardship (Karki et al., 2017).
- **Lack of Policy Implementation:** Even though there are laws like the Forest Rights Act (2006) and Joint Forest Management (JFM) schemes, their execution

at the local level is erratic. Reasons such as slow-moving bureaucracy, poor knowledge of the laws among community members, and little backing from local officials cause these laws to be enforced only half-heartedly. The upshot is that communities do not have the secure rights they need to manage the forests in a way that is both sustainable and good in terms of conserving biodiversity.

4.3 Recommendations for Sustainable Land Management

To effectively integrate indigenous knowledge into modern land management, the following steps are essential:

- **Strengthening Legal Protections:** Indigenous land rights must be safeguarded through the strict enforcement of policies like the Forest Rights Act (2006). Strengthening legal frameworks will help prevent encroachment, deforestation, and exploitation of indigenous lands. Additionally, promoting sustainable forest governance by involving local communities in decision-making processes can enhance conservation efforts and ensure long-term ecological balance (Food and Agriculture Organization of the United Nations, 2018b).
- **Community Participation in Conservation Programs**: Indigenous communities must take part in decision-making and governance roles in biocultural conservation and natural resource management. Closing the governance gap, particularly in the places where indigenous groups lack formal recognition, and supporting community-led initiatives such as Joint Forest Management and community governance of forests ensure that traditional ecological knowledge is part of the sustainable management of those spaces. Financial and technical support can empower these groups not only to govern these biocultural landscapes effectively but also to reclaim authority over them.
- **Blending Traditional and Modern Approaches:** Combining ancient ways of knowing with contemporary technologies like GIS, remote sensing, and ecological research provides a powerful opportunity to upscale sustainable land management. This "tech and tradition" approach not only enhances resource monitoring and provides a real-time view of the places we work but also gives a shot in the arm to climate resilience and biodiversity conservation efforts. Why? Because it respects and preserves cultural heritage. And the big payoff is that these strategies—executed by teams of scientists and working closely with indigenous communities—lead to balancing acts that manage resources, satisfy the climate imperative, and minimize biodiversity losses.
- **Education and Awareness:** It is essential to promote the interaction of knowledge between indigenous elders and their younger counterparts so that traditional wisdom of sustainable environmental management can be carried forward. Community-directed educational programs, storytelling initiatives, and workshops serve both to document and to transmit indigenous practices. Not only is this framing of traditional ecological knowledge (TEK) critical for the generation

of Tejano (the Tejano educators' term for the next generation) to carry forward, but also the integration of this knowledge into formal education (such as at Texas A&M University) serves critical purposes. These include the following:

- Awareness generation
- Interdisciplinary exploration of solution sets
- Appreciation for and shoring up of documented TEK as a part of the folk record

5 Conclusion

Traditional ecological knowledge offers time-tested, sustainable land management practices that secure the biodiversity of the areas in which indigenous peoples live. These practices not only enhance local biodiversity but also contribute to the planet's health as a whole and increase resilience to changes in the environment. Yet, the very practices that enhance the biodiversity of a region and help local communities adapt to environmental changes are under increasing threat from those very same changes.

References

Ajayi, O. C., & Mafongoya, P. L. (Eds.). (2017). *Indigenous knowledge systems and climate change management in Africa*. CTA.

Bijoy, C. R., Gopalakrishnan, S., & Khanna, S. (2010). India and the rights of indigenous peoples: Constitutional, Legislative and Administrative Provisions concerning indigenous and tribal peoples in India and their relation to International Law on indigenous peoples.

Eslamian, S., & Eslamian, F. (2021). *Handbook of water harvesting and conservation: Case studies and application examples*. Wiley.

Filho, W. L., Luetz, J., & Ayal, D. Y. (2020). *Handbook of climate change management: Research, leadership, transformation*. Springer.

Food and Agriculture Organization of the United Nations. (2018a). *Sustainable land management (SLM) in practice in the Kagera Basin: Lessons learned for scaling up at landscape level*. Food & Agriculture Org.

Food and Agriculture Organization of the United Nations. (2018b). *The state of the world's forests 2018: Forest pathways to sustainable development*. Food & Agriculture Org.

Food and Agriculture Organization of the United Nations. (2019). *The state of the world's biodiversity for food and agriculture: FAO commission on genetic resources for food and agriculture assessments 2019*. Food & Agriculture Org.

Hegde, N. G. (2020). Agroforestry: A way forward for sustainable agriculture in India. *Current Science, 119*(7), 1150–1157.

Hunter, D., Borelli, T., & Gee, E. (2020). *Biodiversity, food and nutrition: A new agenda for sustainable food systems*. Routledge.

Karki, M., Hill, R., Xue, D., Alangui, W., Ichikawa, K., & Bridgewater, P. (2017). *Knowing our lands and resources: Indigenous and local knowledge and practices related to biodiversity and ecosystem services in Asia*. UNESCO Publishing.

Kumar, B. M., & Nair, P. K. R. (2007). *Tropical Homegardens: A time-tested example of sustainable agroforestry*. Springer Science & Business Media.
Kumar, V., Singh, J., & Kumar, P. (2020). *Environmental degradation: Causes and remediation strategies*. Agro Environ Media, Publication Cell of AESA, Agriculture and Environmental Science Academy.
Miralles-Wilhelm, F. (2021). *Nature-based solutions in agriculture: Sustainable management and conservation of land, water and biodiversity*. Food & Agriculture Org.
Nakashima, D., Krupnik, I., & Rubis, J. (2018). *Indigenous knowledge for climate change assessment and adaptation*. UNESCO Publishing.
Roué, M., & Molnar, Z. (2017). *Knowing our lands and resources: Indigenous and local knowledge of biodiversity and ecosystem services in Europe and Central Asia*. UNESCO Publishing.
Saito, O., Subramanian, S. M., Hashimoto, S., & Takeuchi, K. (2020). *Managing socio-ecological production landscapes and seascapes for sustainable communities in Asia: Mapping and navigating stakeholders, policy and action*. Springer Nature.
Sillitoe, P. (2017). *Indigenous knowledge: Enhancing its contribution to natural resources management*. CABI.
Singh, P., Singh, R. P., & Srivastava, V. (2019). *Contemporary environmental issues and challenges in era of climate change*. Springer Nature.
Sukartiko, A. C., Nuringtyas, T. R., Marliana, S. N., & Isnansetyo, A. (2018). *Proceeding of the 2nd international conference on tropical agriculture*. Springer.
Tietenberg, T. H., & Lewis, L. (2018). *Environmental and natural resource economics*. Routledge.
Verschuuren, B., & Brown, S. (2018). *Cultural and spiritual significance of nature in protected areas: Governance, management and policy*. Routledge.
Wester, P., Mishra, A., Mukherji, A., & Shrestha, A. B. (2019). *The Hindu Kush Himalaya assessment: Mountains, climate change, sustainability and people*. Springer.
WWAP. (2018). *The United Nations world water development report 2018: Nature-based solutions for water*. UNESCO Publishing.

Sustainable Consumption of Life on Land: Impact of "Cruelty-Free" and "Ethically Certified" Cosmetics Labels

Amrutha Rose J. Valavi, Manjeri Subin Sunder Raj, and Anto Sebastain

1 Introduction

Cruelty-free products entail consumption of products which do not involve any animal for experimentation or apprehension to life on land. Consumption of cruelty-free products adheres to a process of manufacturing and processing consumables without harming animals and environment (Davies & Gutsche, 2016). It is distinct from plant based or vegan, which boycott all animal derivatives. People opt for these cosmetics for several reasons, which include health, environmental concern, compassion, and egoistic satisfaction. In recent years, lifestyle choices of people have inclined toward vegetarianism, vegan, and no-cruelty products across Asia, Europe, and the USA (Wescombe, 2019). Recent trends indicate an increase in vegan and cruelty-free consumables (Beardsworth & Bryman, 2004; Díaz, 2019; Dietz et al., 2010; Kalof et al., 2009; Phillips et al., 2010; Spencer et al., 2007). The consumption of animal-derived products has considerably reduced (Izmirli & Phillips, 2011). It can be noted that global vegan and "no-cruelty" consumption has increased to 23.13 billion US dollars in 2020 and study projected an increase to 61.35 billion US dollars in 2028. Personal preferences and environmental concerns were the major driving factors for these consumer preferences (Fortune Business Insights, 2022). Factors and indicators include environmental concerns, ethical, religious and also animal welfare (Deckers, 2009; Gruzalski, 2002; Rosenfeld, 2018). Some of the other factors and indicators revolved around ecological considerations, social, political, religious, moral, and ethical indicators to (Fox, 2013). These statements were dismissed, and some studies showed that animal welfare and

A. R. J. Valavi (✉) · A. Sebastain
School of Law, CHRIST (Deemed to be University), Pune Lavasa, Maharashtra, India
e-mail: amrutha.valavi@res.christuniversity.in

M. S. S. Raj
School of Law, CHRIST University, Bangalore, Karnataka, India

J. J. Puthenkalam S.J et al. (eds.), *Terra Thrive*,
https://doi.org/10.1007/978-3-032-02738-2_16

environmental sustainability correlate with consumer preference (Matheny & Chan, 2005).

Consumer attitudes can influence and predict the behavior of consumers (Ajzen, 2011). Comprehending consumer behavior patterns toward certain labels or lifestyle choice-based products becomes vital (Zablocki et al., 2019). Researchers indicate that consumers opt for ethical consumption as opposed to traditional ones (Ryan & Casidy, 2018).

It was found the consumer not only showed an inclination toward brands adhering to products labeled as ethically manufactured, they made an effort to avoid brands that were involved in unethical policies (Lim & Shim, 2019). Owing to this behavior, enterprises were forced to invest in ethical manufacturing practices. Despite increased inclination toward labels such as cruelty-free and availability of products that match the descriptions, the market for these products is exclusive and niche. Some researchers are skeptical whether consumers' attitude toward products manufactured ethically and labeled "cruelty-free" actualizes to purchase intentions (Govind et al., 2017; Kushwah et al., 2019).

Additionally, purchase intentions of consumers who opt for cruelty-free products or other ethically produced products may result in sustainable consumption. Sustainable consumption has been defined as the practice of optimizing consumption ensuring minimal impact on the environment and all that encompasses the environment. While internalizing sustainable consumption, consumers mindfully seek to minimize adverse effects on ecosystems. This behavior may be attributed to rationalizing, optimizing, and opting for ethically certified products.

Implications of incorporating sustainability quotients and acknowledging the social issues into business strategies were examined theoretically and practically in recent studies. Value-based business strategies revolve around aspects like ethical standards, environmental responsibilities, consumer perceptions and interest, and need to uphold sustainability (Banerjee, 2002). Recent studies tend to purport that sustainability has been viewed as an end of climate resilient policies to protect life on land (De Chiara, 2016; Neamtu & Dragos, 2015), a prerequisite to widen consumer acceptance toward the goodwill of the management of enterprise. It was also viewed as a means to generate value-based consumables and ensure organizational survival (Gomez-Trujillo et al., 2020). Lifestyle-related attitudes and behavior that claims to contribute sustainability have been evolving in the present times. These lifestyle choices and sustainability trend have been more and more popular in cities in China as well as other regions of Asia, that includes India, Indonesia, Hong Kong, Taiwan, and Philippines (Gupta, 2014; Gupta & Racherla, 2018; Khare, 2011; Posegga et al., 2014). Studies have been conducted to evaluate the common grounds or factors leading to this influence; however, they have failed to draw a correlation among consumer preference, lifestyle choices, consumer attitude, and purchase intention leading to sustainable consumption while consumables labeled as cruelty free, let alone any ethical certification (Joshi & Rahman, 2019).

This study intends to focus on lifestyles choices, consumer preferences, consumer attitudes, purchase intentions, and sustainable consumption by opting for cruelty-free cosmetics and cosmetics with ethical certification. The article has been arranged in the following manner: Sect. 2 has evaluated the literature available on lifestyle choices, consumer preferences, consumer attitudes, consumer preferences, purchase intention, and sustainable consumption. Sections 3 and 4 state the methods and results, lastly Sect. 5 concludes by acknowledging scope for future research and any limitations.

2 Literature Review

2.1 Theoretical Background

2.1.1 Cruelty Free

Cruelty free is a term that has been used to describe several marketing strategies that claim to have refrained from animal-derived products. It may be appended to an enterprise’s conscious avoidance to experimenting or testing on animals, which may coincide with an animal rights path which propagates that people should consider animals as sentient beings. Having said one perspective, there is another side, where animals are treated ethically, for instance, they are not confined to cages or forced to produce anything. The consumables can be merely what is in excess of their needs and cruelty-free eggs or milk from local farmers. There the term cruelty free can be employed to cause a wave in marketing a product. In one of the studies conducted by Coalition Consumer Information, where Leaping Bunny symbol resonated to ensuring cruelty-free process in manufacturing or procuring the good, more than 750 respondents among 1000 responded that the companies should refrain from animal testing or experimentation for cosmetics or home/personal care products. In a study conducted and later published in AV magazine (2015), they found that consumers were inclined toward animal welfare and their rights. People’s attitude toward were also found to be because of their multiple roles in a human’s life, for instance companion, pet or of therapeutic aid (Belk, 1988).

This study will examine the implication of cruelty free on cosmetics and among consumers across the Indian subcontinent on their attitudes and purchase intention leading to sustainable consumption. Companies like Kay, Kama Ayurveda, Lotus, and FAE beauty are some of companies that use the term on their cosmetic products to communicate that they have not tested or experimented on animals to obtain their product.

While examining legal checks on advocating cruelty-free cosmetics, it can be found that in China, testing cosmetics on animals was mandatory till January 2023 (Hirsh, 2023), whereas countries like India have laws that restrict any experimentation on animals for personal care. As far as household products are concerned, there have been deliberations and attempts to ban or restrict products that employ testing on animals. “People for the Ethical Treatment of Animals” (*PETA*) has a guide to navigate and identify enterprises that test and refrain from animal testing or experimentation. Here, their widespread influence has compelled certain brands to opt or reconsider alternatives for the label "Cruelty-free." However, till date, there has been no definite meaning to the term "cruelty free," except a widely accepted interpretation that it would be free from animal test or experimentation and any cruelty upon them while procuring or manufacturing the product. This leads to a more wider and inclusive term "ethically certified."

2.1.2 Ethically Certified

Ethically certified derives from an ardent choice of the consumer to opt for ethical consumption. An act of purchasing a good or service is based on a company's value and refraining from a corporation that produces or manufactures goods in an unethical manner. For instance, opting to avoid products with an ill implication on the environment or society or particular class of beings (Deng, 2013). As opposed to earlier days, where ethical consumption was an elite person's game, currently it has wide spread itself in multiple folds. It not only amplifies the values of the seller or manufactures and other enterprises associated with them but check out a social commitment (Sudbury-Riley & Kohlbacher, 2016). The reasons for consumers to choose ethically manufactured products may be due to environmental concerns, animal welfare, labor conditions, well-being, guilt, self-satisfaction, or merely a habit (Davis et al., 2016; Sebastiani et al., 2013). Consumption of ethically certified products can also be because of lifestyle choices, brand, price, and quality; this was determined in earlier studies by employing the Theory of Planned Behavior (Davies & Gutsche, 2016; Liobikienė & Bernatonienė, 2017). Therefore as stated earlier, ethical consumption may or may not include cruelty-free label, it has much wider acceptance as opposed to cruelty-free products. In this research, we have examined both perspectives among respondents in a collective as the core question is to understand whether they contribute to sustainable consumption.

2.1.3 Labels

Few researchers in the personal care field considered the issue of animal welfare (Liobikienė & Bernatonienė, 2017). However, marketers provide customers with a plethora of labels (e.g., bunny logos) and statements, including negatively worded messaging (e.g., "cruelty-free," "not tested on animals"), highlighting the absence of risky or problematic components or methods due to public outrage This seeks to convert consumers' emotional responses into actual purchases Animal-friendly statements and labeling help to improve brand views (Sheehan & Lee, 2014). Labels and statements referring to the "cruelty-free" concept frequently elicit emotional thoughts, as can items utilizing pastoral imagery or even cutesy lures through a visual interplay (Wang et al., 2017). These literatures depict the underplay in employing a label or calling out for claim (Grappe et al., 2021).

2.2 *Lifestyle Choice, Consumer Attitude, and Sustainable Consumption*

Earlier, only a certain class of people could afford the cost of consuming and participating in a consumer society which dictates a person's choices based on their lifestyle. A consumer society accommodates a large amount of goods and services that are offered to satisfy the needs of the community at large. Role of consumption has elevated because of the emergence of commodities in the online as well as offline market. People's actions are inclined toward what gives them pleasure and utility. Products and services consumed provide a sense of purpose, however it can be temporary. This induces a feeling of wanting to purchase more products to satisfy and maintain the sense of purpose. Another quality that lingers among consumers is the ethos that they support by purchasing a good or service. We may call it a "psychological norm." It validates their moral obligation. These norms are assigned to products and services through trends. These affiliations not only ascribes the consumers and to ease the social status but also allows them a title, cult or societal stratum. Additionally, people take a stand by opting a product or service (Borowiec, 2015). According to a recent study, consumer behavior is an attitude toward reality that includes the pursuit of quick gratification as well as the purpose and meaning of life through the procurement and consumption of commodities and services.

Lifestyle choices are related to the notion encapsulated in the acronym LOHAS (Ray & Anderson, 2001), which stands for Lifestyle or Health and Sustainability. It has been discovered that it is environmentally conscious and exhibits environmentally friendly behaviors. The concept encompasses both products and consumers favors health and environmental sustainability. They care about the environment, as seen by their purchases of organic food, natural personal care items, cruelty-free products, and energy-efficient home equipment. Furthermore, the correlation between environmental performance and service quality has been stressed. As this demonstrates, shifting reasons lead to changing behavior, which indicates a shift in the orientation values of consumers and producers in the free market. It is obvious that lifestyle choices are important drivers of sustainable consumption practices. This deduction, however, is not directed to consumption of cruelty-free cosmetics or cosmetics that possess any ethical certificate.

Consumer attitudes can be shaped by lifestyle. Lifestyle is considered to be intertwined with consumption. While the traditional separation of social classes is based on production relations, it is considered that a consumption-driven lifestyle is a cause of many social divides (Cockerham, 2005). Lifestyle was used to define social status by Max Weber. According to him, lifestyle was used to define social status using the category of "lifestyle." In his opinion, lifestyle not only defines status but it also forms them. Various studies concluded that lifestyle is a consequence of individual choices, and it determines a person's stance in the social hierarchy. This includes anything from dietary habits, choice of sports, hobbies, personal products, or apparels (Abel & Cockerham, 1993; Cockerham, 2005).

Giddens (2002) purports "healthy lifestyle" as a pattern that swirls out health-based choices available in the market. These habits may consist of alcohol consumption, dealing with uncertainties, sports, hygiene, and personal preferences. Here a healthy lifestyle is a pattern of choices that ensure longevity and everything that entails the same. People's choices for a certain lifestyle associate with how they wish to portray their identity. The indication is that elements of a healthy lifestyle should be viewed as meaningful.

Especially among the newly wealthy upper class, who perceive eating as an "immediate pleasure," Parra-Medina et al. (2010) claim that India's growing modernization and consumer society have molded a certain attitude toward food that contributes greatly to obesity and may eventually affect cardiovascular conditions.

This being said, another study suggests that people pay attention to the source where they are consumable, especially if it is animal-derived product. There are instances where consumers have been vocal of their opposition to ill standards of treating animals (Fernqvist & Ekelund, 2014). In order to maintain the demand, enterprises would go out of their way to ensure humane conditions and health of animal and their needs, for instance, a company procuring animal-derived raw materials would choose to procure from local farmer as opposed to a breeder (De-Magistris & Gracia, 2016). Recent studies show that around 21% consumers across the globe are willing to consume for a higher price. Here humane treatment of animals was used as leverage to attract consumers (Zander & Hamm, 2010), here these products are assumed to have certain standards of quality. There are certain assumptions that ethical certification or "no cruelty free" label or painting a humane treatment merely create purchase appeal and cannot tune any connection between lifestyle and consumer attitude. However, these claims have not been substantiated with adequate proof to establish the statement of this correlation as far as "cruelty free product" (Carlson et al., 1993). Most studies and discussion till date revolve around lifestyle choices, consumer attitude, and sustainable consumption in relation to food and dietary habits. Therefore, there is a need to evaluate the correlation between lifestyle choices and consumer attitude while purchasing cruelty-free cosmetics.

H1: Lifestyle choices influence consumer attitudes while purchasing cruelty free and ethically certified cosmetics.

2.3 Consumer Preferences and Consumer Attitudes

Building powerful brands, cultivating a favorable brand image, and growing brand equity are all crucial to organizational effectiveness and inducing preference among consumers. Brand image may influence consumer behavior, influencing a variety of outcomes such as brand preference, satisfaction, and loyalty (Jamal & Goode, 2001). A brand is a distinguishing term and/or signifier used to identify and set one seller's or group of sellers' services and products against those of competitors. Brands with such a substantial reputation have been demonstrated to generate higher revenues (Schuiling & Kapferer, 2004). Brands with more brand preference usually have more brand equity. Brand identification, associations, value perception, loyalty, trademarks, and proprietary information, among many other perspective and cognitive elements of brand equity, have all been recognized. While brand image is not the existing study's focus, it is significantly related to brand preference, which affects customer behavior. One of the most common errors for brand managers is concentrating excessively on brand qualities. Rather, they should focus more on organizational characteristics. It was reiterated that the brand-as-organization strategy relies on organizational qualities that are created by the organization's human resources, ethos, values, and initiatives, such as being inventive, quality-driven, client, or ecologically sustainable (Aaker & Biel, 2013). Improving and refining an organization's behavior are more likely to have a beneficial influence on consumer attitudes than product-related initiatives. Earlier studies have concentrated on organizational behavior, which should assist managers in (a) overcoming the trap of brand attribute absorption and (b) developing strong interpersonal links with consumers. Sustainable acts can attract young customers who are interested in environmental health, and reporting about them can boost brand preference.

Because brand preference is an attitude, analyzing brand preference after consumers have been exposed to a message makes logical sense. Organizational actions are not completely to blame for the relationships that form between a brand and its consumers. Consumer environmental considerations are a crucial component of the problem. Most studies are perplexed as to why consumers' worries and good intentions about environmental health do not translate into pro-environmental behavior. Stronger attitudes typically produce behaviors that are consistent with such attitudes. Other lines of study, however, imply that attitude alone may not be sufficient to elicit commensurate action. This concept appears to be applicable to environmental behavior. According to the current study, attitude structure, and function, consumer understanding of green products and the demand for self-categorization all influence behavior. Organizations may affect how customers view their brands and encourage long-term connections by participating in sustainable behaviors that are aligned with customer attitudes. This strengthens customer loyalty and brand preference (Bagozzi, 1992). For instance, limiting plastic bag usage as a green method should not only assist firms save costs associated with making, procuring, and/or disposing of plastic bags, but it may also help them create relationships with their customers. Anti-disposable shopping bag initiatives have been adopted efficiently in the United States (Rynarzewska, 2019). The above studies have vastly discussed that preferences incline to customer attitude when it comes to consumables other than cruelty free and/or ethically certified cosmetics.

HI 2: Consumer preference influences consumer attitudes while purchasing cruelty free and ethically certified cosmetics.

2.4 Purchase Intention Vis-a-Vis Consumer Attitude and Lifestyle Choice and Consumer Preference

As Ajzen (1991) stated "*Intentions are assumed to capture the motivational factors that influence a behavior; they are indications of how hard people are willing to try, of how much of an effort they are planning to exert, in order to perform the behavior.*"

Customers use his effort, time, and money to buy the goods, thus purchasing intent is very important in his life. Customers' tastes and perceptions are continually influencing the purchase process. The phrase "consumer's judgment based on the whole set of basic as well as exterior characteristics of the product or service" is used to define perceived quality of a product (Grunert, 2005).

Perception on a product directly affects the purchase intention. Customers generally have an idea of the product quality, price, and style before purchasing the same.

Because it has direct relationships that impact each other, purchasing intention grows as well as lowers after utilizing the product. When the quality is excellent, the customer's purchase intention increases, it was (Oliver, 1999) proposed that there are differences between perceived quality and satisfaction. Purchase intentions refer to a customer's opinion of a product after using it. The behavior is seen as crucial in predicting consumers' purchase habits as well as their intentions. According to (Aaker, 1996), perceived quality determines the degree of overall product quality. Through the mediating variable of customer satisfaction, prior studies have discovered an indirect link between perceived product quality and purchase intentions (Bou-Llusar et al., 2001).

Customers' attitudes can be associated with their concern for the effects of their purchases, as well as altruistic impulses, where both would encourage consumers to influence the market to become more ethical. Consumer knowledge is likely to influence corporate tactics. Environmental knowledge had a direct influence on purchase intention of cruelty-free cosmetics, implying that environmental and ethical concerns are important to customers who buy cruelty-free cosmetics (Sebastiani et al., 2012). Environmental knowledge had proportional influence on purchase intention of cruelty-free cosmetics, implying that environmental and ethical concerns are important to customers who buy cruelty-free cosmetics. Marketers tend to give information on environmental safety since it can build a favorable attitude toward businesses. Understanding the characteristics examined in this research can be beneficial to marketers and product creators in the cruelty-free cosmetics business (Mostafa, 2006).

These studies partly discuss purchase intention in relation to consumer attitude and preferences of multiple consumables which briefly included cruelty-free cosmetics; however, it does not draw inferences about lifestyle choices or indulge in a heavy study on these variables. Therefore, in the present, it becomes relevant to test the below hypothesis.

H3: Consumer attitude leads to purchase intention.
H4: Lifestyle choice leads to purchase intention.
H5: Consumer preference leads to purchase intention.

2.5 Sustainable Consumption Derived from Consumer Attitude and Purchase Intention

It has been demonstrated that having a favorable attitude toward a particular conduct increases the likelihood of carrying out that behavior (Ajzen, 1991). According to research on green purchasing, attitudes and beliefs are strong translators of sustainable consumption. Consumers' good attitudes about green hotel operations stimulate eco-friendly consumption intentions. Furthermore, customers' attitudes toward sustainable purchasing are favorably associated with their intents to book green hotels (Tanner & Wölfing Kast, 2003).

It has been demonstrated that having a favorable attitude toward a specific behavior increases the probability of engaging in that behavior (Ajzen, 1991). According to research on green purchasing, attitudes and beliefs are strong mediators of sustainable consumption (Tanner & Wölfing Kast, 2003). Consumers' good attitudes about green hotel operations promote eco-friendly consumption intentions. Furthermore, customers' attitudes toward sustainable purchasing are favorably associated with their intentions to book green hotels. Consumers with a favorable attitude toward the environment will prefer ecologically friendly beverage packaging. In LOHAS Way of Life, Consumer Perception, Subjective Perceived Norm, Behavioral Control Intention to Consume, in a Sustainable Way Consumption making it environmentally friendly A favorable attitude toward sustainable products has been suggested as a motivator for sustainable consumption. Previous study has also found that environmental citizenship motivates consumer behavior toward sustainable purchase (Han & Yoon, 2015).

It has been discovered that consumer decision-making is a multidimensional process, with various contextual aspects like Planned Behavioral Consumption and attitude assisting in decision-making. Planned Behavioral Consumption describes an individual's sensation of ease or difficulty when completing specific actions (Ajzen, 1991). People's behavioral control is influenced by their general attitude, which is mediated by particular control beliefs. This reflects prior customer experience and anticipated inconveniencies.

Green purchasing behavior has been identified as a crucial predictor, demonstrating that a consumer's PI affects his or her likelihood of choosing to purchase green goods (Beckford et al., 2010). PIs and sustainable customers' purchasing patterns for green products and services have been demonstrated to be positively correlated in the Indian context (Yadav & Pathak, 2017). For instance, individuals who have a strong desire to eat organic food are more likely to be oriented toward a diet that is both nourishing and balanced and to maintain a well-balanced lifestyle. According to the research, consumers that have LOHAS impulses are the first to respond favorably to organic produce, green products, and sustainable lifestyle choices. Additionally, they develop a devoted following of buyers for those ecologically friendly items and encourage sustainable practices (Shin, 2011).

In the above literature, they have multiple means of sustainable consumption, however not the correlation of sustainable consumption with consumer attitude and purchase intention.

H6: Consumer attitude leads to sustainable consumption.
H7: Purchasing intention leads to sustainable consumption.

3 Methodology

3.1 *Research Tool*

The proposed model is shown in Fig. 1. Five items from the variable lifestyle choice section of the questionnaire, as shown in Table 1, are included from Boyd Thomas and Lee Okleshen Peters (2009), consumer preference from Yadav and Pathak (2017), consumer attitude from Lovelace and Brickman (2013), purchase intention from Chen and Chang (2012) and Beldad and Hegner (2018), and lastly sustainable consumption from Xiao and Li (2010) and Lombart and Louis (2012). Some of the variables have used a 5-point Likert scale, where 1 = Strongly disagree/Never/Most Unlikely, 2 = Disagree/Rarely/Unlikely, 3 = Neutral/Somewhat agree, 4 = Agree/ Often/ Likely, and 5 = Strongly agree/Always/Most Likely.

3.2 *Sampling Technique and Sample Size*

The research revolves around any consumer with a purchasing power; hence, it is a simple random sampling technique for collecting data. Purpose of this study is to explore the possibility whether lifestyle choices, consumer preference, consumer attitudes, and purchase intention lead to purchase of *cruelty-free* and *ethically certified* cosmetics leading to sustainable consumption. The attitudes of the respondents were measured by using a 5-point Likert scale, the same scale was used to gauge the lifestyle choices, preferences, attitude toward cruelty-free products, or products

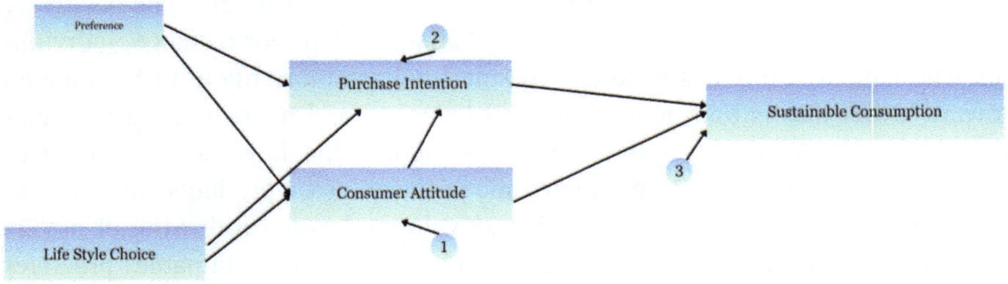

Fig. 1 Proposed Research Model

Table 1 Measurement construct

Items	Variables	Authors
1	*Lifestyle choice*	Boyd Thomas and Lee Okleshen Peters (2009)
	A vegetarian	
	A vegan or plant based consumer	
	Health conscious (physical and mental)	
2	*Consumer preference*	Yadav and Pathak (2017)
	Would you prefer a cosmetic certified as cruelty free?	
	Would you prefer a cosmetic certified as ethical?	
	Would you prefer to avoid cosmetics without cruelty-free certification?	
	Would you prefer to avoid cosmetics without any ethical certification?	
3	*Consumer attitude*	Lovelace and Brickman (2013)
	How often do choose cosmetics with a cruelty-free certification?	
	How often do choose cosmetics with at least one ethical certification?	
4	*Purchase intention*	Chen and Chang (2012) and Beldad and Hegner (2018)
	Would you buy cosmetics without cruelty-free labels?	
	Would you buy cosmetics without any ethical certifications?	
	Would you buy cosmetics which have complied to cruelty-free standards over a cheaper alternative without any certifications?	
	Would you buy cosmetics which have complied to ethical standards over a cheaper alternative without any certifications?	
	Would you buy cosmetics which has complied to ethical standards over a cheaper alternative without any certifications	
5	*Sustainable consumption*	Xiao and Li (2010) and Lombart and Louis (2012)
	Sustainable consumption is avoiding cosmetics without a cruelty-free label	
	Sustainable consumption is avoiding cosmetics without any ethical certification	
	Sustainable consumption is boycotting all cosmetics adhering to unethical standards and animal derived materials	

with any ethical certification and purchase intention. This was to find a correlation between attitude and purchase intention. On determining the purchase intention, the questionnaire further chose to understand whether it contributes to sustainable consumption. Therefore the sample size is randomized by choosing any member of the society who has a purchasing power and those that use any cosmetic product. The questions in the first part of the questionnaire act as a filter to ensure those that use cosmetics and have a purchasing power.

The technique was employed to reduce sample bias and higher-quality data, it had to accommodate the diverse population and ensure adequate representation so that data is not polluted in one or a few demographics.

3.3 Data Collection

The sample research has 466 respondents. The study's sample is composed of consumers with spending capacity. The data was collected using random sampling techniques to ensure maximum exposure, less bias, and open to a diverse population. This study focuses on the assertions whether consumer attitude of cruelty-free cosmetics and their purchase intention can result in sustainable consumption. To guarantee diversity in the data collection for this survey, questionnaires were being used, and web-based surveys are simple and cost-effective.

3.4 Demographics

The age of the majority respondents (57%) were people aged between 18 and 25 and followed by people aged between 26 and 30 and 56 and above (13.9%). The education qualification construct has six categories ranging from Class XII or below to PhD, the questionnaire was extended to ensure the participation of people who opted for skill based education. The highest of 36.3% belongs to the respondents pursuing graduation, followed by post-graduates who held 33.2%. The respondents largely belonged to the range of annual households earning an annual income from 5.1 Lakhs to 10 Lakhs (In rupees) making 22.4% followed by respondents who earned annual household income above 15 Lakh (In rupees) who were 19.7%. These demographics as shown in Table 2, provides insights into the data gathered and helps in further analysis and understanding.

Table 2 Demographics

S. no	Classification	Frequency	Percent
1	Gender		
	Male	200	44.84304933
	Female	242	54.26008969
	Prefer not to say	2	0.448430493
	Other	2	0.448430493
2	Age	Frequency	Percent
	18–25	258	57.8
	26–30	62	13.9
	31–45	34	7.6

Table 2 (continued)

S. no	Classification	Frequency	Percent
	46–55	30	6.7
	56 and above	62	13.9
3	Employment status	Frequency	Percent
	Student/scholar	246	55.2
	Employer	118	26.5
	Unemployed	40	9
	Business	42	9.4
	Total	446	100
4	Educational qualification	Frequency	Percent
	Class 12 or below	8	1.8
	Pursuing undergraduation	162	36.3
	Graduate	84	18.8
	Post-graduate	148	33.2
	Doctoral scholar	16	3.6
	PhD	12	2.7
	Other	16	3.6
	Total	446	100
5	Annual household income	Frequency	Percent
	Upto 3 lakhs	86	19.3
	3.1 lakhs to 5 lakhs	86	19.3
	5.1 lakhs to 10 lakhs	100	22.4
	10.1 lakhs to 15 lakhs	86	19.3
	Above 15 lakhs	88	19.7
	Total	446	100

4 Results

Testing and analyzing variables have been done using IBM SPSS and AMOS. Results from the study have been presented to support and refute the research methodology's hypothesis. The data analysis includes confirmatory factor analysis, descriptive analysis, Cronbach's alpha, regression coefficients, and structural equation modeling. The analysis studies the mediating effect of the variable between consumer attitude and purchase intention, making the model a study analyzing three variables (*Consumer attitude, Purchase Intention, and Sustainable Consumption)*.

4.1 Construct Reliability

Cronbach's alpha reliability test was carried out to measure the internal dependability (Bonett & Wright, 2014). The results above 0.7 can be reliably interpreted as satisfactory. For the constructs, scores above 0.7 are rated satisfactory (Ursachi

Table 3 Construct reliability

Constructs	Mean	SD	Factor analysis	Alpha	AVE	CR
Lifestyle choice						
LC1	3.1973	1.13636	0.78	0.579	0.629	0.773
LC2	2.2646	1.10848	0.807			
LC3	3.5516	1.02805	0.76			
Preference						
PRE1	3.6637	1.16074	0.831	0.832	0.601	0.857
PRE2	3.7848	1.07195	0.795			
PRE3	3.5605	1.09531	0.767			
PRE4	3.4036	1.19865	0.703			
Consumer attitude						
CA1	3.2422	1.1552	0.816	0.809	0.696	0.822
CA2	3.157	1.12481	0.854			
Purchase intention						
PI1	2.5919	1.06816	0.834	0.731	0.761	0.926
PI2	2.426	1.04404	0.803			
PI3	2.8341	1.182	0.934			
PI4	2.8744	1.15045	0.912			
Sustainable consumption						
SC1	3.6996	0.96889	0.897	0.893	0.774	0.911
SC2	3.6547	0.9998	0.907			
SC3	3.6233	1.14236	0.833			

et al., 2015; Nunnally, 1978). All of the constructs have satisfactory Cronbach alpha values, assuring their internal reliability, as shown in Table 3 of the ratings for each construct. As shown in Table 3, when evaluating the Composite Reliability of all components, it is higher than 0.7. A composite reliability higher than 0.7 indicates proven construct reliability (Shrestha, 2021). As indicated in Table 3, average variance extracted ranged between 0.6 and 0.774, which contributes to confirming construct reliability. AVE higher than 0.5 helps to establish a convergent validity (Fornell & Larcker, 1981). Mean and Standard Deviation have been calculated for descriptive statistical analysis. Table 3 displays the factor loadings, AVE, composite reliability, and Cronbach's alpha.

4.2 Kaiser–Meyer–Olkin (KMO) Test

For the purpose of factor analysis, the significance and validity of the data are examined using the Kaiser–Meyer–Olkin (KMO) test. The proportion of common variance is measured by KMO to identify any underlying reasons. KMO values higher than 0.6 indicate the relevance and significance of factor analysis. KMO test values are indicated in Table 4, all values are higher than 0.6, proving relevance and usefulness of factor analysis.

Table 4 Discriminant validity

Correlations	KMO	LC	PRE	CA	PI	SC
LC	0.553	*(0.794)*				
PRE	0.648	0.201**	*(0.775)*			
CA	0.5	0.320**	0.422**	*(0.835)*		
PI	0.549	−0.061	0.129**	0.011	*(0.872)*	
SC	0.705	0.100*	0.458**	0.267**	0.157**	*(0.88)*

**Correlation is significant at the 0.01 level (2-tailed)
*Correlation is significant at the 0.05 level (2-tailed)

The purpose of ascertaining the discriminatory validity was to test if each variable could draw a different perspective. Covariance correlations or standardized linear correlations were used. The indices assume a value anomalous from 1. Dependability of the report was investigated, and it was found that no pair of constructs equaled 1, as indicated in Table 4. This difference indicates that the variables indicate different perspectives. To evaluate discriminant validity, the square root of average (AVE) for each item and its squared correlations with all components are compared. The value of AVE should be bigger than the correlations (Everitt & Yates, 1989). Table 4 demonstrates that the square root of AVE is larger than the cross-construct correlations.

4.3 Confirmatory Factor Analysis

To ensure that the primary variables or constructs were consistent with the theoretical framework of the study, confirmatory factor analysis was conducted on all of the constructs. A two-step model construction process with a measurement model and structural model was suggested by Anderson and Gerbing (1988). It also demonstrates how a concept that cannot be evaluated directly is described by the interconnections between the variables of the measure. The values of the CFA model are not acceptable as far as the normed chi-squared was concerned, however in study conducted by Alavi et al. (2020), they stated that merely viewing at one or more won't hold the model unacceptable or unfit. Additionally, RMSEA also indicates a weak model which may be viewed as a weakness among the variables correlated and further refined. This can be clear from Table 5.

4.4 Regression Coefficient

To determine whether the aforementioned variables contribute to sustainable consumption by purchasing cruelty-free products, logistic regression is used to test their predictive power. The regression's R^2 value is 0.227, indicating a 22.% relationship between the independent and dependent variables, as shown in Table 6.

Table 5 Goodness of fit model

Measure	Symbol	Range	Value	Author
Normed chi-squared	Cmin/df	1.0 < Cmin/df < 3.0	29.443	Alavi et al. (2020)
Goodness of fit index	GFI	GFI > or = 0.9 means satisfactory fit 0.8 < GFI < 0.9 means acceptable fit	0.931	
Normed fit model	NFI	NFI > or = 0.9 means satisfactory fit 0.8 < NFI < 0.9 means acceptable fit	0.665	
Comparative fit model	CFI	CFI > or = 0.9 means satisfactory fit 0.8 < CFI < 0.9 means acceptable fit	0.664	
Root mean square error of approximation	RMSEA	RMSEA <0.08 means good fit 0.08 < RMSEA <0.1 moderate fit RMSEA >0.1 means poor fit	0.253	Olutende et al. (2019)

Table 6 Regression coefficient

		Unstandardized coefficients		Standardized coefficients		
	Value	*B*	Std. error	Beta	*t*	Sig.
(Constant)		1.595	0.237		6.728	<0.001
Lifestyle choice		−0.007	0.054	−0.006	−0.14	0.889
Preference		0.414	0.048	0.405	8.65	<0.001
Consumer attitude		0.087	0.043	0.097	2.016	0.044
Purchase intention		0.118	0.048	0.104	2.443	0.015
R	0.467a					
R^2	0.227					
Adjusted R^2	0.22					
Stand. error of estimate	0.833					

Dependent variable: Sustainable consumption

Despite the R^2 value of the regression less than 0.5, the present model stands strong as the study tries to understand human behavior, which in itself is unpredictable (Ballard, 2019). The values imply that purchase intention and consumer attitude lead sustainable consumption, consumer preference and consumer attitude albeit lifestyle choice lead purchase intention, and lifestyle choice and preference lead consumer attitude. Thus hypothesis has been supported and rejected accordingly as evident in Table 7.

4.5 SEM and Hypothesis Testing

Structural Equation Model is an analytical tool to evaluate the impact of independent variables on the dependent variables as seen in Fig. 2. The proposed model proves the relationship between sustainable consumption as against other

Table 7 Structural equation model results

	Estimate	S.E.	C.R.	*P*	Label
Lifestyle choice → consumer attitude	0.327	0.055	5.91	***	Supported
Preference → consumer attitude	0.422	0.047	9.006	***	Supported
Consumer attitude → purchase intention	−0.023	0.042	−0.547	0.585	Not supported
Preference → purchase intention	0.141	0.046	3.102	0.002	Supported
Lifestyle choice → purchase intention	−0.089	0.051	−1.723	0.085	Not supported
Purchase intention → sustainable consumption	0.176	0.051	3.431	***	Supported
Consumer attitude → sustainable consumption	0.239	0.041	5.768	***	Supported

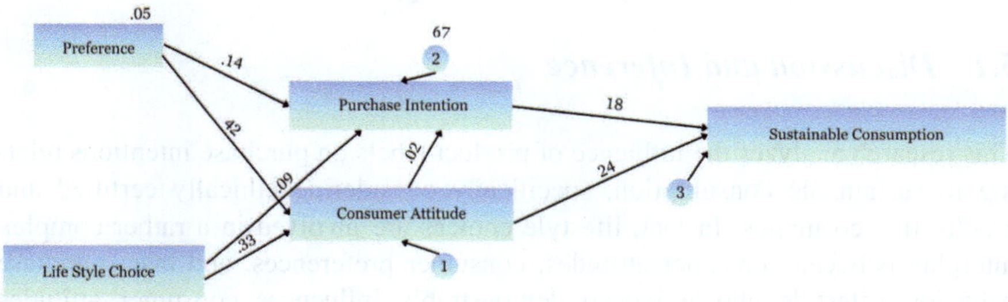

Fig. 2 SEM model

independent variables. Based on SEM and Table 7, hypothesis testing has been done. When the *p* values less than 0.05 support the hypothesis, whereas values more than 0.05 result in the hypothesis being rejected; with a value less than 0.01 and a *p* value of ***, a supported and strong hypothesis is indicated. In Table 7, by looking at the *p* values, it can deduce that the hypothesis between consumer attitude and lifestyle choices, consumer attitude and consumer preference, purchase intention and consumer preference, sustainable consumption and consumer attitude, and sustainable consumption and purchase intention is supported because of their values of less than 0.05. In contrast, there is no evidence to support the ideas linking purchase intention to customer attitude or lifestyle choice. The supported hypotheses demonstrate that the hypothesis formulated during the literature review is accurate; H1, H2, H5, H6, and H7 are supported and proven to be true, whereas H3 and H4 are not.

4.6 Mediating Effect

Additionally, to study the direct implications on sustainable consumption and purchase intention of cruelty free and ethically certified cosmetics, correlation with lifestyle choices, preferences, and consumer attitude was also studied as evident for

Fig. 2. The first relationship was that with lifestyle choices and purchase intention, and consumer preferences and purchase intention. In the former correlation, the correlation did not hold good, as a consequence H4 was not supported, whereas the latter, that is purchase intention and consumer preference did establish a significant relationship holding H5 as proved and supported.

The second relationship that was tested was between sustainable consumption and consumer attitude. It held a significant relationship holding H6 as proved and supported. This indicated the mediating effect of consumer attitude and consumer preference on sustainable consumption as well as purchase intention.

5 Discussion, Inference, Future Study, and Limitations

5.1 Discussion and Inference

This research analyzes the influence of product labels on purchase intentions relating to sustainable consumption, specifically considering ethically certified and cruelty-free cosmetics. In fact, lifestyle choices are involved in a rather complex interplay between consumer attitudes, consumer preferences, and even purchase behavior. Lifestyle choice indeed demonstrably influences consumer attitudes toward products that carry an ethical or cruelty-free label, yet it surprisingly fails to translate powerfully into actual purchase intention. This implies that, at the point of purchase, other factors may dominate the effect of lifestyle considerations, such as price, availability, brand awareness, or even perceived product efficacy. Future research could examine these intervening factors in order to bridge the gap between attitude and action.

A stronger relationship was found between consumer preference, consumer attitudes, and purchase intention. Consumers who express a preference for ethically sourced or cruelty-free products tend to hold positive attitudes toward such products, and this in turn drives purchase behavior. This underlines the importance of informed consumer choice. When consumers are knowledgeable about the ethical implications of their purchases, they are more likely to align their actions with their values. The importance of providing more transparent information regarding the source and production process of products was made evident.

Interestingly, although lifestyle choice showed only weak correlations with purchase intention, consumer preference was clearly the critical driver of purchasing ethically certified and cruelty-free cosmetics. This means that targeted appeals to consumer preferences rather than lifestyle appeals should be the focus for marketing strategies. Companies should focus on specific benefits and values of their product offerings, such as high-quality ingredients, sustainable packaging, or commitment to fair labor practices.

The study also confirmed that there is a significant correlation between consumer attitude and purchase intention in the context of sustainable consumption. Consumers with positive attitudes toward sustainable products are likely to buy them. However, lifestyle choice seems to have little impact on this relationship, especially when it comes to ethically certified and cruelty-free cosmetics. This further fortifies the idea that factors other than lifestyle, including pricing, availability, and brand image, determine purchase behavior.

Therefore, marketers searching for a means to introduce ethically labeled products need to make brand equity robust and target specific consumer preferences. Instead of lifestyle marketing, companies should invest in communication of the unique value proposition, ethical credentials, and positive brand image. Future research might be directed to the impact of locally sourced or substituted goods that may contribute to sustainable consumption without necessarily bearing a specific label. These often appeal to consumers who are looking for local economies and reduced environmental impact.

5.2 *Limitations and Future Scope*

While the study provides evidence of increased conscience among consumers, especially among younger consumers, for healthy choices in products, it also identified significant limitations. There are no standard definitions of terms such as "cruelty-free" and "sustainability." This may cause confusion and inconsistency in labeling. In addition, generational differences were shown in adopting ethical labeling. Older consumers, while often empathetic and mindful in their consumption, do not seek out ethical labels, suggesting that habit and established brand loyalty may be stronger drivers of purchasing behavior than newly introduced labeling conventions. Affordability was also identified as a concern, particularly among older consumers, and accessible pricing strategies for ethically labeled products are needed.

The fact that the study focuses on India also limits its generalizability. Future studies should expand geographically to study cultural and regional variations in consumer attitudes and purchase intentions. Further variables, including religious affiliation and egoistic preferences, could add a more detailed understanding of what drives sustainable consumption. In-depth interviews and focus groups could bring even more depth of understanding of reasons for, or barriers to buying ethically labeled products among people of various age groups. Marketers and policy makers will benefit from that deep understanding and further develop appropriate effective strategies in favor of promoting consumption.

References

Aaker, D. A. (1996). Measuring brand equity across products and markets. *California Management Review, 38*(3), 102–120. https://doi.org/10.2307/41165845

Aaker, D. A., & Biel, A. L. (2013). *Brand equity & advertising*. Psychology Press. https://doi.org/10.4324/9781315799537

Abel, T., & Cockerham, W. C. (1993). Lifestyle or Lebensführung? Critical remarks on the mistranslation of weber's "class, status, party". *The Sociological Quarterly, 34*(3), 551–556. https://doi.org/10.1111/j.1533-8525.1993.tb00126.x

Ajzen, I. (1991). The theory of planned behavior. *Organizational Behavior and Human Decision Processes, 50*(2), 179–211. https://doi.org/10.1016/0749-5978(91)90020-t

Ajzen, I. (2011). Consumer attitudes and behavior. In *Handbook of Consumer Psychology*. https://doi.org/10.4324/9780203809570.ch20

Alavi, M., Visentin, D. C., Thapa, D. K., Hunt, G. E., Watson, R., & Cleary, M. (2020). CHI-square for model fit in confirmatory factor analysis. *Journal of Advanced Nursing, 76*(9), 2209–2211. https://doi.org/10.1111/jan.14399

Anderson, J. C., & Gerbing, D. W. (1988). Structural equation modeling in practice: A review and recommended two-step approach. *Psychological Bulletin, 103*(3), 411–423. https://doi.org/10.1037/0033-2909.103.3.411

AV Magazine Issue 3. (2015). *Issuu*. Retrieved from https://issuu.com/aavs/docs/aavs_av-magazine_2016_cruelty-free-.

Bagozzi, R. P. (1992). The self-regulation of attitudes, intentions, and behavior. *Social Psychology Quarterly, 55*(2), 178. https://doi.org/10.2307/2786945

Ballard, C. (2019, July 8). An ode to R-squared. *Medium*. Retrieved from https://towardsdatascience.com/an-ode-to-r-squared-804d8d0ed22c.

Banerjee, S. B. (2002). Organisational strategies for sustainable development: Developing a research agenda for the new millennium. *Australian Journal of Management, 27*(1_suppl), 105–117. https://doi.org/10.1177/031289620202701s11

Beardsworth, A., & Bryman, A. (2004). Meat consumption and meat avoidance among young people. *British Food Journal, 106*(4), 313–327. https://doi.org/10.1108/00070700410529573

Beckford, C. L., Jacobs, C., Williams, N., & Nahdee, R. (2010). Aboriginal environmental wisdom, stewardship, and sustainability: Lessons from the Walpole Island first nations, Ontario, Canada. *The Journal of Environmental Education, 41*(4), 239–248. https://doi.org/10.1080/00958961003676314

Beldad, A., & Hegner, S. (2018). Determinants of fair trade product purchase intention of Dutch consumers according to the extended theory of planned behaviour. *Journal of Consumer Policy, 41*(3), 191–210. https://doi.org/10.1007/s10603-018-9384-1

Belk, R. W. (1988). Possessions and the extended self. *Journal of Consumer Research, 15*(2), 139. https://doi.org/10.1086/209154

Bonett, D. G., & Wright, T. A. (2014). Cronbach's alpha reliability: Interval estimation, hypothesis testing, and sample size planning. *Journal of Organizational Behavior, 36*(1), 3–15. https://doi.org/10.1002/job.1960

Borowiec, A. A. (2015). Healthy lifestyle and consumer orientation amongst the Warsaw middle class. In *Opuscula Sociologica* (pp. 81–97). https://doi.org/10.18276/os.2015.4-06

Bou-Llusar, J. C., Camisón-Zornoza, C., & Escrig-Tena, A. B. (2001). Measuring the relationship between firm perceived quality and customer satisfaction and its influence on purchase intentions. *Total Quality Management, 12*(6), 719–734. https://doi.org/10.1080/09544120120075334

Boyd Thomas, J., & Lee Okleshen Peters, C. (2009). Silver seniors. *International Journal of Retail & Distribution Management, 37*(12), 1018–1040. https://doi.org/10.1108/09590550911005001

Carlson, L., Grove, S. J., & Kangun, N. (1993). A content analysis of environmental advertising claims: A matrix method approach. *Journal of Advertising, 22*(3), 27–39. https://doi.org/10.1080/00913367.1993.10673409

Chen, Y., & Chang, C. (2012). Enhance green purchase intentions. *Management Decision, 50*(3), 502–520. https://doi.org/10.1108/00251741211216250

Cockerham, W. C. (2005). Health lifestyle theory and the convergence of agency and structure. *Journal of Health and Social Behavior, 46*(1), 51–67. https://doi.org/10.1177/002214650504600105

Davies, I. A., & Gutsche, S. (2016). Consumer motivations for mainstream “ethical” consumption. *European Journal of Marketing, 50*(7/8), 1326–1347. https://doi.org/10.1108/ejm-11-2015-0795

De Chiara, A. (2016). Eco-labeled products: Trend or tools for sustainability strategies? *Journal of Business Ethics, 137*(1), 161–172. Retrieved from http://www.jstor.org/stable/24755753

Deckers, J. (2009). Vegetarianism, sentimental or ethical? *Journal of Agricultural and Environmental Ethics, 22*(6), 573–597. https://doi.org/10.1007/s10806-009-9176-3

De-Magistris, T., & Gracia, A. (2016). Consumers' willingness-to-pay for sustainable food products: The case of organically and locally grown almonds in Spain. *Journal of Cleaner Production, 118*, 97–104. https://doi.org/10.1016/j.jclepro.2016.01.050

Deng, X. (2013). Factors influencing ethical purchase intentions of consumers in China. *Social Behavior and Personality: An International Journal, 41*(10), 1693–1703. https://doi.org/10.2224/sbp.2013.41.10.1693

Díaz, E. M. (2019). Emerging attitudes towards nonhuman animals among Spanish university students. *Society and Animals, 27*(2), 197–223. https://doi.org/10.1163/15685306-12341490

Dietz, T., Frisch, A. S., Kalof, L., Stern, P. C., & Guagnano, G. A. (2010). Values and vegetarianism: An exploratory Analysis. *Rural Sociology, 60*(3), 533–542. https://doi.org/10.1111/j.1549-0831.1995.tb00589.x

Everitt, B. S., & Yates, A. (1989). Multivariate exploratory data analysis: A perspective on exploratory factor analysis. *Biometrics, 45*(1), 342. https://doi.org/10.2307/2532065

Fernqvist, F., & Ekelund, L. (2014). Credence and the effect on consumer liking of food—A review. *Food Quality and Preference, 32*, 340–353. https://doi.org/10.1016/j.foodqual.2013.10.005

Fornell, C., & Larcker, D. F. (1981). Structural equation models with unobservable variables and measurement error: Algebra and statistics. *Journal of Marketing Research, 18*(3), 382. https://doi.org/10.2307/3150980

Fortune Business Insights. (2022). *Vegan food market size, share and growth analysis [2028]* (FBI106421). Retrieved from https://www.fortunebusinessinsights.com/vegan-food-market-106421.

Fox, M. A. (2013). Vegetarianism and veganism. *International Encyclopedia of Ethics*. https://doi.org/10.1002/9781444367072.wbiee009

Giddens, A. (2002). Runaway world: How globalization is reshaping our lives. *Rugman Reviews*, 133–134. https://doi.org/10.1007/978-1-137-28787-8_45

Gomez-Trujillo, A. M., Velez-Ocampo, J., & Gonzalez-Perez, M. A. (2020). A literature review on the causality between sustainability and corporate reputation. *Management of Environmental Quality: An International Journal, 31*(2), 406–430. https://doi.org/10.1108/meq-09-2019-0207

Govind, R., Singh, J. J., Garg, N., & D'Silva, S. (2017). Not walking the walk: How dual attitudes influence behavioral outcomes in ethical consumption. *Journal of Business Ethics, 155*(4), 1195–1214. https://doi.org/10.1007/s10551-017-3545-z

Grappe, C. G., Lombart, C., Louis, D., & Durif, F. (2021). “Not tested on animals”: How consumers react to cruelty-free cosmetics proposed by manufacturers and retailers? *International Journal of Retail & Distribution Management, 49*(11), 1532–1553. https://doi.org/10.1108/ijrdm-12-2020-0489

Grunert, K. G. (2005). Food quality and safety: Consumer perception and demand. *European Review of Agricultural Economics, 32*(3), 369–391. https://doi.org/10.1093/eurrag/jbi011

Gruzalski, B. (2002). Religious vegetarianism: From Hesiod to the Dalai Lama. *Choice Reviews Online, 39*(05), 39-2754. https://doi.org/10.5860/choice.39-2754

Gupta, R. (2014). India @ 2030: The future of the Indian consumer. *Futures, 56*, 81–86. https://doi.org/10.1016/j.futures.2013.10.012

Gupta, S. K., & Racherla, U. S. (2018). Interdependence among dimensions of sustainability. *Management of Environmental Quality: An International Journal, 29*(3), 406–415. https://doi.org/10.1108/meq-06-2017-0051

Han, H., & Yoon, H. J. (2015). Hotel customers' environmentally responsible behavioral intention: Impact of key constructs on decision in green consumerism. *International Journal of Hospitality Management, 45*, 22–33. https://doi.org/10.1016/j.ijhm.2014.11.004

Hirsh, S. (2023, January 5). China removes another layer of animal testing requirements, meaning more brands can become cruelty-free. *Green Matters*. Retrieved from https://www.greenmatters.com/news/why-does-china-require-animal-testing

Izmirli, S., & Phillips, C. J. (2011). The relationship between student consumption of animal products and attitudes to animals in Europe and Asia. *British Food Journal, 113*(3), 436–450. https://doi.org/10.1108/00070701111116482

Jamal, A., & Goode, M. M. (2001). Consumers and brands: A study of the impact of self-image congruence on brand preference and satisfaction. *Marketing Intelligence & Planning, 19*(7), 482–492. https://doi.org/10.1108/02634500110408286

Joshi, Y., & Rahman, Z. (2019). Consumers' sustainable purchase behaviour: Modeling the impact of psychological factors. *Ecological Economics, 159*, 235–243. https://doi.org/10.1016/j.ecolecon.2019.01.025

Kalof, L., Dietz, T., Stern, P. C., & Guagnano, G. A. (2009). Social psychological and structural influences on vegetarian beliefs. *Rural Sociology, 64*(3), 500–511. https://doi.org/10.1111/j.1549-0831.1999.tb00364.x

Khare, A. (2011). Impact of Indian cultural values and lifestyles on meaning of branded products: Study on university students in India. *Journal of International Consumer Marketing, 23*(5), 365–379. https://doi.org/10.1080/08961530.2011.602953

Kushwah, S., Dhir, A., & Sagar, M. (2019). Ethical consumption intentions and choice behavior towards organic food. Moderation role of buying and environmental concerns. *Journal of Cleaner Production, 236*, 117519. https://doi.org/10.1016/j.jclepro.2019.06.350

Lim, J. S., & Shim, K. (2019). Corporate social responsibility beyond borders: U.S. consumer boycotts of a global company over sweatshop issues in supplier factories overseas. *American Behavioral Scientist, 63*(12), 1643–1664. https://doi.org/10.1177/0002764219835241

Liobikienė, G., & Bernatonienė, J. (2017). Why determinants of green purchase cannot be treated equally? The case of green cosmetics: Literature review. *Journal of Cleaner Production, 162*, 109–120. https://doi.org/10.1016/j.jclepro.2017.05.204

Lombart, C., & Louis, D. (2012). Consumer satisfaction and loyalty: Two main consequences of retailer personality. *Journal of Retailing and Consumer Services, 19*(6), 644–652. https://doi.org/10.1016/j.jretconser.2012.08.007

Lovelace, M., & Brickman, P. (2013). Best practices for measuring students' attitudes toward learning science. *CBE Life Sciences Education, 12*(4), 606–617. https://doi.org/10.1187/cbe.12-11-0197

Matheny, G., & Chan, K. M. (2005). Human diets and animal welfare: The illogic of the larder. *Journal of Agricultural and Environmental Ethics, 18*(6), 579–594. https://doi.org/10.1007/s10806-005-1805-x

Mostafa, M. M. (2006). Antecedents of Egyptian consumers' green purchase intentions. *Journal of International Consumer Marketing, 19*(2), 97–126. https://doi.org/10.1300/j046v19n02_06

Neamtu, B., & Dragos, D. C. (2015). Sustainable public procurement: The use of eco-labels. *European Procurement & Public Private Partnership Law Review, 10*(2), 92–101. Retrieved from https://www.jstor.org/stable/26639578

Nunnally, J. C. (1978). *Psychometric theory*. McGraw-Hill Companies.

Oliver, R. L. (1999). Whence consumer loyalty? *Journal of Marketing, 63*, 33. https://doi.org/10.2307/1252099

Olutende, O. M., Wanzala, M. N., & Wamukoya, E. K. (2019). Psychometric properties of the health-promoting lifestyle profile II: Validation of the self-realization and health responsibility constructs among Kenyan university students. *Saudi Journal of Nursing and Health Care, 02*(11), 365–373. https://doi.org/10.36348/sjnhc.2019.v02i11.004

Parra-Medina, D., Wilcox, S., Wilson, D. K., Addy, C. L., Felton, G., & Poston, M. B. (2010). Heart healthy and ethnically relevant (HHER) lifestyle trial for improving diet and physical activity in underserved African American women. *Contemporary Clinical Trials, 31*(1), 92–104. https://doi.org/10.1016/j.cct.2009.09.006

Phillips, C., Izmirli, S., Aldavood, J., Alonso, M., Choe, B., Hanlon, A., Handziska, A., Illmann, G., Keeling, L., Kennedy, M., Lee, G., Lund, V., Mejdell, C., Pelagic, V., & Rehn, T. (2010). An international comparison of female and male students' attitudes to the use of animals. *Animals, 1*(1), 7–26. https://doi.org/10.3390/ani1010007

Posegga, O., Zylka, M. P., Putzke, J., Fischbach, K., & Schoder, D. (2014). Understanding the lifestyle of health and sustainability—An exploratory study. *International Journal of Organisational Design and Engineering, 3*(3/4), 338. https://doi.org/10.1504/ijode.2014.065071

Ray, P. H., & Anderson, S. R. (2001). *The cultural creatives: How 50 million people are changing the world.* Crown.

Rosenfeld, D. L. (2018). Why some choose the vegetarian option: Are all ethical motivations the same? *Motivation and Emotion, 43*(3), 400–411. https://doi.org/10.1007/s11031-018-9747-6

Ryan, J., & Casidy, R. (2018). The role of brand reputation in organic food consumption: A behavioral reasoning perspective. *Journal of Retailing and Consumer Services, 41*, 239–247. https://doi.org/10.1016/j.jretconser.2018.01.002

Rynarzewska, A. I. (2019). Plastic bag fee: Would consumers be willing to pay for plastic grocery bags to 'go green'? *Journal of Managerial Issues, 31*(3), 312–330. Retrieved from https://www.jstor.org/stable/45206627

Schuiling, I., & Kapferer, J. (2004). Executive insights: Real differences between local and international brands: Strategic implications for international marketers. *Journal of International Marketing, 12*(4), 97–112. https://doi.org/10.1509/jimk.12.4.97.53217

Sebastiani, R., Montagnini, F., & Dalli, D. (2012). Ethical consumption and new business models in the food industry. Evidence from the Eataly case. *Journal of Business Ethics, 114*(3), 473–488. https://doi.org/10.1007/s10551-012-1343-1

Sheehan, K. B., & Lee, J. (2014). What's cruel about cruelty free: An exploration of consumers, moral heuristics, and public policy. *Journal of Animal Ethics, 4*(2), 1–15. https://doi.org/10.5406/janimalethics.4.2.0001

Shin, D. (2011). Effects on LOHAS image of family restaurant, customer value, customer satisfaction and intention of relation. *Journal of Korea Service Management Society, 12*(1), 189–216. https://doi.org/10.15706/jksms.2011.12.1.008

Shrestha, N. (2021). Factor analysis as a tool for survey analysis. *American Journal of Applied Mathematics and Statistics, 9*(1), 4–11. https://doi.org/10.12691/ajams-9-1-2

Spencer, E. H., Elon, L. K., & Frank, E. (2007). Personal and professional correlates of US medical students' vegetarianism. *Journal of the American Dietetic Association, 107*(1), 72–78. https://doi.org/10.1016/j.jada.2006.10.034

Sudbury-Riley, L., & Kohlbacher, F. (2016). Ethically minded consumer behavior: Scale review, development, and validation. *Journal of Business Research, 69*(8), 2697–2710. https://doi.org/10.1016/j.jbusres.2015.11.005

Tanner, C., & Wölfing Kast, S. (2003). Promoting sustainable consumption: Determinants of green purchases by Swiss consumers. *Psychology & Marketing, 20*(10), 883–902. https://doi.org/10.1002/mar.10101

Ursachi, G., Horodnic, I. A., & Zait, A. (2015). How reliable are measurement scales? External factors with indirect influence on reliability estimators. *Procedia Economics and Finance, 20*, 679–686. https://doi.org/10.1016/s2212-5671(15)00123-9

Wang, T., Mukhopadhyay, A., & Patrick, V. M. (2017). Getting consumers to Recycle*NOW*! When and why cuteness appeals influence prosocial and sustainable behavior. *Journal of Public Policy & Marketing, 36*(2), 269–283. https://doi.org/10.1509/jppm.16.089

Wescombe, N. J. (2019). Communicating veganism: Evolving theoretical challenges to mainstreaming ideas. *Studies in Media and Communication, 7*(2), 1. https://doi.org/10.11114/smc.v7i2.4367

Xiao, J. J., & Li, H. (2010). Sustainable consumption and life satisfaction. *Social Indicators Research, 104*(2), 323–329. https://doi.org/10.1007/s11205-010-9746-9

Yadav, R., & Pathak, G. S. (2017). Determinants of consumers' green purchase behavior in a developing nation: Applying and extending the theory of planned behavior. *Ecological Economics, 134*, 114–122. https://doi.org/10.1016/j.ecolecon.2016.12.019
Zablocki, A., Makri, K., Schlegelmilch, B. B., & Houston, M. J. (2019). Corrigendum to 'emotions within online reviews and their influence on product attitudes in Austria, USA and Thailand' [journal of interactive marketing, volume 46, may 2019, pages 20–39]. *Journal of Interactive Marketing, 46*, 113. https://doi.org/10.1016/j.intmar.2019.04.001
Zander, K., & Hamm, U. (2010). Consumer preferences for additional ethical attributes of organic food. *Food Quality and Preference, 21*(5), 495–503. https://doi.org/10.1016/j.foodqual.2010.01.006

Index

J. J. Puthenkalam S.J et al. (eds.), *Terra Thrive*,
https://doi.org/10.1007/978-3-032-02738-2

MIX
Papier aus verantwortungsvollen Quellen
Paper from responsible sources
FSC® C105338
FSC
www.fsc.org

If you have any concerns about our products,
you can contact us on
ProductSafety@springernature.com

In case Publisher is established outside the EU,
the EU authorized representative is:
Springer Nature Customer Service Center GmbH
Europaplatz 3, 69115 Heidelberg, Germany

Printed by Libri Plureos GmbH
in Hamburg, Germany